New Perspectives on Calcium Antagonists

The positions taken by the authors of the chapters are their own and not necessarily those of the American [illegible]

Preface

The advent of a new class of therapeutically effective agents is always an event of considerable interest. This is especially so for newer types of Ca^{2+} antagonists—potentially important groups of compounds that act by specifically blocking one or more of the diverse physiological actions of Ca^{2+}.

Although a wide variety of compounds have been identified as directly or indirectly altering certain actions of Ca^{2+}, increasing attention has been focused on those newer groups of agents that have specific inhibitory actions on the inward Ca^{2+} current in the heart and on the voltage-sensitive Ca^{2+} channels of smooth muscle. *New Perspectives on Calcium Antagonists* reflects these interests: many of the chapters primarily deal with the delineation of the actions of inhibitors of Ca^{2+} uptake. But as the title indicates, the broader range of actions of other types of Ca^{2+} antagonists is also considered.

The terminology for this general field of research and especially for the newer types of Ca^{2+} antagonists is imprecise. In the heart, where the slow-channel Ca^{2+} current is blocked, these agents have appropriately been called Ca^{2+}-blocking agents, Ca^{2+}-influx blockers, or Ca^{2+}-channel blockers. In smooth muscle, where more than one type of Ca^{2+} entry channel apparently exists, this group of drugs may have an even more specific mode of action in blocking only voltage-operated (voltage-sensitive) channels. Perhaps as our knowledge of the basic mechanisms and ranges of action of the various Ca^{2+} antagonists in smooth muscle is increased, the terminology will be redefined to delineate categories of agents capable of inhibiting voltage-sensitive Ca^{2+} channels in different types of smooth muscle.

There are two approaches to studying newer types of Ca^{2+} antagonists. First, especially in heart and various types of smooth muscle, their effects on muscle contractility and associated parameters related to Ca^{2+} mobilization can be examined. This yields information about the mechanisms and specificity of action of various agents. Second, these drugs can be used to block subsequent contractile or secretory events. The increasing precision of this type of blockade conveys a considerable amount of information about important Ca^{2+}-related cellular events in a number of systems.

This book presents a sequence of perspectives on the various actions and effects of Ca^{2+} antagonists. The chapters are grouped into four sections.

- The first five chapters present different aspects of the basic framework important for understanding how and where Ca^{2+} antagonists act. The types of antagonists, the nature of the Ca^{2+} channel, the subcellular regulatory mechanisms for contraction in different types of muscle and muscle cell structures, and the cellular actions of different antagonists in cardiac muscle are outlined.
- The second group of five chapters focuses on the mechanisms important for the actions and effects of Ca^{2+} antagonists in different types of vascular smooth muscle. Possible models for actions of agents affecting Ca^{2+} uptake, techniques used to delineate actions of different types of Ca^{2+} antagonists, and the heterogeneity of vascular systems and the resulting divergent effects of different Ca^{2+} antagonists are highlighted.
- The third section contains four chapters on the actions of Ca^{2+} and the selectivity of various Ca^{2+} antagonists in secretory systems. Generally the concentrations of the agents at which Ca^{2+}-channel blockade is attained are considerably higher in secretory systems than in smooth or cardiac muscle. Thus in these systems the newer Ca^{2+} antagonists may be valuable tools for understanding physiological function.
- The final four chapters emphasize different aspects of the physiological actions of Ca^{2+} antagonists with real or potential therapeutic effects. The Ca^{2+}-uptake (channel) blockers clearly have useful actions on ischemic myocardial tissue, and some may also be effective antiasthmatic agents. Eventually Ca^{2+} antagonists may even be used in the central nervous system, where a number of Ca^{2+}-dependent pharmacological effects have been described.

Thus the objective of *New Perspectives on Calcium Antagonists* is not only to summarize current knowledge of the actions of Ca^{2+} antagonists but also to indicate the direction of future advances in both knowledge and therapeutics arising from the rapid expansion of the types and numbers of Ca^{2+} antagonists. The emerging rationale for progress is that documented differences in cellular and molecular patterns of Ca^{2+} mobilization in the heart, smooth muscle, secretory systems, and other tissues provide a scientific basis for the development of an equally diverse group of Ca^{2+} antagonists of increasing specificity and efficacy.

The Editor

Contents

Molecular and Cellular Parameters for Activity of Calcium Antagonists

Effects of Calcium Antagonists on Excitation-Contraction Coupling

1

Calcium Antagonists: Basic Chemical and Pharmacological Aspects

D. J. TRIGGLE

Department of Biochemical Pharmacology, School of Pharmacy, State University of New York, Buffalo, New York

A LARGE COLLECTION OF MOLECULES, including di- and trivalent cations (Mn^{2+}, La^{3+}), local anesthetics (procaine), antibiotics (neomycin C), tricyclics (trifluoperazine), and antiepileptics (dilantin), have been designated as Ca^{2+} antagonists (95). The ability to inhibit Ca^{2+}-dependent processes with varying potencies and specificities (Fig. 1*A*; 95) has been attributed to all these agents. This structural diversity is testimony to the multiple roles of both extracellular and intracellular Ca^{2+} in regulating cellular activity (Fig. 2). It is expected, by analogy with the role of receptor antagonists in providing classification of receptors, that a corresponding classification will be available for Ca^{2+} sites, but this has not yet been achieved. For certain structures, however, some definition of site of action is becoming apparent.

Trifluoperazine and structurally related antipsychotic agents appear to exert their antagonism through binding to calmodulin (60), the Ca^{2+}-dependent regulator protein or intracellular Ca^{2+} receptor that subserves multiple Ca^{2+}-dependent functions including contractile, secretory, and transport processes (70, 102). Certain naphthalene derivatives (Fig. 1*A*) appear to function similarly. Particularly interesting is a group of compounds, termed *Ca^{2+}-channel antagonists*, that includes verapamil and methoxyverapamil (D 600), nifedipine and related 1,4-dihydropyridines, diltiazem and cinnarizine, and related diphenylmethylalkylamines (Fig. 1*B*). These structures are the main focus of this chapter.

CALCIUM CHANNELS AND CALCIUM-CHANNEL ANTAGONISTS

Calcium channels responsible for the entry of Ca^{2+} into the cell in response to a membrane signal, which may be chemical or electrical, have been classified into two major types (Fig. 3; 1, 11, 56, 78, 81, 86): potential-dependent channels (PDC), which are activated by membrane depolarization, and receptor-operated channels (ROC), which are associated with and activated by specific

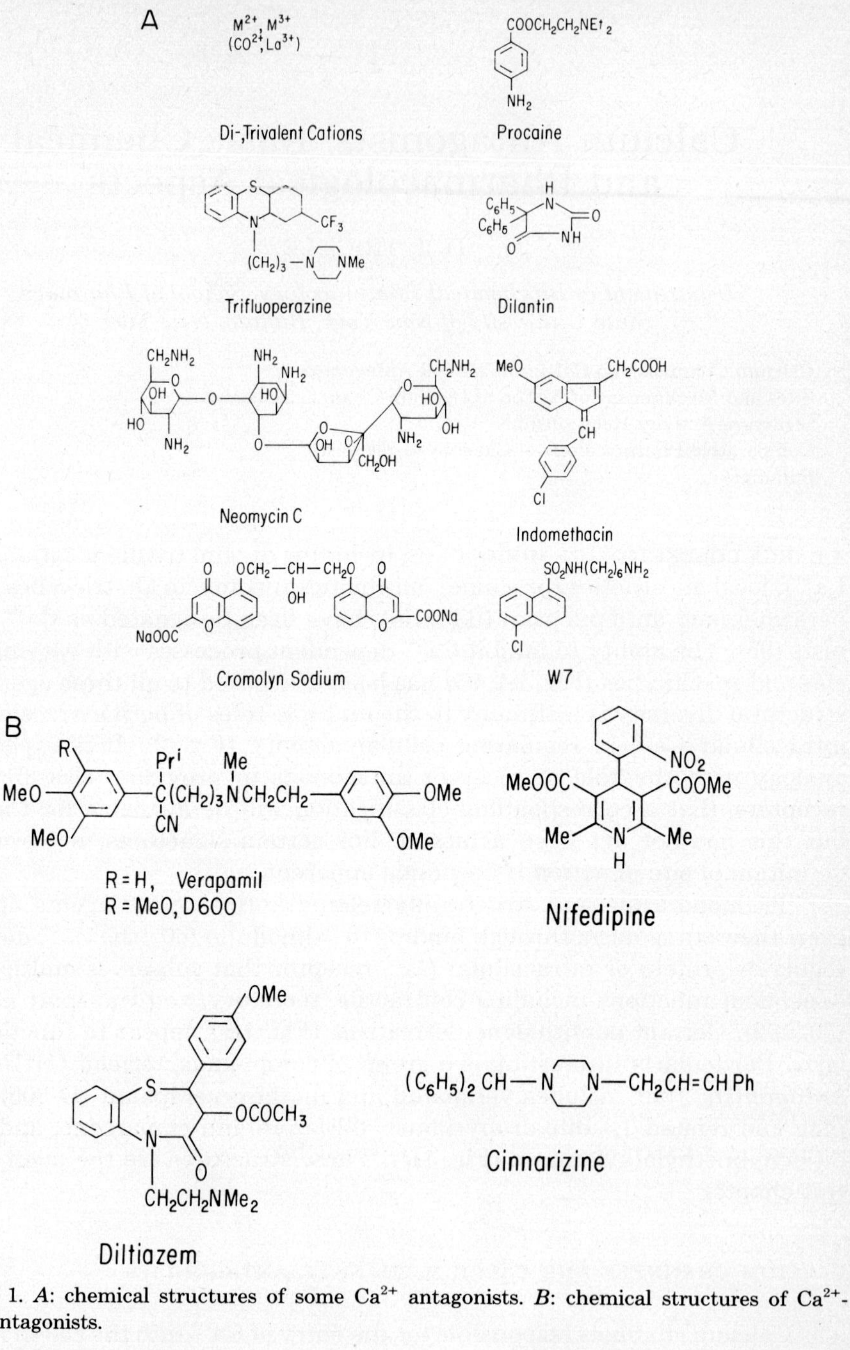

FIG. 1. *A*: chemical structures of some Ca^{2+} antagonists. *B*: chemical structures of Ca^{2+}-channel antagonists.

receptors independent of membrane potential. It is unlikely that either of these channels are totally specific for Ca^{2+}. Thus an agonist-induced secretory or mechanical event may use to a greater or lesser extent either channel population depending on both the extent of dependence of the response on extracel-

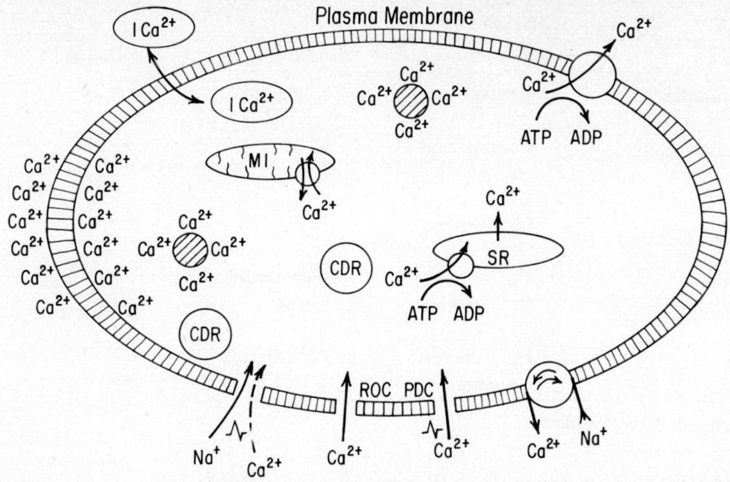

FIG. 2. Schematic representation of Ca^{2+} regulation. Intracellular Ca^{2+}, with resting free ionized concentration $< 10^{-7}$ M (57), is bound to internal membrane surface and to intracellular proteins and is sequestered in mitochondrion and sarcoplasmic reticulum (13). The Ca^{2+}-dependent regulator protein calmodulin (CDR) is shown as intracellular Ca^{2+} receptor (70). Calcium pumping across plasma membrane is depicted by a Ca^{2+}-ATPase (13, 87) and an Na^{+}-Ca^{2+} exchange mechanism that can operate in either direction (10). Specific entry processes for Ca^{2+} include the Na^{+} (minor) and Ca^{2+} (major) channels. A nonphysiological bypass mechanism is depicted by ionophore-mediated entry (I Ca^{2+}; 86). ROC, receptor-operated channel; PDC, potential-dependent channel, SR, sarcoplasmic reticulum, MI, mitochondrion.

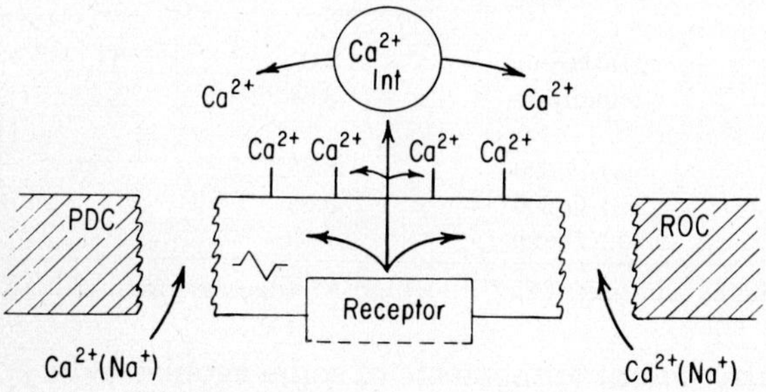

FIG. 3. Schematic representation of membrane-mediated Ca^{2+}-mobilization processes. Shown are Ca^{2+} influx through receptor-operated (ROC) and potential-dependent (PDC) channels as well as mobilization of intracellular Ca^{2+}.

lular Ca^{2+} and the association of membrane depolarization with the receptor-activation process (11, 14, 33, 78). It must be emphasized that intracellular Ca^{2+} is also mobilized to sustain these Ca^{2+}-dependent processes and that the relative extent of use of intracellular and extracellular sources is both tissue and stimulant dependent (11, 14, 17, 78, 97).

The agents designated as Ca^{2+}-channel antagonists apparently exert their actions primarily at the potential-dependent Ca^{2+} channels (11, 24, 25, 33, 78, 86), a conclusion supported by several lines of evidence. These agents inhibit the slow inward Ca^{2+} current in cardiac tissue in the concentration range at which they produce a negative inotropic effect (Table 1). They serve as competitive antagonists against extracellular Ca^{2+} in smooth muscle and are potent inhibitors of smooth muscle contractile activity produced by K^{+} depo-

TABLE 1. *Ca^{2+} antagonists in cardiac tissue*

Preparation	Activity	Antagonist	ED_{50}, M	Ref.
Guinea pig papillary	Contraction	Verapamil	5×10^{-7}	24, 52
Guinea pig atria	Contraction	D 600	$< 10^{-7}$	34
		Nifedipine	10^{-7}	
Cat papillary	Ca^{2+} current	D 600	2×10^{-6}	96
Cat ventricular trabeculae	Ca^{2+} current	Verapamil	$< 10^{-6}$	52–54
	Sr^{2+} current	D 600	$< 10^{-6}$	
Cat papillary	Ca^{2+} current	Nifedipine	6×10^{-7}	4, 8
	Contraction	Nifedipine	6×10^{-7}	

ED_{50}, mean effective dose.

TABLE 2. *Competitive activities of Ca^{2+} antagonists*

Preparation	Antagonist	pA_2	Ref.
Guinea pig taenia coli	Prenylamine	6.44	37, 41, 80
	Diltiazem	6.90	
	Verapamil	7.36	
	Nifedipine	9.43	
Guinea pig atria	Prenylamine	4.98	37, 41
	Diltiazem	6.57	
	Verapamil	6.27	
	Nifedipine	8.24	
Canine coronary	Diltiazem	7.0	41, 73
	Nifedipine	8.84	
Guinea pig ileum	(±)-D 600	8.85	44
	(−)-D 600	8.84	
	(+)-D 600	7.89	

Contractile responses were induced by addition of Ca^{2+} to a high-K^+ (depolarizing) solution.

larization (Table 2). They are also potent antagonists of some agonist-induced mechanical events, and there is a reasonably good correlation between this activity and the dependence of the response on extracellular Ca^{2+} (Table 3). Additionally these agents inhibit Ca^{2+} uptake into smooth muscles, including guinea pig taenia coli and ileum (66, 85), rabbit aorta (45, 58), and rat portal vein (71), at concentrations where they potently inhibit mechanical response.

These lines of evidence are consistent with the thesis that agents such as verapamil, D 600, nifedipine, diltiazem, and cinnarizine are potent antagonists of Ca^{2+} movement through certain Ca^{2+} channels. The sites and mechanisms of actions of these agents remain to be defined.

SITES AND MECHANISMS OF ACTION OF CALCIUM-CHANNEL ANTAGONISTS

Even within the limited group of compounds described as Ca^{2+}-channel antagonists, the obvious structural heterogeneity suggests that more than one site and mechanism of action exist. The inorganic cations Mn^{2+}, Co^{2+}, and La^{3+} apparently function as general antagonists and are effective in a wide variety of Ca^{2+}-dependent processes (11, 71, 75, 81, 86, 95). This nonselectivity of action probably arises from a general ability of these cations to substitute for Ca^{2+} at Ca^{2+}-binding sites. Thus they modify membrane excitability by either screening or binding to surface negative charges on plasma membranes (42). These cations probably occupy the cation coordination site of Ca^{2+} channels and either block channel function or actually enter the cell (28, 35), where they can substitute for or block Ca^{2+} binding at the intracellular Ca^{2+} receptors.

Although the organic Ca^{2+}-channel antagonists are generally competitive against Ca^{2+}, it is clear that they lack the widespread Ca^{2+}-blocking properties of the inorganic cations, and it is not likely that they compete directly with Ca^{2+} at the cation coordination site of the Ca^{2+} channel. Furthermore important differences in activity can be seen among the various structural classes of these compounds.

There is limited evidence regarding whether the organic antagonists have intracellular or extracellular sites of action. D 600 is ineffective in skinned smooth muscle and myocardial preparations, which is consistent with a membrane locus of action (24, 26). Accordingly intracellular D 600 is ineffective in inhibiting Ca^{2+} currents in barnacle muscle (83), and very high concentrations of D 600 or verapamil are needed to block Ca^{2+} movements in intracellular organelles such as sarcoplasmic reticulum and mitochondria (16, 20, 27). The quaternary methyl bromide derivative of D 600, however, is ineffective in both myocardial and smooth muscle preparations (65, 86), the activity of D 600 ($pK_a \sim 8.5$) is enhanced at alkaline pH (93), and the activity of quaternary ammo-

TABLE 3. *Ca^{2+} antagonists: effects on smooth muscle excitation-contraction coupling*

Preparation	Agonist	Antagonist	ED_{50}, M	Ref.
Rabbit aorta	NE	Verapamil	1.2×10^{-4}	88
	K^+	Verapamil	2.7×10^{-8}	
Rabbit mesenteric artery	NE	Verapamil	4.5×10^{-5}	88
	K^+	Verapamil	5.4×10^{-8}	
Canine trachea	ACh	Verapamil	$> 10^{-4}$	23
	K^+	Verapamil	$\ll 10^{-4}$	
Rat portal vein	NE	D 600	5×10^{-8}	36, 43
	K^+	D 600	5×10^{-8}	
Guinea pig ileum	ACh	D 600	2.7×10^{-8}	85
	K^+	D 600	3.4×10^{-8}	
	ACh	Nifedipine	5.1×10^{-9}	
	K^+	Nifedipine	3.4×10^{-9}	
Dog cerebral artery	$PGF_{2\alpha}$	Verapamil	2×10^{-7}	90
	K^+	Verapamil	2×10^{-7}	

NE, norepinephrine; ACh, acetylcholine; $PGF_{2\alpha}$, prostaglandin $F_{2\alpha}$.

nium analogues of nifedipine is also greatly reduced (86). Since quaternary ammonium compounds should have reduced ability to cross the membrane, their diminished activity suggests an intracellular locus of action, presumably at the internal surface of the plasma membrane. Alternatively it is possible that the sites of action of D 600 and nifedipine are located in the membrane interior and that some permeation, hindered by net positive charge, is needed to reach these sites.

This last suggestion may be consistent with the well-documented observation (4, 5, 7, 19, 74) that the negative inotropic and Ca^{2+} current–blocking properties of verapamil and D 600 depend strongly on frequency and voltage features that are difficult to reconcile with views of these agents as "simple channel plugs." The actions of both verapamil and D 600 increase with increasing frequency of stimulation such that there is an actual inversion of the normal force-frequency relationship (5). This might indicate that these antagonists, in addition to blocking slow inward current, also impair the availability of Ca^{2+} perhaps at a membrane-bound locus or at an intracellular site of Ca^{2+} recycling (4, 5, 7, 47, 48). The frequency-dependent changes are also modified by the level of the membrane potential, however, such that inhibition increases with depolarization (19). This suggests that verapamil and D 600 may exert an effect on the myocardial slow inward Ca^{2+} current by modifying the voltage-dependent activation and/or inactivation parameters of this channel. Verapamil and D 600 can inhibit Ca^{2+}-channel recovery from inactivation (55), and McDonald et al. (67) have shown that recovery from channel block after administration of D 600 is characterized by two voltage-dependent time constants, both of which increase as membrane potential becomes less negative. The finding that inhibition increases with increasing frequency of stimulation and decreasing membrane potential is consistent with a state-dependent block where the antagonist interacts preferentially with the depolarized, inactivated state of the Ca^{2+} channel (61) and where recovery occurs more rapidly from the repolarized resting state. This situation is potentially analogous to the better investigated frequency-dependent inhibition of voltage-dependent Na^+ channels by local anesthetics and some antiarrhythmic drugs (39, 40, 51).

Whether this postulated mechanism of action is valid for other Ca^{2+}-channel antagonists is not certain. However, the actions of nifedipine on both contractile and inward current processes have been reported to depend less on frequency of stimulation (3, 4, 8) than the actions of verapamil and D 600, possibly suggesting that nifedipine has more "pure" channel-blocking action. More extensive studies are needed to fully define this point.

Few studies are available that define the site of action of diltiazem or the cinnarizine class of compounds, although some pharmacological evidence indicates that cinnarizine has a different spectrum of action than verapamil or nifedipine (95, 99). However, [^{3}H]cinnarizine has been reported to show a component of saturable binding in vascular smooth muscle that is blocked by D 600 (32).

STRUCTURE-ACTIVITY RELATIONSHIPS

The diversity of molecular structures that are Ca^{2+} antagonists is consistent with many modes and sites of action. It is therefore unlikely that a single structure-activity correlation can be described. Within specific series of compounds, however, defined structure-activity correlations are becoming available and are consistent with antagonist action at specific sites rather than action through nonspecific membrane perturbation.

Stereoselectivity of action is an important factor in defining a drug action as specific. Thus the apparently selective action of the (*S*)-(−)-enantiomers of verapamil and D 600 in reducing Ca^{2+} current in cardiac action potentials is very significant (6, 7). Ludwig and Nawrath (63) have reported a similar stereoselectivity in cardiac contractility studies where (−)-D 600 was a more potent negative inotropic agent. Furthermore (−)-D 600 is more potent than (+)-D 600 as an antagonist of responses induced by extracellular Ca^{2+}–dependent K^+ depolarization (Ca^{2+}/K^+) in guinea pig ileal smooth muscle (44), and a similar stereoselectivity for both D 600 and verapamil has been reported for Ca^{2+}/K^+–induced contractions of frog slow skeletal muscle (49).

Mannhold et al. (65) have analyzed a rather limited series of verapamil analogues to correlate structure with negative inotropic potency. Their data (Table 4) reveal a good correlation among activity, the electronic parameter σ, and the steric parameter [molar volume (MV)]

$$\log \frac{1}{ED_{50}} = 0.96\,\sigma + 0.626\mathrm{MV} \qquad (1)$$

Thus activity increases with decreasing electron density of the substituted aromatic ring and also depends on the steric bulk of the substituent, indicating that optimum interaction depends on specific substituent properties. Consistent with this was a lack of correlation between negative inotropic activity and the octanol:water partition coefficient.

Rather more extensive structure-activity correlations are available for the 1,4-dihydropyridines of the nifedipine series. Rodenkirchen et al. (82) have described correlations between the negative inotropic activity and aryl substituent characteristics in a series of 2,6-dimethyl 3, 5-di(carbomethoxy)-4-aryl 1,4-dihydropyridines (Fig. 4). As noted previously with in vivo studies of the vasodilating properties of these compounds (42, 62) and their effects as inhibitors of guinea pig ileal smooth mechanical responses (86), activity lies in the sequence X = ortho ≥ meta ≫ para. Although this dihydropyridine series is more potent in the smooth muscle preparation, possibly due to pharmacokinetic factors, there is a good correlation between activities in cardiac and smooth muscles (Fig. 5), consistent with a fundamentally similar mode of interaction in both preparations.

Since the effects of these substituents on activity seem relatively independent of electronic character (electron-releasing or electron-withdrawing properties), steric factors are likely to be important. This is consistent with the

TABLE 4. *Negative inotropic potency of verapamil analogues*

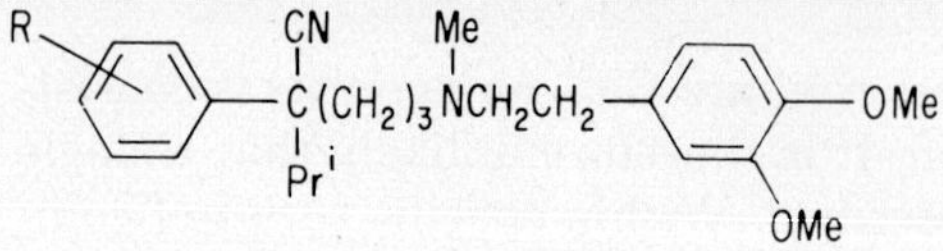

R	ED_{50}, M
3,4,5-$(MeO)_3$	1.2×10^{-6}
4-MeO	1.6×10^{-5}
3,4-Cl_2	7.9×10^{-7}
3-CF_3	2.6×10^{-6}
3-MeO, 4-OH	1.07×10^{-5}
3,4-Me	8.1×10^{-6}
3,4-$(MeO)_2$	3.5×10^{-6}

From Mannhold et al. (65).

good correlation noted (82) between negative inotropic activity and the substituent constant B_1, a measure of minimum width of the substituted derivatives (Fig. 4)

$$\log \frac{1}{EC_{50}} = 5.06 + 0.80B_1 \tag{2}$$

where EC_{50} is mean effective concentration. This relationship suggests that insertion of the ortho substituent into a specific cavity of the receptor site is important in determining optimum binding and activity.

The importance of steric factors is also indicated by considerations of the solid-state structures of nifedipine and related compounds (94). In these compounds the 4-aryl ring is oriented approximately perpendicular to the plane of the 1,4-dihydropyridine ring (Fig. 6) with limited rotational flexibility. Twisting of the 4-aryl ring from this perpendicular orientation causes increased puckering of the 1,4-dihydropyridine ring and a reduction in activity (Fig. 7). These findings are consistent with a rather specific interaction of the nifedipine-type structure with its site of action. It is also clear, however, that there must be elements of considerable molecular flexibility in this interaction because substantial variation is possible (12, 42) in the substituent character at the 3-position of the 1,4-dihydropyridine ring with full maintenance of activity (Table 5).

Although it is clear that Ca^{2+} antagonists, notably verapamil, D 600,

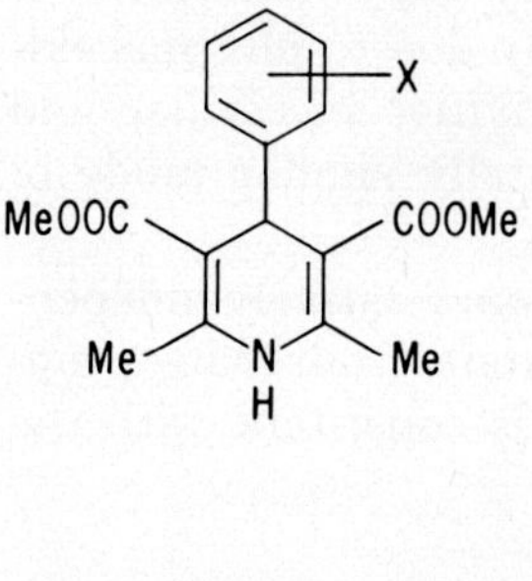

FIG. 4. Structural formula of 2,6-dimethyl 3,5-di(carbomethoxy)-4-aryl 1,4-dihydropyridine.

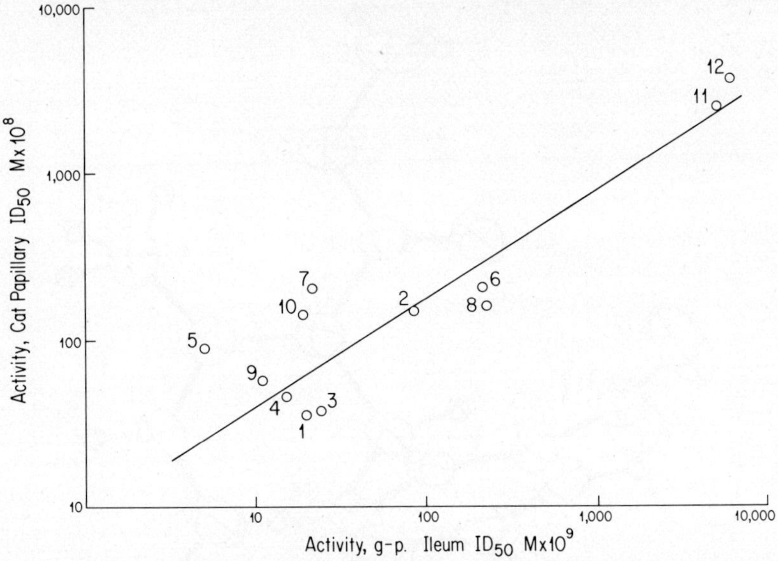

FIG. 5. Correlation between activities of a series of 2,6-dimethyl 3,5-di(carbomethoxy)-4-aryl 1,4-dihydropyridines as negative inotropic agents (82) and as inhibitors in guinea pig ileal longitudinal smooth muscle mechanical responses (86). Substituent position in aryl ring: *1*, 2-NO_2; *2*, 2-MeO; *3*, 2-Me; *4*, 2-Cl; *5*, 2-CF_3; *6*, H; *7*, 3-Cl; *8*, 3-MeO; *9*, 3-NO_2; *10*, 3-N_3; *11*, 4-Me; *12*, 4-Cl.

nifedipine, and diltiazem, inhibit an inward Ca^{2+} current measured either electrophysiologically or mechanically and that this property is certainly a very important determinant of their inhibitory effects on smooth and cardiac muscle, it is also clear that these agents may have significant effects on other ionic and receptor-mediated processes. It is important to determine these effects and their relationship to the Ca^{2+} antagonistic properties of the agents, with respect to both the usefulness of these agents as pharmacological tools (86, 95) and their several clinical uses (79, 91).

Both verapamil and D 600 reduce the tetrodotoxin-sensitive Na^+ current at the high concentrations (10^{-4} M) needed to inhibit the inward Ca^{2+} current in the squid axon (2). There are several indications that a similar effect on Na^+ current can also be exerted in cardiac muscle (84, 89); the stereoselectivity of the action of verapamil and D 600, in which negative inotropic activity is associated with the (−)-enantiomer, is particularly interesting in this respect. Bayer et al. (6) reported that the (+)-enantiomers had a remarkably different activity, namely reduction of the rapid Na^+ current (Table 6). If this stereoselectivity of action does exist for the Na^+ channels, it is consistent with a specific, rather than a nonspecific, mechanism of action. Kass and Tsien (46) have also described multiple effects of D 600 on cardiac plateau currents that indicate that although 10^{-6} M D 600 does not inhibit the pacemaker K^+ current (I_{K2}), it does inhibit the K^+-dependent, slowly activating current, I_x. Verapamil (10^{-6}–5×10^{-6} M) also inhibits K^+ influx and efflux in cardiac Purkinje fibers in both active and quiescent states (77). Some Na^+-dependent systems that

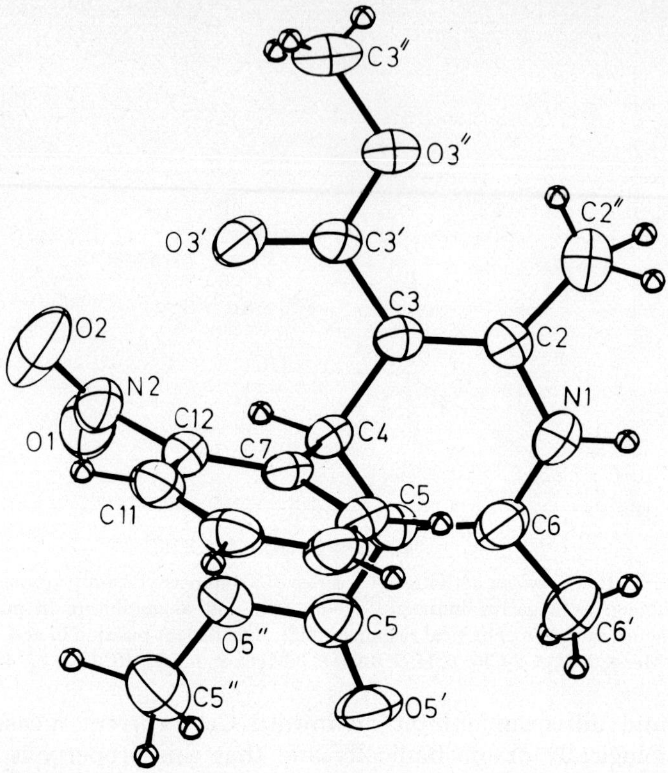

FIG. 6. Solid-state structure of 2,6-dimethyl 3,5-di(carbomethoxy)-4-(2-nitro)phenyl 1,4-dihydropyridine (nifedipine). [From Triggle et al. (94).]

are inhibited by verapamil or D 600 are listed in Table 6. More detailed analysis of such systems is required to determine to what extent the observed inhibition may be secondary to inhibition of Ca^{2+} movements.

Besides effects on ionic processes, several of the Ca^{2+} antagonists inhibit specific ligand binding to membrane receptors, notably muscarinic and α-adrenergic receptors (Table 7). Of interest is the competitive nature of the interactions exhibited, although the K_i (inhibition constant) values are greater than those for Ca^{2+} antagonism. Furthermore, for the action of D 600 in guinea pig ileum, where a direct comparison can be made (Table 2), muscarinic receptor binding is nonstereoselective, whereas Ca^{2+} antagonism is stereoselective.

The contribution of these several "non–Ca^{2+} antagonist" properties to the total pharmacological profile of these agents is uncertain. The preceding data indicate, however, that these antagonists do not have absolute specificity for Ca^{2+} channels and that caution should be used in interpreting results, particularly when relatively high antagonist concentrations are utilized. Thus the use (as reported in the literature: 11, 75, 86, 95) of D 600 or verapamil at concentrations exceeding 10^{-5} M may well indicate action at sites other than Ca^{2+} channels.

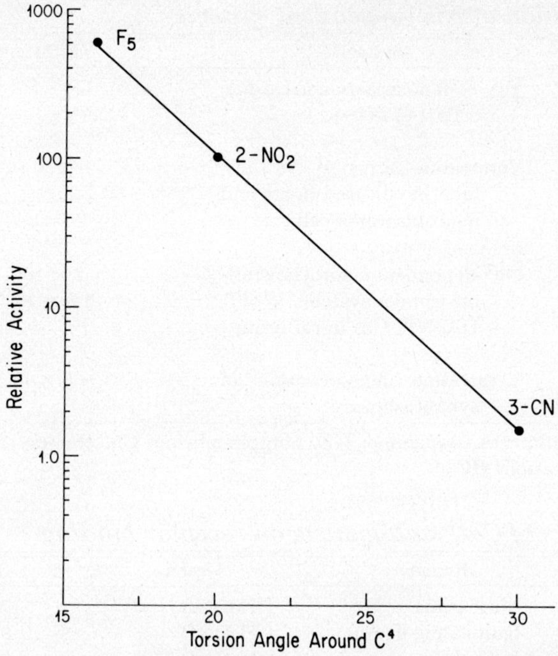

FIG. 7. Correlation between biological activity (inhibition of guinea pig ileal longitudinal smooth muscle mechanical responses) and deviation of planarity of 1,4-dihydropyridine ring in three 1,4-dihydropyridine antagonists. [From Triggle et al. (94).]

TABLE 5. *Coronary vasodilating properties of nifedipine analogues*

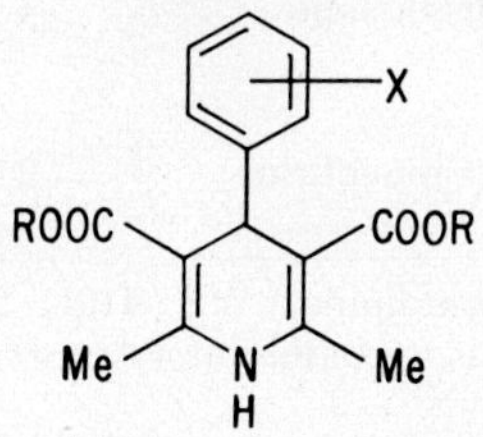

X	R	Relative Activity
2-NO_2	Me (nifedipine)	xxxx
2-NO_2	Et	xxx
2-NO_2	Pr^i	xxx
2-NO_2	$(CH_2)_2$—OMe	xxx
2-NO_2	$CH_2C{\equiv}CH$	xxx
3-NO_2	Me	xx
3-NO_2	Et	xxx
3-NO_2	Pr^i	xxx
3-NO_2	$(CH_2)_2$—OMe	xxx
3-NO_2	$CH_2C{\equiv}CH$	xxx
3-NO_2	$(CH_2)_2OC_3H_7^i$	xxxx

xxxx, Most potent. [From Bossert et al. (12).]

TABLE 6. *Inhibition of Na^+-dependent systems*

Agent	System	K_i, M	Ref.
Verapamil D 600	V_{max} of depolarization in cardiac cells, (+) > (−)		6
D 600	Veratridine-activated Na^+ uptake in cultured heart and neuroblastoma cells	$\sim 3 \times 10^{-6}$	30
Verapamil	Na^+-dependent neurotransmitter uptake systems (5-HT, DA, NE, Ch) in rat brain	2×10^{-6}–1.5×10^{-4}*	68
Verapamil D 600	Veratridine depolarization of synaptosomes	$\sim 5 \times 10^{-5}$*	72

5-HT, serotonin; DA, dopamine; NE, norepinephrine; Ch, choline. *Values are mean inhibitory concentrations (IC_{50}).

TABLE 7. *Effects of Ca^{2+} antagonists on receptor binding*

Agent	Receptor	Ligand	K_i, M	Ref.
(±)-D 600	Muscarinic	[^{3}H]QNB	7×10^{-6}	44
(−)-D 600	(guinea pig ileum)	[^{3}H]QNB	8×10^{-6}	
(+)-D 600		[^{3}H]QNB	7×10^{-6}	
YC-93		[^{3}H]QNB	1.3×10^{-6}	
(±)-D 600	Muscarinic (rat brain)	[^{3}H]QNB	8.8×10^{-6}	22
(±)-D 600	α-Adrenergic (rat brain)	[^{3}H]WB 4101	1.1×10^{-6}	22
(±)-D 600	Opiate (rat brain)	[^{3}H]Naloxone	4×10^{-6}	22
Verapamil	α-Adrenergic (rat liver)	[^{3}H]Epinephrine	2×10^{-5}*	9

QNB, 3-quinuclidinyl benzilate; YC-93, nicardipine; WB 4101, 2,6-(dimethoxyphenoxyethyl)methylamino-1,4-benzodioxan. *Value is mean inhibitory concentration (IC_{50}).

COMPARATIVE PHARMACOLOGY OF CALCIUM-CHANNEL ANTAGONISTS

Tissue and agonist selectivity of Ca^{2+} antagonists is of major pharmacological and therapeutic interest. In a broad sense selectivity depends on the presence and extent of activation of the Ca^{2+} channels sensitive to verapamil, nifedipine, and other members of this putative class of antagonists. Table 3 summarizes some data for smooth muscle and shows that although responses to K^+ depolarization are uniformly sensitive, considerable variation exists for agonist sensitivity. Thus in rabbit aorta contractile responses to norepinephrine are sensitive to D 600 only at significantly higher concentrations, where nonspecific spasmolytic effects are likely operative. This is consistent with the

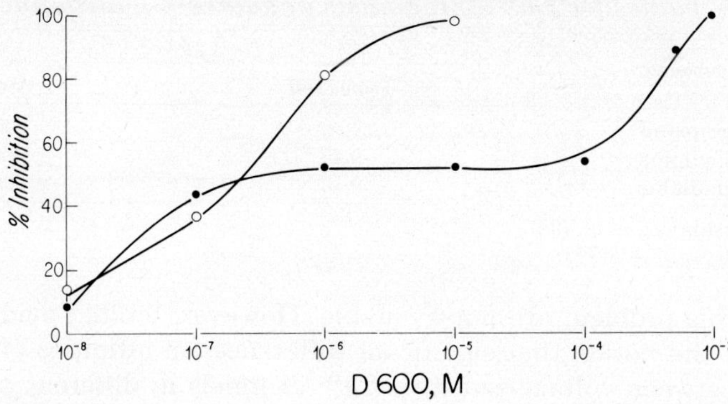

FIG. 8. Inhibition of responses of rat aorta to maximum concentrations of norepinephrine (10^{-4} M; ●) or K^+ (80 mM; ○).

known use of intracellular Ca^{2+} by norepinephrine in this preparation (97). A simple interpretation of such data is that systems highly sensitive to the Ca^{2+} antagonists use potential-dependent Ca^{2+} channels to satisfy excitation-contraction coupling requirements and that insensitive systems use receptor-operated Ca^{2+} channels or intracellular sources. Additional complexity arises, however, from observations that agonists may use more than one source of Ca^{2+}, depending on the agonist, its concentration, and the component of mechanical response being measured. Figure 8 shows the sensitivity of norepinephrine- and K^+-induced responses in rat aorta to D 600. Clearly the response to norepinephrine has at least two mechanisms of Ca^{2+} mobilization, one of which, as judged by D 600 sensitivity, is identical to that activated by K^+ depolarization. Similar observations have been made for other tissues (23, 33). Thus a simple comparison of the mean effective dose (ED_{50}) values for Ca^{2+} antagonist inhibition may be very misleading, since the existence of multiple mechanisms of Ca^{2+} translocation is not indicated.

Of more subtle importance is whether, by analogy with pharmacological receptors, subclasses of Ca^{2+} channels exist. The quantitative data necessary

TABLE 8. *Effects of Ca^{2+} antagonists on stimulus-secretion coupling*

Preparation	Activity	Antagonist	ID_{50}, M	Ref.
Pituitary	ACTH, GH, TSH release			
	induced by K^+	Verapamil	$\sim 10^{-5}$	21
	induced by releasing	Verapamil		
	factor	D 600	$\sim 10^{-5}$	
Pancreatic	Insulin release induced	D 600	$\sim 2 \times 10^{-6}$	
islets	by glucose	Diltiazem	$\sim 10^{-5}$	15, 64
		Nifedipine	$\sim 10^{-6}$	
Rat brain	^{45}Ca uptake evoked by	Verapamil, D 600	$\sim 5 \times 10^{-5}$	72
synaptosomes	K^+ depolarization	Nifedipine	$\gg 3 \times 10^{-5}$	

ACTH, adrenocorticotropin; GH, growth hormone; TSH, thyroid stimulating hormone.

TABLE 9. *pA_2 values for Ca^{2+} antagonists in guinea pig atria and taenia coli*

Compound	pA_2	
	Taenia coli	Atria
Verapamil	7.36	6.67
Nifedipine	9.43	8.24
Niludipine	9.16	6.32

From Hashimoto et al. (38).

to resolve this problem are not available. However, limited evidence is consistent with the notion that significant differences in affinities of antagonists may exist between voltage-sensitive Ca^{2+} channels in different preparations. Thus voltage-sensitive Ca^{2+} channels in *Paramecium* are insensitive to D 600 (18). A number of stimulus-secretion systems show relatively low sensitivity to verapamil, nifedipine, and diltiazem (Table 8). Uptake of ^{45}Ca evoked by K^+ depolarization of rat brain synaptosomes is not very sensitive to verapamil or D 600 and, interestingly, is insensitive to nifedipine but sensitive to the inorganic antagonist Mn^{2+} (72).

Comparison of pA_2 values of verapamil, nifedipine, and niludipine for Ca^{2+} antagonism in guinea pig atria and taenia coli reveals the considerable smooth muscle selectivity of niludipine (Table 9). Similarly nimodipine has been reported to select for cerebral vascular tissue (50). These observations are especially interesting because the structural differences between nifedipine, niludipine, and nimodipine are small (Fig. 9), indicating that these side chain substitutions may be used to probe subtle variations in the proposed binding site for these Ca^{2+} antagonists.

Important differences in the selectivity of action also seem apparent with cinnarizine, flunarizine, and related compounds (31, 98, 99). These agents are more effective on smooth muscle than they are on cardiac tissue (Table 10) and differ significantly from verapamil in this respect. Furthermore significant differences have been noted among different vascular tissues with respect to

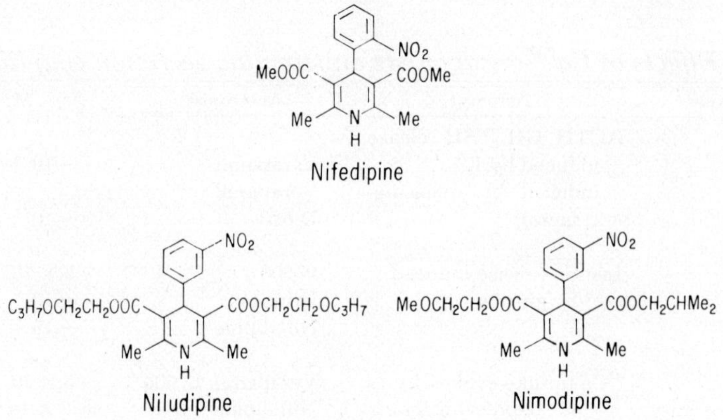

FIG. 9. Structural formulas of nifedipine, niludipine, and nimodipine.

TABLE 10. *Relative activities of cinnarizine and flunarizine in cardiac and vascular tissues*

Antagonist	Activity Ratio	
	Rat caudal artery: cat papillary	Rat caudal artery: rat portal vein
Flunarizine	> 200	> 200
Cinnarizine	> 70	9
Verapamil	3.3	0.4

From Van Hueten (99).

inhibitory activity against either agonists or K^+ depolarization. Sensitivity to inhibition tends to increase with decreasing vessel size.

Related to this is the question of whether changes in Ca^{2+} channels may occur during development and pathological conditions. Galper and Catterall (29) have described changes in the sensitivity of embryonic chick hearts in ovo to D 600 and tetrodotoxin such that during development a change occurs in the slow Ca^{2+} and Na^+ channels that reduces their sensitivity to D 600. Vascular tissue from spontaneously hypertensive rats shows marked and well-known differences in reactivity compared to normotensive control tissue, and suggestions have been made that these differences at least partly relate to differences in Ca^{2+} binding (59, 101). Several reports indicate differences in sensitivity to verapamil, D 600, and nifedipine in the normotensive and hypertensive states (76, 92), and although more quantitative data are needed, they suggest alterations in the antagonist-sensitive Ca^{2+}-mobilization machinery.

SUMMARY

A remarkable range of chemical structures serve as Ca^{2+} antagonists. It is improbable that they share a common mechanism of action. However, some agents, including verapamil, D 600, nifedipine, diltiazem, and cinnarizine, have been classified as Ca^{2+}-channel antagonists that inhibit that class of Ca^{2+} channels activated by membrane potential. Their mechanisms of action at this site remain to be defined, however. It is clear that they also can exert an inhibitory effect on other ionic processes and on ligand-receptor binding processes. It is likely that these compounds are not uniformly effective against all voltage-sensitive Ca^{2+} channels. This selectivity arises from varying extents of activation of these channels in a given response and probably also because these antagonists have different affinities for different subclasses of Ca^{2+} channels. Continued investigation of these compounds will further elucidate their usefulness as pharmacological tools and may indicate new therapeutic specificities.

REFERENCES

1. BAKER, P. F. Transport and metabolism of calcium ions in nerve. *Prog. Biophys. Mol. Biol.* 24: 177–223, 1972.
2. BAKER, P. F., AND H. G. GLITSCH. Voltage-dependent changes in the permeability of nerve membranes to cal-

cium and other divalent cations. *Philos. Trans. R. Soc. London Ser. B* 270: 389–409, 1975.

3. BAUMANN, K. On the action of nifedipine under conditions of variable stimulation patterns and $[Ca^{2+}]_o$ in guinea-pig atrium. *Naunyn-Schmiedebergs Arch. Pharmacol.* 294: 161–168, 1976.
4. BAYER, R., AND T. EHARA. Comparative studies on calcium antagonists. *Prog. Pharmacol.* 2: 31–37, 1978.
5. BAYER, R., R. HENNEKES, R. KAUFMANN, AND R. MANNHOLD. Inotropic and electrophysiological actions of verapamil and D 600 in mammalian myocardium. I. Pattern of inotropic effects of the racemic compounds. *Naunyn-Schmiedebergs Arch. Pharmacol.* 290: 49–68, 1975.
6. BAYER, R., D. KLAUSCHE, R. KAUFMANN, AND R. MANNHOLD. Inotropic and electrophysiological actions of verapamil and D 600 in mammalian myocardium. III. Effects of the optical isomers on transmembrane action potentials. *Naunyn-Schmiedebergs Arch. Pharmacol.* 290: 81–97, 1975.
7. BAYER, R., R. KAUFMANN, AND R. MANNHOLD. Inotropic and electrophysiological actions of verapamil and D 600 in mammalian myocardium. II. Pattern of inotropic effects of the optical isomers. *Naunyn-Schmiedebergs Arch. Pharmacol.* 290: 69–80, 1975.
8. BAYER, R., R. RODENKIRCHEN, R. KAUFMANN, J. H. LEE, AND R. HENNEKES. The effects of nifedipine on contraction and monophasic action potentials of isolated cat myocardium. *Naunyn-Schmiedebergs Arch. Pharmacol.* 30: 29–37, 1977.
9. BLACKMORE, P. F., M. F. EL-REFAI, AND J. H. EXTON. α-Adrenergic blockade and inhibition of A 23187 mediated Ca^{2+} uptake by the calcium antagonist verapamil in rat liver cells. *Mol. Pharmacol.* 15: 598–606, 1979.
10. BLAUSTEIN, M. P. The interrelationship between sodium and calcium fluxes across cell membranes. *Rev. Physiol. Pharmacol. Biochem.* 70: 33–82, 1974.
11. BOLTON, T. B. Mechanisms of action of transmitters and other substances on smooth muscle. *Physiol. Rev.* 59: 606–718, 1979.
12. BOSSERT, F., H. HORSTMANN, H. MEYER, AND W. VATER. Einfluss der esterfunktion auf die vasodilatierenden eigenschaften von 1,4-dihydro-2,6-dimethyl-4-nitrophenyl-pyridine-3,5-dicarbonsaureestern. *Arzneim. Forsch.* 29: 226–229, 1979.
13. CARAFOLI, E., AND M. CROMPTON. The regulation of intracellular calcium. *Curr. Top. Membr. Transp.* 10: 151–216, 1978.
14. CASTEELS, R. Electro- and pharmacomechanical coupling in vascular smooth muscle. *Chest* 78, Suppl.: 150–156, 1980.
15. CONN, P. M., J. MARIAN, M. McMILLAN, AND D. ROGERS. Evidence for calcium mediation of gonadotrophin releasing hormone action in the pituitary. *Cell Calcium* 1: 7–20, 1980.
16. CRANKSHAW, D. J., R. A. JANIS, AND E. E. DANIEL. The effects of Ca^{2+} antagonists on Ca^{2+} accumulation by subcellular fractions of rat myometrium. *Can. J. Physiol. Pharmacol.* 55: 1028–1032, 1977.
17. DANIEL, E. E., D. J. CRANKSHAW, AND C.-Y. KWAN. Intracellular sources of calcium for activation of smooth muscle. In: *Topics in Autonomic Pharmacology,* edited by S. Kalsner. Baltimore: Urban and Schwarzenburg, 1979, vol. 1, p. 421–484.
18. ECKERT, R., AND P. BREHM. Ionic mechanisms of excitation in Paramecium. *Annu. Rev. Biophys. Bioeng.* 8: 353–384, 1979.
19. EHARA, T., AND R. KAUFMANN. The voltage- and time-dependence effects of (−)-verapamil on the slow inward current in isolated cat ventricular myocardium. *J. Pharmacol. Exp. Ther.* 207: 49–55, 1978.
20. ENTMAN, M. L., J. C. ALLEN, E. P. BORNET, P. C. GILLETTE, E. T. WALLICK, AND A. SCHWARTZ. Mechanisms of calcium accumulation and transport in cardiac relaxing system (sarcoplasmic reticulum membrane): effects of verapamil, D 600, X 537A and A 23187. *J. Mol. Cell. Cardiol.* 4: 681–687, 1972.
21. ETO, S., J. WOOD, M. HUTCHINS, AND N. FLEISCHER. Pituitary $^{45}Ca^{++}$ uptake and release of ACTH, GH and TSH: effect of verapamil. *Am. J. Physiol.* 226: 1315–1320, 1974.
22. FAIRHURST, A. S., M. L. WHITTAKER, AND F. J. EHLERT. Interactions of D 600 (methoxyverapamil) and local anesthetics with rat brain α-adrenergic and muscarinic receptors. *Biochem. Pharmacol.* 29: 155–162, 1980.
23. FARLEY, J. M., AND P. R. MILES. The sources of calcium for acetylcholine-induced contractions of dog tracheal smooth muscle. *J. Pharmacol. Exp. Ther.* 207: 340–346, 1978.
24. FLECKENSTEIN, A. Specific pharmacology of calcium in myocardium, cardiac pacemakers and vascular smooth muscle. *Annu. Rev. Pharmacol. Toxicol.* 17: 149–166, 1977.
25. FLECKENSTEIN, A. Specific inhibitors and promoters of calcium action in the excitation-contraction coupling of heart muscle and their role in the prevention or production of myocardial lesions. In: *Calcium and the Heart,* edited by P. Harris and L. Opie. London: Academic, 1971, p. 135–188.
26. FLECKENSTEIN, A., K. NAKAYAMA, G. FLECKENSTEIN-GRUN, AND Y. K. BYON. Interactions of H ions, Ca-antagonistic drugs and cardiac glycosides with excitation-contraction coupling of vascular smooth muscle. In: *Ionic Actions on Vascular Smooth Muscle,* edited by E. Betz. Berlin: Springer-Verlag, 1976, p. 117–123.
27. FREY, M., AND J. JANKE. The effect of organic Ca-antagonists (verapamil, prenylamine) on the calcium transport system in isolated mitochondria of rat cardiac muscle. *Pfluegers Arch.* 356: R26, 1975.
28. FUKUDA J., AND K. KAWA. Permeation of manganese, cadmium, zinc and beryllium through Ca^{2+}-channels of an insect muscle membrane. *Science* 196: 309–311, 1977.
29. GALPER, J. B., AND W. A. CATTERALL. Developmental changes in the sensitivity of embryonic heart cells to tetrodotoxin and D 600. *Dev. Biol.* 65: 216–227, 1978.
30. GALPER, J. B., AND W. A. CATTERALL. Inhibition of sodium channels by D 600. *Mol. Pharmacol.* 15: 174–178, 1979.
31. GODFRAIND, T., A. KABA, AND P. POLSTER. Differences in sensitivity of arterial smooth muscles to inhibition of their contractile response to depolarization by potassium. *Arch. Int. Pharmacodyn. Ther.* 172: 235–239, 1968.
32. GODFRAIND T., AND N. MOREL. An upper limit to the number of calcium channels in smooth muscle as estimated with cinnarizine. *Eur. J. Pharmacol.* 41: 245–246, 1977.
33. GOLENHOFEN, K. Theory of P and T systems for calcium activation in smooth muscle. In: *Physiology of Smooth Muscle,* edited by E. Bülbring and M. F. Shuba. New York: Raven, 1976, p. 197–202.
34. GORLITZ, B. D., J. WAGNER, AND H. SCHÜMANN. Functional antagonism between calcium antagonists and noradrenaline in isolated guinea-pig atria. *Naunyn-Schmiedebergs Arch. Pharmacol.* 238: 311–317, 1973.
35. HAGIWARA, S. Ca-dependent action potential. In: *Dynamic Properties of Lipid Membranes and Bilayers,* edited by G. Eisenman. New York: Dekker, 1975, p. 359–381.
36. HARRIS, A., V. C. SWAMY, D. J. TRIGGLE, AND D. H.

WATERS. Responses to noradrenaline of portal vein strips from normotensive and spontaneously hypertensive rats. *J. Autonom. Pharmacol.* 1: 61–66, 1980.

37. HASHIMOTO, K., Y. NAKAGAWA, H. NABATA, AND S. IMAI. In vitro analysis of Ca-antagonistic effects of prenylamine as mechanisms for its cardiac actions. *Arch. Int. Pharmacodyn. Ther.* 231: 212–221, 1978.
38. HASHIMOTO, K., K. TAKEDA, Y. KATANO, Y. NAKAGAWA, T. TSUKADA, T. HASHIMOTO, N. SHIMAMOTO, K. SAKAI, T. OTORRII, AND S. IMAI. Effects of Niludipine (Bay a 7168) on the cardiovascular system. With a note on its calcium-antagonistic effects. *Arzneim. Forsch.* 29: 1368–1373, 1979.
39. HILLE, B. Local anesthetics: hydrophilic and hydrophobic pathways for the drug-receptor reaction. *J. Gen. Physiol.* 69: 497–515, 1977.
40. HONDEGHEM, L. M., AND B. G. KATZUNG. Time- and voltage-dependent interactions of antiarrhythmic drugs with cardiac sodium channels. *Biochim. Biophys. Acta* 472: 373–398, 1977.
41. IMAI, S. Anti-anginal agents—revival of interests in vasodilators. *Trends Pharmacol. Sci.* 2: 87–89, 1979.
42. IWANAMI, M., T. SHIBANUMA, M. FUJIMOTO, R. KAWAI, K. TAMAZAWA, T. TAKENAKA, K. TAKAHASHI, AND M. MURAKAMI. Synthesis of new water-soluble dihydropyridine vasodilators. *Chem. Pharm. Bull.* 27: 1426–1440, 1979.
43. JETLEY, M., AND A. H. WESTON. Some effects of sodium nitroprusside, methoxyverapamil and nifedipine on rat portal vein. *Br. J. Pharmacol.* 68: 311–320, 1980.
44. JIM, K. *Characterization of the Processes of Ca^{2+}-Utilization, Desensitization and Muscarinic Receptor Activation in Guinea-Pig Ileal Longitudinal Smooth Muscle* (PhD Thesis). Buffalo: State Univ. of New York, 1980.
45. KARAKI, H., AND G. B. WEISS. Alterations in high and low affinity binding of ^{45}Ca in rabbit aortic smooth muscle by norepinephrine and potassium after exposure to lanthanum and low temperature. *J. Pharmacol. Exp. Ther.* 211: 86–92, 1979.
46. KASS, R. S., AND R. W. TSIEN. Multiple effects of calcium antagonists on plateau currents in cardiac Purkinje fibers. *J. Gen. Physiol.* 66: 169–192, 1975.
47. KAUFMANN, R. Calcium movements controlling cardiac contractility: cellular control systems and sites of action. *Prog. Pharmacol.* 2: 25–37, 1978.
48. KAUFMANN, R., R. BAYER, T. FÜRNISS, H. KRAUSE, AND H. TRITTHART. Calcium-movement controlling cardiac contractility II. Analog computation of cardiac excitation-contraction coupling on the basis of calcium kinetics in a multi-compartment model. *J. Mol. Cell. Cardiol.* 6: 543–559, 1974.
49. KAUMANN, A. J., AND O. D. UCHITEL. Reversible inhibition of potassium contractures by optical isomers of verapamil and D 600 on slow muscle fibres of the frog. *Naunyn-Schmiedebergs Arch. Pharmacol.* 229: 21–27, 1976.
50. KAZDA, S., F. HOFFMEISTER, B. GARTHOFF, AND R. TOWART. Prevention of the postischaemic impaired reperfusion of the brain by nimodipine (Bay e 9736). *Acta Neurol. Scand. Suppl.* 72: 302–303, 1979.
51. KENDIS, J. J., K. R. COURTNEY, AND E. N. COHEN. Anesthetics: molecular correlates of voltage- and frequency-dependent sodium channel block in nerve. *J. Pharmacol. Exp. Ther.* 210: 446–452, 1979.
52. KOHLHARDT, M., B. BAUER, H. KRAUSE, AND A. FLECKENSTEIN. Differentiation of the transmembrane Na and Ca channels in mammalian cardiac fibres by the use of specific inhibitors. *Pfluegers Arch.* 335: 309–322, 1972.
53. KOHLHARDT, M., B. BAUER, H. KRAUSE, AND A. FLECKENSTEIN. Selective inhibition of the transmembrane Ca conductivity of mammalian myocardial fibers by Ni, Co and Mn ions. *Pfluegers Arch.* 338: 115–124, 1973.
54. KOHLHARDT, M., A. HERDEY, AND M. KÜBLER. Interchangeability of Ca ions and Sr ions as charge carriers of the slow inward current in mammalian myocardial fibres. *Pfluegers Arch.* 344: 149–158, 1973.
55. KOHLHARDT, M., AND Z. MNICH. Studies on the inhibitory effect of verapamil on the slow inward current in mammalian ventricular myocardium. *J. Mol. Cell. Cardiol.* 10: 1037–1052, 1978.
56. KOSTYUK, P. G. Calcium ionic channels in electrically excitable membranes. *Neuroscience* 5: 945–959, 1980.
57. KRETSINGER, R. H. The informational role of calcium in the cytoplasm. *Adv. Cyclic Nucleotide Res.* 11: 1–26, 1979.
58. KREYE, V. A. W., G. D. BARON, J. B. LÜTH, AND H. SCHMIDT-GAYK. Mode of action of sodium nitroprusside on vascular smooth muscle. *Naunyn-Schmiedebergs Arch. Pharmacol.* 288: 381–402, 1975.
59. KWAN, C.-Y., L. BELBECK, AND E. E. DANIEL. Abnormal biochemistry of vascular smooth muscle plasma membrane as an important factor in the initiation and maintenance of hypertension in rats. *Blood Vessels* 16: 259–268, 1979.
60. LEVIN, R. M., AND B. WEISS. Selective binding of antipsychotics and other psychoactive agents to the calcium-dependent activator of cyclic nucleotide phosphodiesterase. *J. Pharmacol. Exp. Ther.* 208: 454–459, 1979.
61. LINDEN, J., AND G. BROOKER. The influence of resting membrane potential on the effect of verapamil on atria. *J. Mol. Cell. Cardiol.* 12: 325–331, 1980
62. LOEV, B., M. M. GOODMAN, K. M. SNADER, R. TEDESCHI, AND E. MACKO. "Hantzsch-type" dihydropyridine hypotensive agents. *J. Med. Chem.* 17: 956–965, 1974.
63. LUDWIG, C., AND H. NAWRATH. Effects of D 600 and its optical isomers on force of contraction in cat papillary muscles and guinea-pig auricles. *Br. J. Pharmacol.* 59: 411–417, 1977.
64. MALAISSE, W., G. DEVIS, D. G. PIPELEERS, AND G. SOMERS. Calcium antagonists and islet function. IV. Effect of D 600. *Diabetologia* 12: 77–81, 1976.
65. MANNHOLD, R., R. STEINER, W. HAAS, AND R. KAUFMANN. Investigations on the structure-activity relationships of verapamil. *Naunyn-Schmiedebergs Arch. Pharmacol.* 302: 217–226, 1978.
66. MAYER, C. J., C. VAN BREEMEN, AND R. C. CASTEELS. The action of lanthanum and D 600 on the calcium exchange in the smooth muscle cells of the guinea-pig taenia coli. *Pfluegers Arch.* 337: 333–350, 1972.
67. McDONALD, T. F., D. PELZER, AND W. TRAUTWEIN. On the mechanism of slow calcium channel block in heart. *Pfluegers Arch.* 385: 175–179, 1980.
68. McGEE, R., AND J. E. SCHNEIDER. Inhibition of high affinity synaptosomal uptake systems by verapamil. *Mol. Pharmacol.* 16: 877–885, 1979.
69. McLAUGHLIN, S. Electrostatic potentials at membrane-solution interfaces. *Curr. Top. Membr. Transp.* 9: 71–144, 1977.
70. MEANS, A. R., AND J. R. DEDMAN. Calmodulin—an intracellular calcium receptor. *Nature London* 285: 73–77, 1980.
71. MIKKELSEN, R. B. Lanthanides as calcium probes in biomembranes. In: *Biological Membranes*, edited by D. Chapman and D. F. H. Wallach. London: Academic, 1976, vol. 3, p. 133–190.
72. NACHSEN, D. A., AND M. P. BLAUSTEIN. The effects of some organic "calcium antagonists" on presynaptic

nerve terminals. *Mol. Pharmacol.* 16: 579–586, 1979.

73. NAGAO, T., T. IKEO, AND M. SATO. Influence of calcium ions on responses to diltiazem in coronary arteries. *Jpn. J. Pharmacol.* 27: 330–332, 1977.
74. NAWRATH, H. Cardiac effects of verapamil and D 600. *Prog. Pharmacol.* 2: 55–68, 1979.
75. PAROD, R. J., AND J. W. PUTNEY, JR. Stimulus-permeability coupling in rat lacrimal gland. *Am. J. Physiol.* 239 (*Gastrointest. Liver Physiol.* 2): G106–G113, 1980.
76. PEDERSEN, O. L., E. MIKKELSEN, AND K.-E. ANDERSSON. Effects of extracellular calcium on potassium and noradrenaline induced contractions in the aorta of spontaneously hypertensive rats—increased sensitivity to nifedipine. *Acta Pharmacol. Toxicol.* 43: 137–144, 1978.
77. POSNER, P., B. L. MILLER, AND C. R. LAMBERT. The effects of verapamil on potassium fluxes in canine cardiac Purkinje fibres. *Eur. J. Pharmacol.* 34: 369–372, 1975.
78. PUTNEY, J. W., JR. Stimulus-permeability coupling: role of calcium in the receptor regulation of membrane permeability. *Pharmacol. Rev.* 30: 209–245, 1978.
79. RAHWAN, R. G., M. F. PIASCIK, AND D. T. WITIAK. The role of calcium antagonism in the therapeutic action of drugs. *Can. J. Physiol. Pharmacol.* 57: 443–460, 1979.
80. REIMER, J., F. DÖRFLER, C.-J. MAYER, AND G. ULBRECHT. Calcium-antagonistic effects on the spontaneous activity of guinea-pig taenia coli. *Pfluegers Arch.* 351: 241–258, 1974.
81. REUTER, H. Divalent cations as charge carriers in excitable membranes. *Prog. Biophys. Mol. Biol.* 26: 1–43, 1973.
82. RODENKIRCHEN, R., R. BAYER, R. STEINER, F. BOSSERT, H. MEYER, AND E. MÖLLER. Structure-activity studies on nifedipine in isolated cardiac muscle. *Naunyn-Schmiedebergs Arch. Pharmacol.* 310: 69–78, 1978.
83. ROJAS, E., AND M. LUXORO. Coupling between ionic conductance changes and contractions in barnacle muscle fibres under membrane potential control. In: *Actualités Neurophysiologiques*, edited by A. M. Monnier. Paris: Masson, 1974, p. 159–169.
84. ROSEN, M., J. P. ILVENTO, H. GELBAND, AND C. MERKER. Effects of verapamil on electrophysiologic properties of canine cardiac Purkinje fibers. *J. Pharmacol. Exp. Ther.* 189: 414–422, 1974.
85. ROSENBERGER, L. B., M. K. TICKU, AND D. J. TRIGGLE. The effects of Ca^{2+} antagonists on mechanical responses and Ca^{2+} movements in guinea-pig ileal longitudinal smooth muscle. *Can. J. Physiol. Pharmacol.* 57: 333–347, 1979.
86. ROSENBERGER, L. B., AND D. J. TRIGGLE. Calcium, calcium translocation and specific calcium antagonists. In: *Calcium in Drug Action*, edited by G. B. Weiss. New York: Plenum, 1978, p. 3.
87. SCHATZMANN, H. J. Active calcium transport and Ca^{2+}-activated ATPase in human red cells. *Curr. Top. Membr. Transp.* 6: 126–168, 1975.
88. SCHÜMANN, H. J., B. D. GÖRLITZ, AND J. WAGNER. Influence of papaverine, D 600 and nifedipine on the effects of noradrenaline and calcium on the isolated aorta and mesenteric artery of the rabbit. *Naunyn-Schmiedebergs Arch. Pharmacol.* 289: 409–418, 1975.
89. SHIGENOBU, K., J. A. SCHNEIDER, AND N. SPERELAKIS. Verapamil blockade of slow Na^{+} and Ca^{2+} responses in myocardial cells. *J. Pharmacol. Exp. Ther.* 190: 280–288, 1974.
90. SHIMIZU, K., T. OHTA, AND N. TODA. Evidence for greater susceptibility of isolated dog cerebral arteries to Ca antagonists than peripheral arteries. *Stroke* 11: 261–266, 1980.
91. SINGH, B. N., G. ELLRODT, AND C. T. PETER. Verapamil: a review of its pharmacological properties and therapeutic uses. *Drugs* 15: 169–197, 1978.
92. SUTTER, M. C., AND C. C. PANG. Differential effects of D 600 on contractile response of aorta and portal vein from spontaneously hypertensive rats (SHR) (abstr.). *Pharmacologist* 21: 312, 1979.
93. TRIGGLE, C. R. Site of action of D 600 in guinea-pig ileal longitudinal smooth muscle (abstr.). *J. Physiol. London* 305: 82P, 1980.
94. TRIGGLE, A. M., E. SHEFTER, AND D. J. TRIGGLE. The crystal structures of calcium channel antagonists: 2,6-dimethyl-3,5-dicarbomethoxy-4-(2-nitro(and 3-cyano, 4-dimethylamino and 2,3,4,5,6-pentafluoro)) phenyl-1,4-dihydropyridine. *J. Med. Chem.* 23: 1442–1445, 1980.
95. TRIGGLE, D. J., AND V. C. SWAMY. Pharmacology of agents that affect calcium: agonists and antagonists. *Chest* 78, Suppl.: 174–180, 1980.
96. TRITTHART, H., R. VOLKMANN, R. WEISS, AND A. FLECKENSTEIN. Calcium-mediated action potentials in mammalian myocardium. *Naunyn-Schmiedebergs Arch. Pharmacol.* 280: 239–252, 1973.
97. VAN BREEMEN, C., P. AARONSON, R. LOUTZENHISER, AND K. MEISHERI. Ca^{2+} movements in smooth muscle. *Chest* 78, Suppl.: 157–165, 1980.
98. VAN NUETEN, J. M., J. VAN BEEK, and P. A. J. JANSSEN. Effects of flunarizine on calcium-induced responses of peripheral vascular smooth muscle. *Arch. Int. Pharmacodyn. Ther.* 232: 42–52, 1978.
99. VAN NUETEN, J. M., J. VAN BEEK, and P. A. J. JANSSEN. The vascular effects of flunarizine as compared with those of other clinically used vasoactive substances. *Arzneim. Forsch.* 28: 2082–2087, 1978.
100. VERLOOP, A., W. HOOGENSTRAATEN, AND J. TIPKER. Development and application of new steric substituent parameters in drug design. In: *Drug Design*, edited by E. J. Ariëns. London: Academic, 1976, vol. 7, p. 165–208.
101. WEBB, R. C., AND R. C. BHALLA. Altered calcium sequestration by subcellular fractions of vascular smooth muscle from spontaneously hypertensive rats. *J. Mol. Cell. Cardiol.* 8: 651–661, 1976.
102. WOLFF, D. J., AND C. O. BROSTROM. Properties and functions of the calcium-dependent regulator protein. *Adv. Cyclic Nucleotide Res.* 11: 27–88, 1979.

2

Calcium Channels in the Barnacle Muscle Membrane

N. LAKSHMINARAYANAIAH

Department of Pharmacology, Thomas Jefferson University Medical College, Philadelphia, Pennsylvania

IT IS RECOGNIZED THAT THERE ARE channels or pores in excitable membranes (3, 4, 31, 36, 37, 57). Some of these polar, water-filled channels are pathways for the selective flow of K^+ (3–5, 34), some allow passage of Na^+ (32, 33, 35, 37), some carry Ca^{2+} (21, 22), and some appear to carry both Na^+ and Ca^{2+} (8, 9, 22, 52). The characteristics of Na^+ and K^+ channels are better understood than the properties of the slow Ca^{2+} channel (3, 5, 31, 35–37). The study of the properties of the slow channel in heart muscle has attracted considerable attention (17, 59, 60, 76). In recent years other biological preparations in which slow Ca^{2+} channels exist have been used in voltage-clamp experiments to study the slow-channel characteristics (2, 22, 49). One such preparation is the giant single barnacle muscle fiber, whose properties were described for the first time in 1963 by Hoyle and Smyth (40). Since then Hagiwara (21, 22) and others (43, 46–48, 55) have used it in several studies to characterize the Ca^{2+} channel; their findings are discussed below. Special emphasis is placed on those studies (some preliminary) in which agents (ions and/or drugs) that act as Ca^{2+} antagonists are used as probes to monitor the structural characteristics of the Ca^{2+} channel in the barnacle muscle membrane.

CALCIUM SPIKE

Under normal conditions an all-or-none action potential is not elicited from the barnacle muscle fiber. However, when the intracellular Ca^{2+} concentration is reduced below 8×10^{-8} M by injecting or internally perfusing Ca^{2+}-binding agents such as potassium fluoride, potassium citrate, ethylenediaminetetraacetic acid (EDTA), or ethylene glycol-bis(β-aminoethylether)-*N*,*N*′-tetraacetic acid (EGTA) (23, 26, 28, 55), an all-or-none action potential (Ca^{2+} spike) can be obtained.

Hagiwara and Takahashi (29) studied the characteristics of the Ca^{2+} spike and found that its rate of rise and amplitude increased as the Ca^{2+} concentration in the external saline was raised. Sodium had little effect on these parameters. Injection of Ca^{2+} buffers to control intracellular Ca^{2+} concentration showed (28) that when $[Ca^{2+}]_i$ was about 10^{-8} M, single spikes were obtained on stimulation. As $[Ca^{2+}]_i$ increased to about 10^{-7} M, the membrane response became oscillatory without all-or-none spikes. The oscillatory potential changed to repetitive all-or-none spikes when Ca^{2+} in the saline increased from 20 to about 40 mM. In addition the nature of the response (single spike, multiple spikes, or oscillatory) was governed by the frequency of stimulation (55). Some of these responses are unpredictable. Figure 1 shows some of the

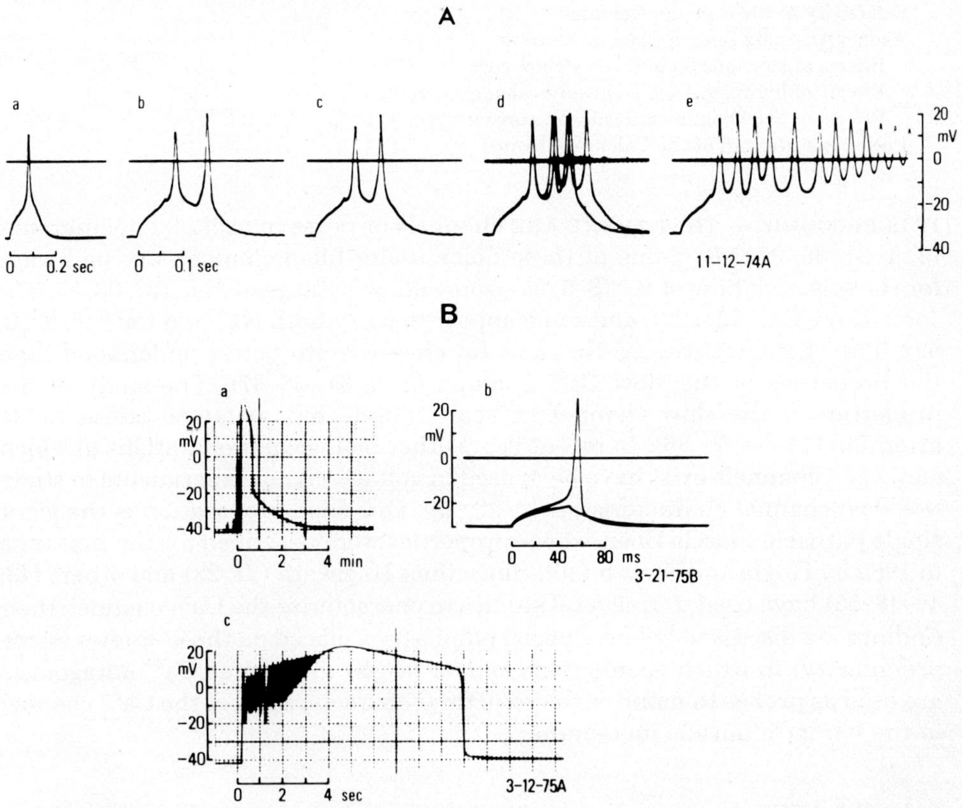

FIG. 1. Nature of action potentials elicited from a barnacle muscle fiber internally perfused with a solution containing (in mM) potassium acetate (180), glycine (740–800), tris(hydroxymethyl)aminomethane (Tris)-Cl (5), and EGTA (3); pH = 7.3. *A*: sequence of action potentials (*a–e*) recorded when muscle fiber was subjected to 5 successive stimulations, 1 every 3.3 s. Fiber was immersed in ASW containing 10 mM Ca^{2+} and 20 mM Mg^{2+}. *B*: *a*: action potential lasting several minutes; *b*: single spike observed in high-Ca^{2+} (60 mM) and zero-Mg^{2+} ASW; *c*: long-lasting action potential obtained just after solution change from high-Ca^{2+} ASW to ASW containing zero Ca^{2+} and 60 mM Mg^{2+}. [From Murayama and Lakshminarayanaiah (55).]

responses elicited in an internally perfused barnacle muscle fiber (*Balanus nubilus*). The muscle fiber subjected to a threshold stimulus lasting approximately 50 ms responded with a single spike (Fig. 1*A*, *a*), and the overshoot was about 15 mV. With successive pulses of the same magnitude applied every 3.3 s, the initial response of a single spike changed to multiple-spike responses (Fig. 1*A*, *b* and *c*), repetitive discharge (Fig. 1*A*, *d*), and finally a plateau lasting from a few seconds (Fig. 1*A*, *e*) to several minutes (Fig. 1*B*, *a*). When the external solution was replaced by a high-Ca^{2+} ASW (artificial seawater, in mM: Na^+, 430; K^+, 10; Mg^{2+}, 0; Ca^{2+}, 60; Cl^-, 565; tris(hydroxymethyl)-aminomethane, 10; sucrose, 75; osmolality = 1,000 and pH = 7.2], the muscle fiber stabilized and upon stimulation gave a single spike (Fig. 1*B*, *b*). When this solution was replaced by ASW containing no added Ca^{2+} and 60 mM Mg^{2+}, however, the fiber gave a long-lasting action potential for the first pulse applied a few minutes after the solution change. This is probably due to the inward movement of residual Ca^{2+}, since subsequent stimulations gave no action potential of any sort. Thus the nature of the membrane response to stimulation depended on $[Ca^{2+}]_i$, which in turn was regulated by the flow of Ca^{2+} from the outside during stimulation. Generally it was observed that when the Ca^{2+} gradient across the membrane was high, stimulation always generated a single spike.

When $[Ca^{2+}]_i$ increased to about 5×10^{-7} M, neither spikes nor oscillatory potentials were observed. As $[Ca^{2+}]_i$ increased to 8×10^{-7} M and above, contractions appeared (28).

Tetrodotoxin (TTX) blocks Na^+ channels at extremely low concentrations (100–150 nM). In contrast, 20–30 μM TTX had no effect on the Ca^{2+} spike (27). Similarly procaine (which eliminates the Na^+ spike at a concentration of 5–10 mM) had no effect on Ca^{2+} permeability but enhanced the amplitude of the Ca^{2+} spike by suppressing the outward current in the K^+ channel (27). When the procaine concentration was further increased, however, the amplitude of the Ca^{2+} spike was reduced.

Measurements of the rate of rise of the Ca^{2+} spike and its overshoot as a function of the extracellular Ca^{2+} concentration showed that both parameters increased with an increase in $[Ca^{2+}]_o$. But for equal increments of $[Ca^{2+}]_o$, the increase in the magnitude of the overshoot decreased and attained saturation. This behavior is characteristic of an adsorption process, and consequently the effects of several divalent cations on generation of the Ca^{2+} spike and its amplitude have been explained on the basis of the Langmuir adsorption equation (21, 22, 29). Divalent ions were considered bound to membrane sites (charged groups of the membrane) in a competitive manner with different affinities that decreased in the order La^{3+}, $UO_2^{2+} > Zn^{2+}$, Co^{2+}, $Fe^{2+} > Mn^{2+} > Ni^{2+} > Ca^{2+} > Mg^{2+}$, Sr^{2+} (29). A concentration of 10–30 mM Co^{2+} or Mn^{2+} eliminated the Ca^{2+} spike. These ions had no affect on the Na^+ spike at a similar concentration (22). Such differences in the actions of pharmacological agents and ions indicate that Na^+ and Ca^{2+} channels have different physical and/or chemical structures in the membrane.

VOLTAGE-CLAMP EXPERIMENTS

The barnacle muscle fiber either microinjected (25) or internally perfused (43, 55) and subjected to voltage clamp gave an inward current due to flow of Ca^{2+} followed by a late outward current due to movement of K^+. These currents overlapped at positive membrane potentials beyond zero (24, 55). Furthermore oscillations in membrane currents (Fig. 2) made the separation of membrane current measured at any given voltage into its components (inward Ca^{2+} current, I_{Ca}; outward K^+ current, I_K) difficult and ambiguous. This problem can be overcome by suppressing one of the currents. Hagiwara et al. (24) found that application of 20 mM Co^{2+} reduced oscillations without affecting I_K but depressed the initial I_{Ca}. Also zero-Ca^{2+} ASW eliminated oscillations without interfering with I_K (43, 55). This procedure completely eliminated I_{Ca} in the barnacle *Megabalanus psittacus* (43, 48), whereas a small inward current, probably due to residual Ca^{2+}, persisted in *Balanus nubilus* (13, 46, 55).

Depression of I_K by tetraethylammonium (TEA) ion applied either inside (43, 55) or outside (24) the fiber reduced oscillations sufficiently and enabled measurement of I_{Ca}. Uniform spacial clamping of potential is a problem with the barnacle muscle fiber, however, because it has an intricate network of clefts and tubules that extend into the interior of the fiber (7, 39, 66). These extensive invaginations make the potential across the membrane nonuniform even though the clamp system as a whole works well (43). Consequently, in studying the effects of several agents on I_{Ca}, one must use some normalizing procedure to overcome the effects due to nonuniform potential control. Without such a procedure quantitative interpretation of results becomes less reliable.

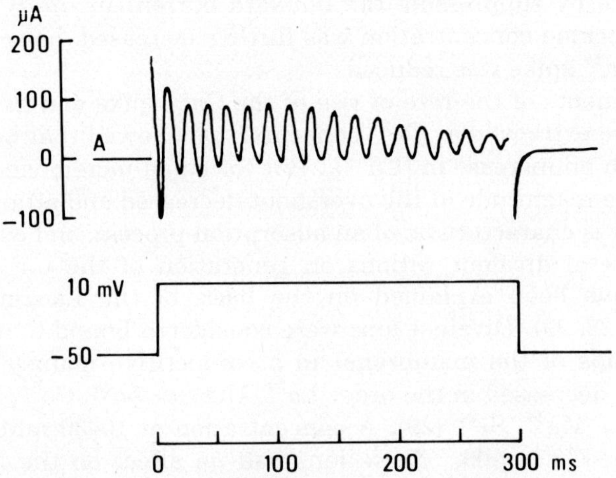

FIG. 2. Oscillations in membrane current associated with application of a 60-mV depolarization to a muscle fiber membrane under voltage clamp. The internal solution contained (in mM) potassium acetate (180), Tris-Cl (5), Tris-EGTA (3), and sucrose (540–620). See text for composition of ASW.

With muscle fibers internally perfused with 180 mM TEA acetate and immersed in ASW containing a known concentration of Ca^{2+}, I_{Ca} values were recorded as functions of time at several voltages. From these records, current voltage (*I*-V) curves were constructed and are shown in Figure 3. The muscle fiber in zero-Ca^{2+} ASW gave a small inward current followed by an outward current. Similar curves were obtained at other Ca^{2+} concentrations. In general, as the step depolarization was increased, membrane current increased, reached a maximum, and then decreased. Increases in $[Ca^{2+}]_o$ also increased the membrane current. Figure 3 shows that the maxima of all the *I*-V curves corresponding to several $[Ca^{2+}]_o$ lie at absolute zero membrane potential. In view of this, I_{Ca} was measured at zero membrane potential as a function of $[Ca^{2+}]_o$. These results, normalized by taking I_{Ca} measured in ASW containing 20 mM Ca^{2+} as 100%, are shown in Figure 4. The curve followed the Langmuir equation (21, 22, 29)

$$I_{Ca}(V) = \frac{(I_{Ca})_{max}(V)\cdot[Ca^{2+}]_o}{[Ca^{2+}]_o + K'_{Ca}(V)} \qquad (1)$$

where $(I_{Ca})_{max}(V)$ is the voltage-dependent maximum current and $K'_{Ca}(V)$ is the voltage-dependent apparent dissociation constant of the site or group binding Ca^{2+}. A double reciprocal plot of $1/I_{Ca}(V)$ versus $1/[Ca^{2+}]_o$ gave a straight line (Fig. 4, *inset*), from which values of 35.7 mM and 279% were derived for $K'_{Ca}(V)$ and $(I_{Ca})_{max}(V)$, respectively.

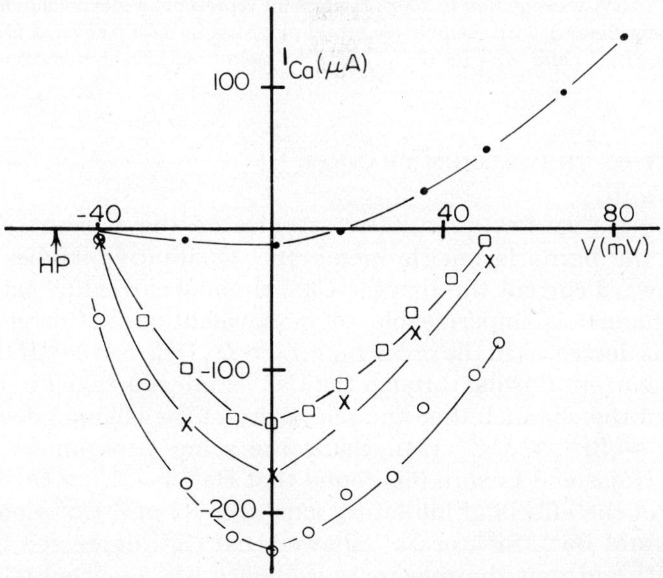

FIG. 3. Current-voltage curves constructed from membrane currents measured in barnacle muscle fibers immersed in several ASW solutions containing zero Ca^{2+} (●), 40 mM Ca^{2+} (□), and 100 mM Ca^{2+} (×, ○). *Curves* □ and × correspond to voltage-clamp experiments performed on same fiber, first immersed in ASW containing 40 mM Ca^{2+} and then in ASW containing 100 mM Ca^{2+}. *Curve* ○ was obtained from a different fiber immersed in 100-mM Ca^{2+} ASW. [Redrawn from Lakshminarayanaiah and Beirao (46).]

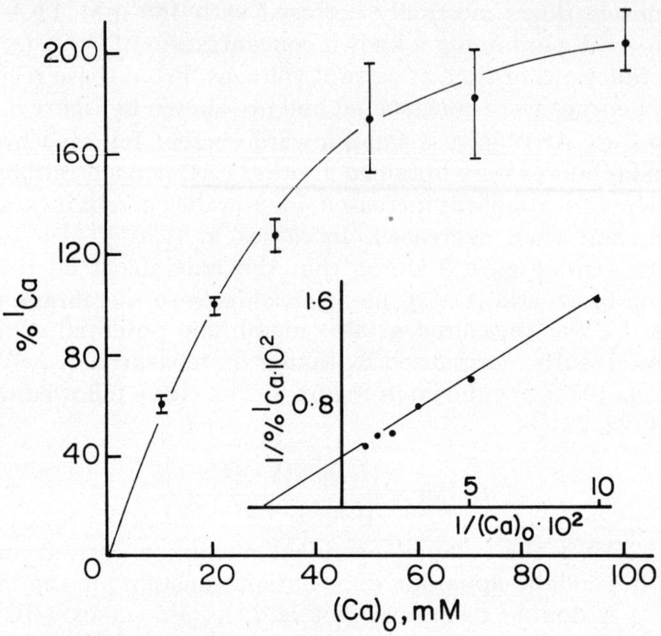

FIG. 4. Inward Ca^{2+} current measured as function of $[Ca^{2+}]_o$ at zero membrane potential by applying a depolarizing pulse of +50 mV when membrane was held at −50 mV. The I_{Ca} measured in 20-mM Ca^{2+} ASW was equated to 100%. Each point represents a mean value for I_{Ca} measured in 3 or 4 different fibers ± 2 SE. Double reciprocal plot of same data (shown in *inset*) gave values of 35.7 mM for $K'_{Ca}(V)$ and 279% for $(I_{Ca})_{max}(V)$. [Redrawn from Lakshminarayanaiah and Beirao (46).]

SELECTIVITY OF THE CALCIUM CHANNEL

There are very few quantitative studies on the selectivity of the Ca^{2+} channel in the barnacle muscle membrane. Qualitative studies of measurements of inward current through the Ca^{2+} channel carried by cations indicate that the channel is impermeable to monovalent ions. Current carried by divalent ions decreases in the order $I_{Ba} > I_{Sr} > I_{Ca}$ (Fig. 5*A*; 24). If the maximum quantity of current flowing through the Ca^{2+} channel is taken to represent the selectivity of the channel, then the selectivity of the channel decreases in the order $Ba^{2+} > Sr^{2+} > Ca^{2+}$ (24). Using the same criterion for determining selectivity, Rojas and Luxoro (63) found that $Ba^{2+} > Ca^{2+} > Sr^{2+} \gg Mg^{2+}$.

Study of the effects of inhibitory ions (e.g., 20 mM Co^{2+}) on I_{Ba}, I_{Sr}, and I_{Ca} with 20 mM Ba^{2+}, Sr^{2+}, or Ca^{2+} showed that Co^{2+} depressed I_{Ba} more than I_{Ca} (Fig. 5*B*) and that the selectivity sequence was reversed ($Ca^{2+} > Sr^{2+} > Ba^{2+}$). This reversal in the selectivity of the channel was attributed to either the low affinity of Ba^{2+} or the high affinity of Ca^{2+} for the negatively charged sites on the membrane (24). Because of this, I_{Ba} was more sensitive to the blocking agent than I_{Ca}.

AGENTS AFFECTING INWARD CALCIUM CURRENT

Although much is known about changes in ionic conductance responsible for excitation, few details exist about the biochemical components triggering these permeability changes. Several approaches for studying this include chemical and/or other modifications, such as variation of hydrogen and divalent ion concentrations (19, 20, 30, 58, 67, 68, 77), use of enzymes (1, 6, 62, 64, 74, 75), and cross-linking or use of group-specific reagents (10, 16, 41, 58, 67–73). Some of these studies suggest that membrane proteins may be involved in gating ionic channels.

Studies of the effects of divalent ions on membrane currents (51), particularly peak Na^+ current (38), show that voltage shifts in Na^+ activation occur after changes in concentration of divalent ions (38). These shifts are explained on the basis of changes in external surface potential initiated by fixed charges near the Na^+ channels after changes in divalent ion concentration. Similar shifts are not observed in the case of the barnacle muscle fiber when $[Ca^{2+}]_o$ is increased or decreased (13). Lakshminarayanaiah (45) has estimated the charge density of the negative groups associated with the barnacle muscle membrane to be 1 *e* per 217 $Å^2$ membrane surface, assuming that the charges are uniformly distributed over the surfaces of the membrane. Calcium ions bind to these charged sites or groups according to the reaction

$$[Ca^{2+}]_o + S \rightleftharpoons CaS \rightarrow S + [Ca^{2+}]_i \qquad (2)$$

where S represents the sites involving two negative charges. The I_{Ca}, given by *Equation 1*, is determined by the magnitude of CaS. When other divalent ions compete with Ca^{2+} for the sites on the membrane, competitive inhibition of I_{Ca} occurs.

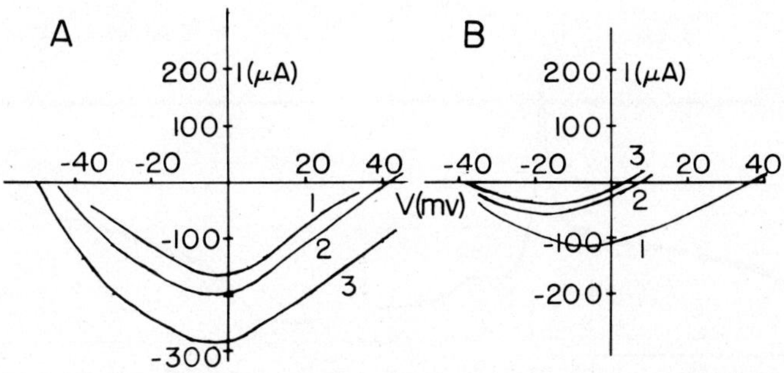

FIG. 5. Current-voltage curves constructed from peak membrane currents obtained by voltage clamp of barnacle muscle fibers. *A*: holding potential was −65 mV; external ASW contained 150 mM TEA plus 20 mM Ca^{2+} (*curve 1*), Sr^{2+} (*curve 2*), or Ba^{2+} (*curve 3*) and no other divalent cations. *B*: curves obtained when ASW contained 20 mM Co^{2+} plus 20 mM Ca^{2+} (*curve 1*), Sr^{2+} (*curve 2*), or Ba^{2+} (*curve 3*). Holding potential was −65 mV. [Redrawn from Hagiwara et al. (24).]

Effects of Zinc Ions on Inward Calcium Current

Although no detailed examination of the effect of extracellularly applied Zn^{2+} on the Ca^{2+} spike was done by Hagiwara and Takahashi (29), they found its capacity to reversibly bind to the sites on the membrane to be greater than that of Ca^{2+}. Murayama and Lakshminarayanaiah (55) studied the effects of Zn^{2+} on the Ca^{2+} spike and I_{Ca} by adding Zn^{2+} (5–10 mM) to the internal perfusion fluid. Internally applied Zn^{2+} had no effect on the resting membrane potential. About 5 mM Zn^{2+} (exact concentration is unknown due to the presence of the ligand EGTA in the internal solution) had little effect on the oscillatory action potential (Fig. 6*A*), whereas at higher concentrations it eliminated the Ca^{2+} spike (Fig. 6*B*).

Membrane currents recorded under voltage clamp when step depolarizing pulses are applied both in the absence and presence of Zn^{2+} are shown in Figure 7. These data were used to construct the *I*-V curves in Figure 8, which show that Zn^{2+} depressed I_{Ca} without affecting the K^+ currents (55).

Effects of Hydrogen Ions on Inward Calcium Current

Several studies with different biological preparations (15, 18, 30, 35, 48, 53, 54, 65, 77) show that reduction of pH has basically two effects: reversible

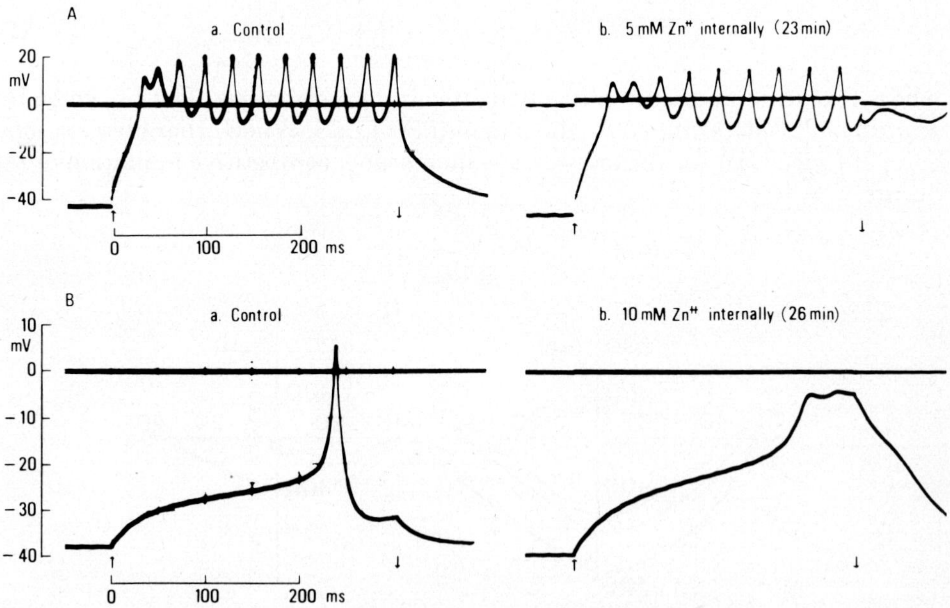

FIG. 6. Action potentials from 2 barnacle muscle fibers immersed in ASW containing 10 mM Ca^{2+} and 20 mM Mg^{2+} and internally perfused with 180 mM potassium acetate and sucrose. *A*: oscillatory action potential (control) is not significantly affected by internal application of 5 mM Zn^{2+}. *B*: single-spike action potential (control) is suppressed by internal application of 10 mM Zn^{2+}.

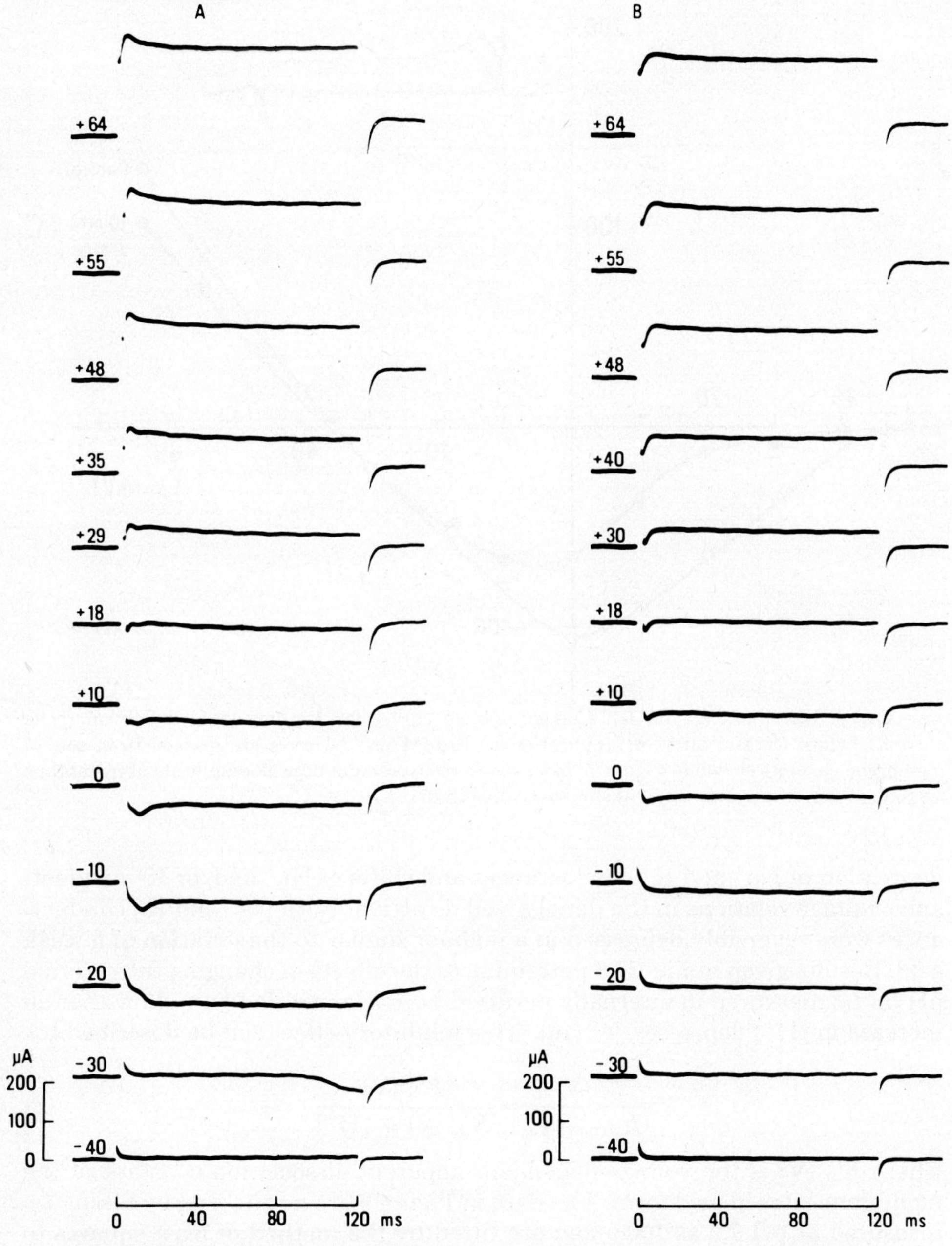

FIG. 7. Effect of 10 mM Zn^{2+} on membrane currents in voltage-clamped barnacle muscle fiber internally perfused with 180 mM potassium acetate containing 60 mM TEA and bathed in ASW containing 60 mM Ca^{2+} and zero Mg^{2+}. *A*: set of membrane currents related to step depolarizing pulses of 10 mV from holding potential of −50 mV before application of Zn^{2+}. *B*: set of membrane currents 30 min after internal application of 10 mM Zn^{2+}. [From Murayama and Lakshminarayaniah (55).]

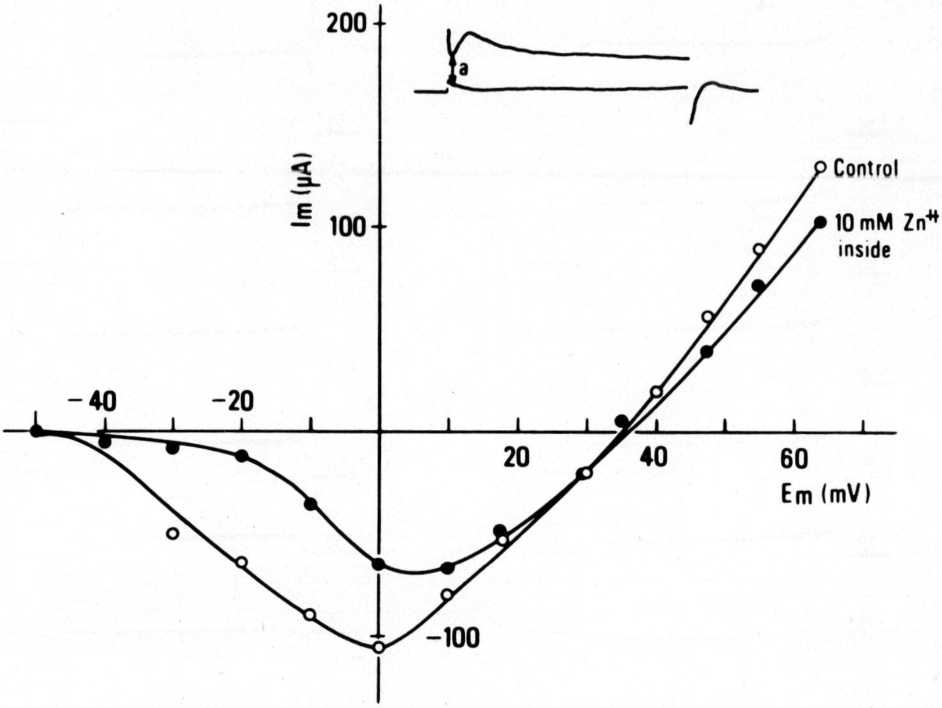

FIG. 8. Effect of Zn^{2+} on I_{Ca}. Current-voltage curves for 1st downward deflection (peak current) before (○) and after (●) application of 10 mM Zn^{2+}. Curves are derived from sets of membrane currents shown in Figure 7. *Inset* shows measurement of peak amplitude of membrane current. [From Murayama and Lakshminarayanaiah (55).]

depression of Na^+ and K^+ conductances and shifts of Na^+ and/or K^+ conductance-voltage relations in the depolarized direction. Both Na^+ and K^+ conductances were reversibly depressed in a manner similar to the titration of a weak acid. Results given in Figure 9 pertaining to the effects of changing the external pH on I_{Ca} measured in internally perfused barnacle muscle fibers show that an increase in $[H^+]$ depresses I_{Ca} (46). This inhibitory effect can be described by

$$\frac{I_{Ca}(V)}{(I_{Ca})_{max}(V)} = \frac{K'_h(V)}{K'_h(V) + [H^+]} \tag{3}$$

where K'_h (V) is the voltage-dependent apparent dissociation constant of the membrane sites in acid form. The data in Figure 9 are normalized by taking I_{Ca} measured at pH 7.2 as 100% and are fitted by the method of least squares to the solid curve by using values of 100% and 3.9 for $(I_{Ca})_{max}(V)$ and $pK'_a = -\log K'_h(V)$, respectively. The apparent pK'_a value after correcting for the presence of 100 mM of Mg^{2+} used in the measurements was 5.2 for the chemical group of the Ca^{2+} channel. An identical value has been derived for the negatively charged carboxyl group of the Na^+ channel (35). Without further work it is difficult to determine the chemical group of the Ca^{2+} channel. It may be the

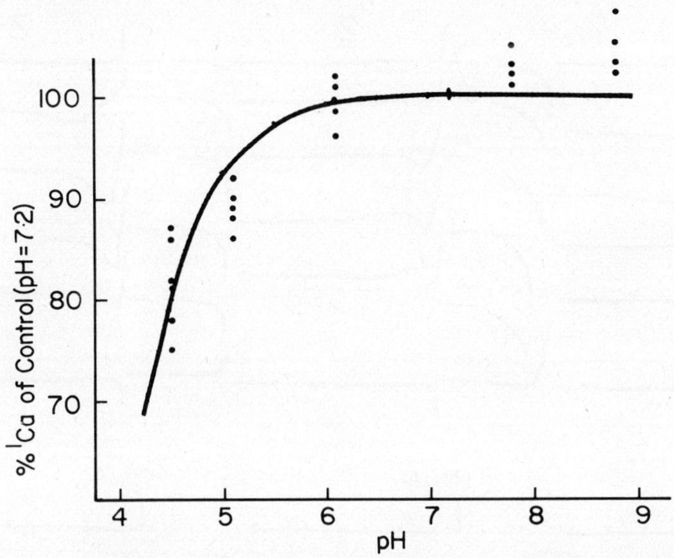

FIG. 9. Effect of external pH on I_{Ca}. Points represent measurements made on different single fibers. I_{Ca} has been normalized by taking its value at pH 7.2 as 100%. Curve is drawn from *Equation 3* using values of 100% and $10^{-3.9}$ for $(I_{Ca})_{max}(V)$ and $K'_h(V)$, respectively. [Redrawn from Lakshminarayanaiah and Beirao (46).]

imidazole group of the histidine residue of a protein associated with the Ca^{2+} channel. This needs to be identified by studying the effects of group-specific reagents on the negative site of the channel.

Effects of D 600 on Inward Calcium Current

Methoxyverapamil (D 600) is a known Ca^{2+} antagonist (11, 12, 42, 44, 50, 56). Applied externally in the perfused barnacle muscle fiber, it reversibly inhibited I_{Ca} (Fig. 10; 63). The exact concentration used is unknown, but it was rather high (~1 mM). Applied internally at a concentration of 2 mM, D 600 exerted no effect on I_{Ca} (63). Similar results were reported for Ba^{2+} and Sr^{2+} currents flowing through the Ca^{2+} channel (63). This asymmetric divalent ion-independent effect of a known Ca^{2+} antagonist seems to indicate a direct effect on a membrane component of the Ca^{2+} channel.

PROBABLE STRUCTURE OF THE CALCIUM CHANNEL

Studying the permeability of the Na^+ and K^+ channels to several inorganic and organic ions of different sizes and shapes has made it possible to estimate channel sizes. Hille (34) determined the relative permeabilities of K^+, Tl^+, Rb^+, and NH_4^+ in the K^+ channel of the myelinated nerve to be 1:2.3:0.91:0.13. Other ions (Li^+, Na^+, Cs^+, methylamine, guanidine, hydrazine, and hydroxylamine) were relatively impermeable. These data were correlated with the minimum

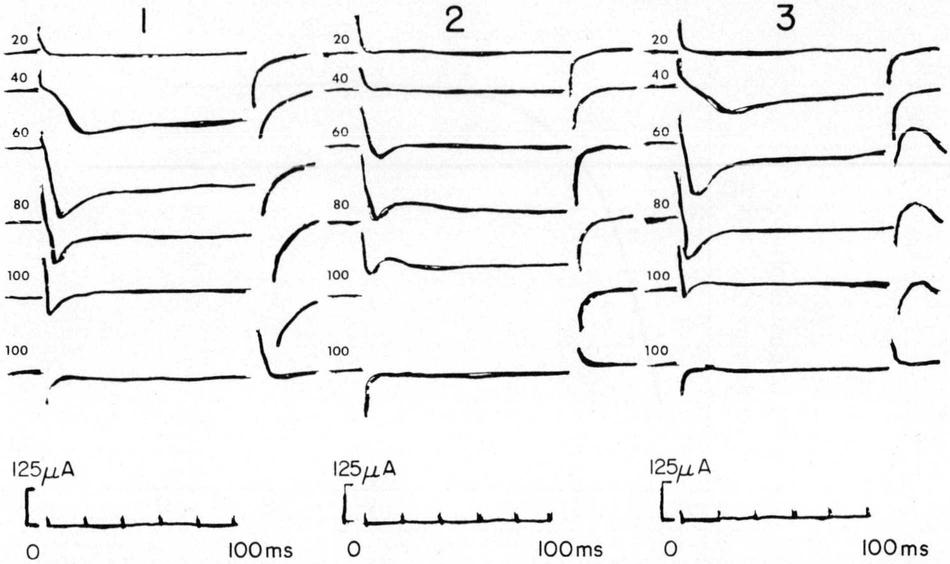

FIG. 10. Reversible depression of I_{Ca} by external application of D 600. *Panel 1*: membrane currents recorded under voltage-clamp conditions in an internally perfused (200 mM cesium acetate and 10 mM EGTA) barnacle muscle fiber immersed in ASW containing 60 mM $CaCl_2$ and 2.5% alcohol, pH 5.5. *Panel 2*: membrane currents with same solution used in *panel 1* plus 1 mM D 600 in ASW. *Panel 3*: recovery of membrane currents with same solution used in *1*. [Redrawn from Rojas and Luxoro (63).]

pore diameter required to allow the ions to pass. On this basis, the K^+ pore diameter was 3.0–3.3 Å. Similar kinds of measurements by Bezanilla and Armstrong (14) showed that ions with crystal diameters from 2.66 Å (K^+) to 2.96 Å (Rb^+, NH_4^+) passed through the K^+ pore, whereas ions with smaller (Na^+ = 1.9 Å, Li^+ = 1.36 Å) and larger (Cs^+ = 3.3 Å, TEA^+ = 8 Å) diameters did not. It is considered (3) that the K^+ channel has a narrow pore (diam 2.6–3 Å) with a mouth 8 Å in diameter to just admit the TEA ion. Ions with diameters larger than 3 Å are excluded sterically from the pore, whereas ions smaller than 2.6 Å are excluded because of unfavorable interaction with the oxygen-lined wall of the pore. Similarly Hille (32, 33) models the Na^+ channel as an oxygen-lined pore 3 × 5 Å in cross section. Hille (36, 37) has presented a hypothetical picture of the Na^+ channel that takes its selectivity, gating, and voltage-dependent properties into account.

A similar quantitative approach to determine the geometric dimensions of the Ca^{2+} channel in the barnacle muscle membrane is hampered because of the inability to uniformly space-clamp the membrane to any given voltage due to the complex structure of the muscle fiber. Qualitatively, however, the physical structure of the Ca^{2+} channel should be similar to that of the Na^+ or K^+ channel in view of its several properties described above. A tentative, highly imaginative picture is shown in Figure 11. It is drawn after Hille's (36, 37) fanciful diagram for an ionic channel like the Na^+ or K^+ channel.

Surface charges on the membrane give rise to the formation of ionic double layers on both the inside and outside of the membrane interfaces. A protein macromolecule in the membrane seems to be involved in channel or pore formation. The open pore is filled with aqueous fluid because the Q_{10} for Ca^{2+} conductance of the membrane (i.e., flow of ions in the pore) is about 1.5 (43), which is very close to the Q_{10} of 1.4 for Ca^{2+} conductance in bulk water when its limiting conductances at 25°C and 35°C are considered (61).

The polar pore has a selectivity filter on the outside. This filter represents the several energy barriers to the flow of ions. The shapes and sizes of these barriers are unknown. Similarly the shape and structure of the gate on the inside of the membrane are unknown. The voltage sensor moves under the influence of the intramembrane electric field and controls the opening and closing of the gate. The voltage sensor may be a dipolar or quadrupolar protein molecule residing in the hydrophobic region of the membrane. It is possible that some polar part of the voltage sensor that protrudes into the aqueous phase of the channel might itself serve as the gate whose movement, under the influence of the electric field, will open and close the channel.

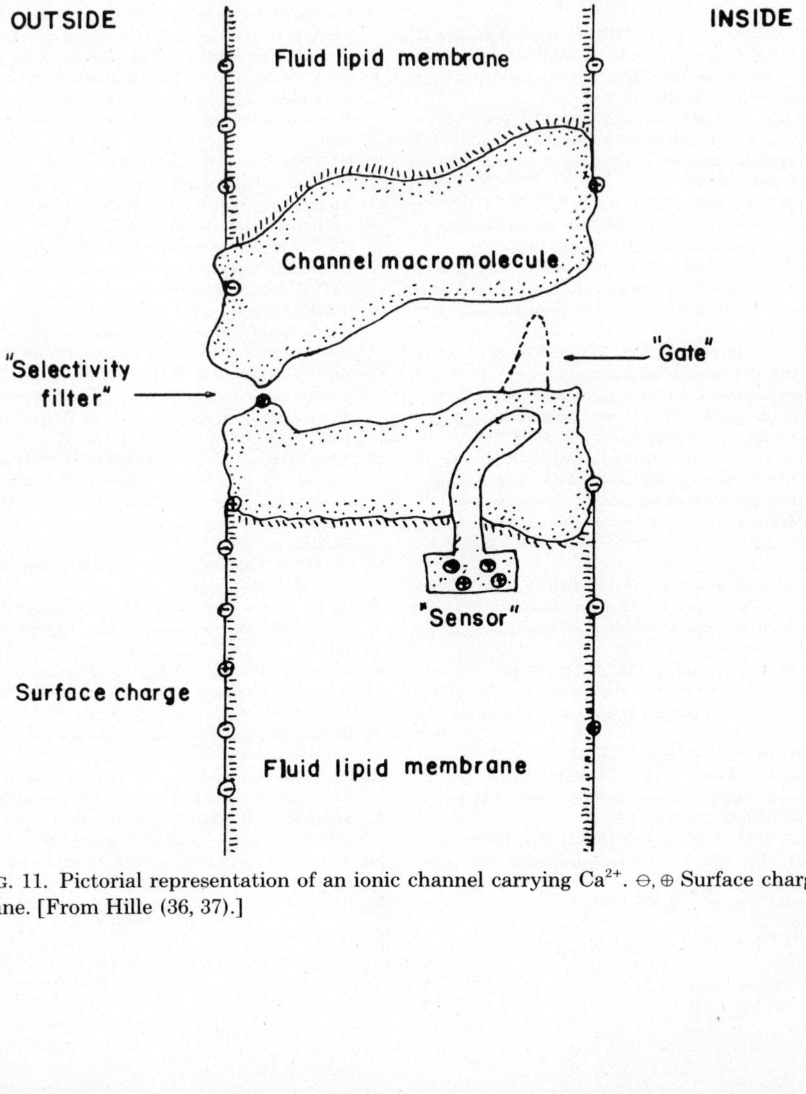

FIG. 11. Pictorial representation of an ionic channel carrying Ca^{2+}. ⊖, ⊕ Surface charges on membrane. [From Hille (36, 37).]

Some of my work has been supported in part by National Institutes of Health Grant HL-15835. I would like to thank Dr. C. Paul Bianchi for his interest and encouragement.

REFERENCES

1. ABBOTT, N. J., T. DEGUCHI, D. T. FRAZIER, K. MURAYAMA, T. NARAHASHI, A. OTTOLENGHI, AND C. M. WONG. The action of phospholipases on the inner and outer surface of the squid giant axon membrane. *J. Physiol. London* 220: 73–86, 1972.
2. ADAMS, D. J., AND P. W. GAGE. Ionic currents in response to membrane depolarization in an *Aplysia* neurone. *J. Physiol. London* 289: 115–141, 1979.
3. ARMSTRONG, C. M. Ionic pores, gates, and gating currents. *Q. Rev. Biophys.* 7: 179–210, 1974.
4. ARMSTRONG, C. M. Evidence for ionic pores in excitable membranes. *Biophys. J.* 15: 932–933, 1975.
5. ARMSTRONG, C. M. Potassium pores of nerve and muscle membranes. In: *Membranes. Lipid Bilayers and Biological Membranes: Dynamic Properties*, edited by G. Eisenman. New York: Dekker, 1975, vol. 3, p. 325–358.
6. ARMSTRONG, C. M., F. BEZANILLA, AND E. ROJAS. Destruction of sodium conductance inactivation in squid axons perfused with pronase. *J. Gen. Physiol.* 62: 375–391, 1973.
7. BACIGALUPO, J., M. LUXORO, S. RISSETTI, AND C. VERGARA. Extracellular space and diffusion barriers in muscle fibers from *Megabalanus psittacus* (Darwin). *J. Physiol. London* 288: 301–312, 1979.
8. BAKER, P. F., H. MEVES, AND E. B. RIDGWAY. Effects of manganese and other agents on the calcium uptake that follows depolarizations of squid axons. *J. Physiol. London* 231: 511–526, 1973.
9. BAKER, P. F., H. MEVES, AND E. B. RIDGWAY. Calcium entry in response to maintained depolarization of squid axons. *J. Physiol. London* 231: 527–548, 1973.
10. BARCHI, R. L., AND J. B. WEIGELE. Characteristics of saxitoxin binding to the sodium channel of sarcolemma isolated from rat skeletal muscle. *J. Physiol. London* 295: 383–396, 1979.
11. BAYER, R., R. HENNEKES, R. KAUFMANN, AND R. MANNOLD. Inotropic and electrophysiological actions of verapamil and D600 in mammalian myocardium. I. Pattern of inotropic effects of the racemic compounds. *Naunyn-Schmiedebergs Arch. Pharmacol.* 290: 49–68, 1975.
12. BAYER, R., D. KALUSCHE, R. KAUFMANN, AND R. MANNOLD. Inotropic and electrophysiological actions of verapamil and D600 in mammalian myocardium. III. Effects of optical isomers on transmembrane action potentials. *Naunyn-Schmiedebergs Arch. Pharmacol.* 290: 81–97, 1975.
13. BEIRAO, P. S., AND N. LAKSHMINARAYANAIAH. Calcium carrying system in the giant muscle fiber of the barnacle species, *Balanus nubilus*. *J. Physiol. London* 293: 319–327, 1979.
14. BEZANILLA, F., AND C. M. ARMSTRONG. Negative conductance caused by entry of sodium and cesium ions into the potassium channel of squid axon. *J. Gen. Physiol.* 60: 588–608, 1972.
15. CARBONE, E., R. FIORAVANTI, G. PRESTIPINO, AND E. WANKE. Action of extracellular pH on Na^+ and K^+ membrane currents in giant axon of *Loligo vulgaris*. *J. Membr. Biol.* 43: 295–315, 1978.
16. COOKE, I. M., J. M. DIAMOND, A. D. GRINNELL, S. HAGIWARA, AND H. SAKATA. Suppression of the action potential in nerve by nitrobenzene derivatives. *Proc. Natl. Acad. Sci. USA* 60: 470–477, 1968.
17. CRANEFIELD, P. F. *The Conduction of the Cardiac Impulse.* Mount Kisco, NY: Futura, 1975.
18. DROUIN, H., AND R. THE. The effect of reducing extracellular pH on the membrane currents of the Ranvier node. *Pfluegers Arch.* 313: 80–88, 1969.
19. FRANKENHAEUSER, B., AND A. L. HODGKIN. The action of calcium on the electrical properties of squid axons. *J. Physiol. London* 137: 218–244, 1957.
20. GILBERT, D. L., AND G. EHRENSTEIN. Effect of divalent cations on potassium conductance of squid axons: determination of surface charge. *Biophys. J.* 9: 447–463, 1969.
21. HAGIWARA, S. Ca spike. In: *Advances in Biophysics*, edited by M. Kotani. Tokyo: Univ. of Tokyo Press, 1973, vol. 4, p. 71–102.
22. HAGIWARA, S. Ca dependent action potential. In: *Membranes. Lipid Bilayers and Biological Membranes: Dynamic Properties*, edited by G. Eisenman. New York: Dekker, 1975, vol. 3, p. 359–381.
23. HAGIWARA, S., S. CHICHIBU, AND K. NAKA. The effects of various ions on the resting and spike potentials of barnacle muscle fibers. *J. Gen. Physiol.* 48: 163–179, 1964.
24. HAGIWARA, S., J. FUKUDA, AND D. C. EATON. Membrane currents carried by Ca, Sr, and Ba in barnacle muscle fiber during voltage clamp. *J. Gen. Physiol.* 63: 564–578, 1974.
25. HAGIWARA, S., H. HAYASHI, AND K. TAKAHASHI. Calcium and potassium currents of the membrane of a barnacle muscle fiber in relation to calcium spike. *J. Physiol. London* 205: 115–129, 1969.
26. HAGIWARA, S., AND K. NAKA. The initiation of spike potential in barnacle muscle fibers under low intracellular Ca^{++}. *J. Gen. Physiol.* 48: 141–162, 1964.
27. HAGIWARA, S., AND S. NAKAJIMA. Differences in Na and Ca spikes as examined by application of TTX, procaine and manganese ions. *J. Gen. Physiol.* 49: 793–806, 1966.
28. HAGIWARA, S., AND S. NAKAJIMA. Effects of intracellular Ca ion concentration upon excitability of the muscle fiber membrane of a barnacle. *J. Gen. Physiol.* 49: 807–818, 1966.
29. HAGIWARA, S., AND K. TAKAHASHI. Surface density of calcium ions and calcium spikes in the barnacle muscle fiber membrane. *J. Gen. Physiol.* 50: 583–601, 1967.
30. HILLE, B. Charges and potentials at the nerve surface. Divalent ions and pH. *J. Gen. Physiol.* 51: 221–236, 1968.
31. HILLE, B. Ionic channels in nerve membrane. *Prog. Biophys. Mol. Biol.* 21: 1–32, 1970.
32. HILLE, B. The permeability of sodium channels to organic cations in myelinated nerves. *J. Gen. Physiol.* 58: 599–619, 1971.
33. HILLE, B. The permeability of sodium channels to metal cations in myelinated nerves. *J. Gen. Physiol.* 59: 637–658, 1972.
34. HILLE, B. Potassium channels in myelinated nerves. *J. Gen. Physiol.* 61: 669–686, 1973.
35. HILLE, B. Ionic selectivity of Na and K channels of nerve membranes. In: *Membranes. Lipid Bilayers and Biological Membranes: Dynamic Properties*, edited by G. Eisenman. New York: Dekker, 1975, vol. 3, p. 255–323.
36. HILLE, B. Ionic channels of nerve: questions for theoretical chemists. *Biosystems* 8: 195–199, 1977.
37. HILLE, B. Ionic channels in excitable membranes. Current

problems and biophysical approaches. *Biophys. J.* 22: 283–294, 1978.

38. HILLE, B., A. M. WOODHULL, AND B. I. SHAPIRO. Negative surface charge near sodium channels of nerve: divalent ions, monovalent ions, and pH. *Philos. Trans. R. Soc. London Ser. B* 270: 301–318, 1975.
39. HOYLE, G., P. A. McNEILL, AND A. I. SELVERSTON. Ultrastructure of barnacle giant muscle fibers. *J. Cell Biol.* 56: 74–91, 1973.
40. HOYLE, G., AND T. SMYTH, JR. Giant muscle fibers in a barnacle *Balanus nubilus*, Darwin. *Science* 139: 49–50, 1963.
41. HUNEEUS-COX, F., H. L. FERNANDEZ, AND B. H. SMITH. Effects of redox and sulfhydryl reagents on the bioelectric properties of the giant axon of the squid. *Biophys. J.* 6: 675–689, 1966.
42. KASS, R. S., AND R. W. TSIEN. Multiple effects of calcium antagonists on plateau currents in cardiac Purkinje fibers. *J. Gen. Physiol.* 66: 169–192, 1975.
43. KEYNES, R. D., E. ROJAS, R. E. TAYLOR, AND J. VERGARA. Calcium and potassium systems of a giant barnacle muscle fiber under membrane potential control. *J. Physiol. London* 229: 409–455, 1973.
44. KOHLHARDT, M., B. BAUER, H. KRAUSE, AND A. FLECKENSTEIN. Differentiation of the transmembrane Na and Ca channels in mammalian cardiac fibers by the use of specific inhibitors. *Pfluegers Arch.* 335: 309–322, 1972.
45. LAKSHMINARAYANAIAH, N. Evaluation of membrane surface charge density: a discussion of some models. *Bull. Math. Biol.* 39: 643–662, 1977.
46. LAKSHMINARAYANAIAH, N., AND P. S. BEIRAO. Calcium system of a giant barnacle muscle fiber. *Proc. W. Pharmacol. Soc.* 22: 301–307, 1979.
47. LAKSHMINARAYANAIAH, N., AND E. ROJAS. Effects of anions and cations on the resting membrane potential of internally perfused barnacle muscle fibers. *J. Physiol. London* 233: 613–634, 1973.
48. LAKSHMINARAYANAIAH, N., AND E. ROJAS. Effects of pH and ionic strength on the potassium system in the internally perfused giant barnacle muscle fiber. *Pfluegers Arch.* 358: 349–366, 1975.
49. LEE, K. S., N. AKAIKE, AND A. M. BROWN. Properties of internally perfused, voltage-clamped, isolated nerve cell bodies. *J. Gen. Physiol.* 71: 489–507, 1978.
50. McDONALD, T. F., D. PELZER, AND W. TRAUTWEIN. On the mechanism of slow calcium channel block in heart. *Pfluegers Arch.* 385: 175–179, 1980.
51. McLAUGHLIN, S. G. A., G. SZABO, AND G. EISENMAN. Divalent ions and the surface potential of charged phospholipid membranes. *J. Gen. Physiol.* 58: 667–687, 1971.
52. MEVES, H., AND W. VOGEL. Calcium inward currents in internally perfused giant axons. *J. Physiol. London* 235: 225–265, 1973.
53. MOZHAYEVA, G. N., AND A. P. NAUMOV. Effect of surface charge on the steady-state potassium conductance of nodal membrane. *Nature London* 228: 164–165, 1970.
54. MOZHAYEVA, G. N., AND A. P. NAUMOV. Effect of the surface charge on the steady potassium conductivity of the membrane of the node of Ranvier—II. Change in ionic strength of the external solution. *Biophysics* 17: 644–649, 1972.
55. MURAYAMA, K., AND N. LAKSHMINARAYANAIAH. Some electrical properties of the membrane of the barnacle muscle fibers under internal perfusion. *J. Membr. Biol.* 35: 257–283, 1977.
56. NAWRATH, H., R. E. TEN EICK, T. F. McDONALD AND W. TRAUTWEIN. On the mechanism underlying the action of D600 on slow inward current and tension in mammalian myocardium. *Circ. Res.* 40: 408–414, 1977.
57. NEWMARK, P. Pores for thought. *Nature London* 284: 659–660, 1980.
58. NONNER, W., B. C. SPALDING, AND B. HILLE. Low intracellular pH and chemical agents slow inactivation gating in sodium channels of muscle. *Nature London* 284: 360–363, 1980.
59. REUTER, H. Divalent cations as charge carriers in excitable membranes. *Prog. Biophys. Mol. Biol.* 26: 1–43, 1973.
60. REUTER, H., AND H. SCHOLZ. A study of the ionic selectivity and the kinetic properties of the calcium dependent slow inward current in mammalian cardiac muscle. *J. Physiol. London* 264: 17–47, 1977.
61. ROBINSON, R. A., AND R. H. STOKES. *Electrolyte Solutions*. London: Butterworths, 1959, p. 465.
62. ROJAS, E. Membrane potentials, resistance and ion permeability in squid giant axons injected or perfused with proteases. *Proc. Natl. Acad. Sci. USA* 53: 306–311, 1965.
63. ROJAS, E., AND M. LUXORO. Coupling between ionic conductance changes and contractions in barnacle muscle fibers under membrane potential control. In: *Actualités Neurophysiologiques*, edited by A. M. Monnier. Paris: Masson, 1974, p. 159–169.
64. ROJAS, E., AND B. RUDY. Destruction of the sodium conductance inactivation by a specific protease in perfused nerve fibers from *Loligo*. *J. Physiol. London* 262: 501–531, 1976.
65. SCHAUF, C. L., AND F. A. DAVIS. Sensitivity of the sodium and potassium channels of *Myxicola* giant axons to changes in external pH. *J. Gen. Physiol.* 67: 185–195, 1976.
66. SELVERSTON, A. Structure and function of the transverse tubular system in crustacean muscle fibers. *Am. Zool.* 7: 515–525, 1967.
67. SHRAGER, P. Ionic conductance changes in voltage clamped crayfish axons at low pH. *J. Gen. Physiol.* 64: 666–690, 1974.
68. SHRAGER, P. Specific channel groups involved in the control of ionic conductance in nerve. *Ann. NY Acad. Sci.* 264: 293–303, 1975.
69. SHRAGER, P. Slow sodium inactivation in nerve after exposure to sulfhydryl blocking agents. *J. Gen. Physiol.* 69: 183–202, 1977.
70. SHRAGER, P., R. I. MACY, AND A. STRICKHOLM. Internal perfusion of crayfish giant axons: action of tannic acid, DDT, and TEA. *J. Cell. Physiol.* 74: 77–90, 1969.
71. SHRAGER, P., AND C. PROFERA. Inhibition of the receptor for tetrodotoxin in nerve membranes by reagents modifying carboxyl groups. *Biochim. Biophys. Acta* 318: 141–146, 1973.
72. SHRAGER, P., A. STRICKHOLM, AND R. I. MACY. Chemical modification of crayfish axons by protein cross linking aldehydes. *J. Cell. Physiol.* 74: 91–99, 1969.
73. SIGWORTH, F. J., AND B. C. SPALDING. Chemical modification reduces the conductance of sodium channels in nerve. *Nature London* 283: 293–295, 1980.
74. TASAKI, I., AND T. TAKENAKA. Effects of various potassium salts and proteases upon excitability of intracellularly perfused squid giant axons. *Proc. Natl. Acad. Sci. USA* 52: 804–810, 1964.
75. TOBIAS, J. M. Effects of phospholipases, collagenase and chymotrypsin on impulse conduction and resting potential in the lobster axon with parallel experiments on frog muscle. *J. Cell. Comp. Physiol.* 46: 183–207, 1955.
76. TRAUTWEIN, W. Membrane currents in cardiac muscle fibers. *Physiol. Rev.* 53: 793–835, 1973.
77. WOODHULL, A. M. Ionic blockage of sodium channels in nerve. *J. Gen. Physiol.* 61: 687–708, 1973.

3

Differences in Skeletal, Cardiac, and Smooth Muscle Contractile Element Regulation by Calcium

JAMES T. STULL AND C. FAGG SANFORD

Moss Heart Center and Department of Pharmacology, University of Texas Health Science Center, Dallas, Texas

THE PRIMARY EVENT IN excitation-contraction coupling of all types of muscles is the release of Ca^{2+} from membrane components, such as the sarcolemma and the sarcoplasmic reticulum, before activation of the contractile mechanism. Although there are many similarities in the biochemical and physiological properties of the membrane components of skeletal, cardiac, and smooth muscle cells, there are also many important distinctions. A comparison of these properties in relation to excitation-contraction coupling is beyond the scope of this chapter. Other chapters in this series, however, present discussions pertinent to this topic.

One of the primary contractile proteins is myosin. Myosin from all types of muscle consists of light-chain and heavy-chain protein subunits (Fig. 1). Myosin is a hexamer with one pair of heavy chains (each of mol wt 200,000) and two distinct pairs of light chains that vary in molecular weight depending on the muscle source (mol wt 15,000–27,000). The total molecular weight of the myosin complex is about 460,000. Thick filaments of the contractile elements are composed of myosin molecules arranged in a parallel array. The heavy chains form an elongated structure in which a tail portion of the myosin molecule is embedded in the thick filament and adds structural rigidity. Projections of the remaining portion of the myosin molecule arise from the thick filaments at regular intervals to form cross bridges. These globular projections from the thick filament contain the myosin ATPase activity and the sites of actin binding. Muscle contraction is thought to consist of a cyclic attachment and detachment of this globular portion of the myosin molecule to the actin filament, and the energy needed for this process is provided by adenosine triphosphate (ATP). It is thought that the cyclical association and dissociation of myosin and actin in relation to hydrolysis of ATP is a key step

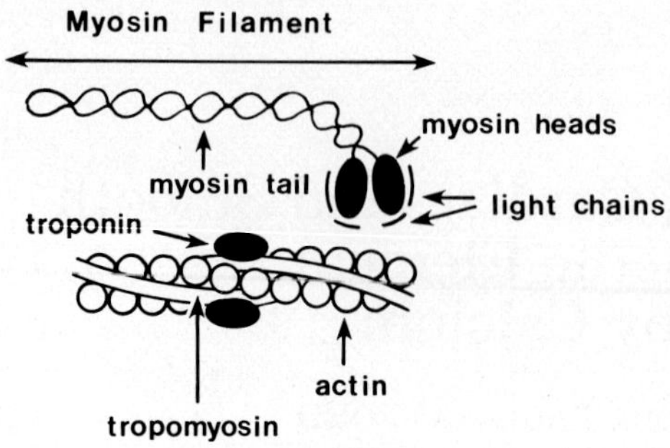

FIG. 1. The primary proteins in contractile elements of muscles. Diagram illustrates relationships between the myofibrillar proteins for cardiac and skeletal muscles. Precise location of myosin light-chain subunits is not known. Smooth muscles are similar except they lack troponin. (Leiotonin of smooth muscle system has not been included for lack of detailed information.)

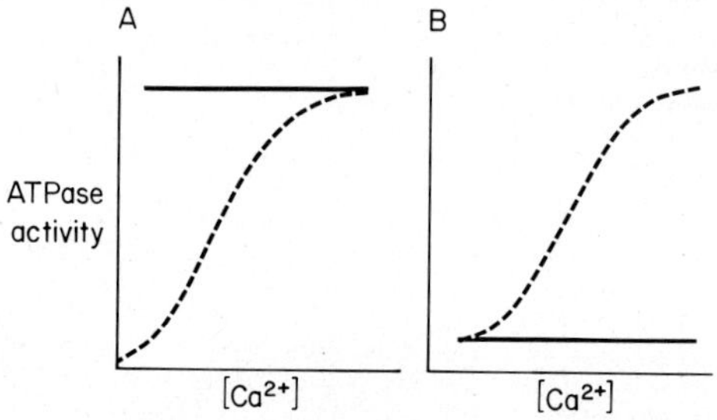

FIG. 2. Calcium regulation of actomyosin ATPase activity in cardiac and skeletal (*A*) and smooth (*B*) muscles. *Solid line* shows actomyosin ATPase activity alone without any significant regulation by Ca^{2+}. *Dashed line* represents actomyosin ATPase activity in presence of regulatory proteins from striated (*A*) and smooth (*B*) muscles.

in the sliding of thin and thick filaments past each other and in the subsequent contraction of muscle.

Actin is a globular protein (mol wt 43,000) that polymerizes to form a double helical filament (Fig. 1). The thin filaments thus formed by actin play an important role in the activation of myosin ATPase activity in the presence of Mg^{2+} (the actin-activated Mg^{2+}-dependent ATPase activity of myosin). Myosin shows ATPase activity under various conditions, but it is the Mg^{2+}-dependent ATPase activity that is thought to be physiologically relevant. Recent reviews provide a more detailed discussion of actin-myosin interactions (1, 31).

The cyclic interaction of myosin with actin is regulated by particular protein components within the contractile elements. In the relaxed state, the concentration of Ca^{2+} in the sarcoplasm is less than 0.1 μM, due to sequestration and storage in specific membrane components. Upon the release of Ca^{2+} into the sarcoplasm, the concentration of Ca^{2+} may rise to 10 μM (3, 9, 21) and activation of the contractile elements follows with subsequent contraction. The primary focus of many investigators has been to identify and characterize the particular protein components that regulate in a Ca^{2+}-dependent manner the actin-myosin interaction. In mammalian muscles this regulation falls into two general classes (1, 10, 31). In cardiac and skeletal muscles, pure myosin and actin hydrolyze ATP at a maximal rate. The addition of particular regulatory proteins results in inhibition of the ATPase activity at concentrations of Ca^{2+} less than 0.1 μM. In the presence of 10 μM Ca^{2+}, the ATPase activity is similar to that observed with myosin and actin alone. Thus the Ca^{2+}

regulation of contractile elements for these two types of muscles is an inhibitory system (Fig. 2). In smooth muscles, on the other hand, pure actin and myosin alone do not show any significant ATPase activity in the presence or absence of Ca^{2+}. Upon the addition of particular regulatory proteins, the ATPase activity is stimulated in the presence of Ca^{2+} but remains low in the absence of Ca^{2+}. Thus the regulation of smooth muscle actomyosin systems by Ca^{2+} has properties of an activation system rather than an inhibitory system (Fig. 2).

SKELETAL AND CARDIAC MUSCLE CONTRACTILE ELEMENTS

Troponin-Tropomyosin and Thin Filament Regulation

The regulation of the actin-myosin interaction in cardiac and skeletal muscles by Ca^{2+} is mediated primarily by the regulatory proteins troponin and tropomyosin. Tropomyosin has a molecular weight of 66,000 and is composed of two helical subunits that coil about each other. Tropomyosin molecules are polymerized end-to-end and form a strand that lies in the grooves of the actin filament (Fig. 1). Troponin, on the other hand, is composed of three distinct protein subunits: troponin I, which inhibits actomyosin ATPase activity; troponin C, which binds Ca^{2+}; and troponin T, which binds to tropomyosin. In the absence of Ca^{2+}, troponin I binds tightly to actin and tropomyosin, maintaining tropomyosin in an inhibitory position for the actin-myosin interaction. Binding of Ca^{2+} to troponin C results in a reduction of the binding affinity of troponin I for actin-tropomyosin and increases the interaction of troponin C with troponin T. These conformational changes favor the dissociation of troponin I from actin and tropomyosin and allow tropomyosin to shift into the actin groove. Thus tropomyosin moves into a noninhibitory position and the actin-myosin interaction may take place in a cyclic fashion. When Ca^{2+} is removed from troponin C in the thin filament, tropomyosin shifts to block the actin-myosin interaction, preventing the hydrolysis of ATP and initiating the relaxed state in muscle.

Although the functional properties of troponin from cardiac and skeletal muscles are the same, there are some important differences in the biochemical properties of these particular proteins. Troponin I from skeletal muscle is smaller than troponin I from cardiac muscle. The major difference is the extra 26 amino acid residues at the N-terminal end of cardiac troponin I, which are not shared by troponin I from red and white skeletal muscles (33). Within this extra series of amino acid residues is a single site that is phosphorylated by cyclic adenosine 3′,5′-monophosphate (cyclic AMP)–dependent protein kinase. This phosphorylation leads to a decrease in the activity of cardiac actomyosin ATPase in the presence of Ca^{2+} (25). It has been postulated that phosphorylation of this site in cardiac muscle in response to agents that increase cyclic AMP may play some role in the inotropic response of cardiac muscle (for further discussion see 26). Troponin T from cardiac muscle is also larger than its analogous subunit from skeletal muscle. Any distinctions in the respective functional or regulatory properties of this protein complex are not evident.

Troponin C from skeletal muscle has a molecular weight of 18,000 and has four Ca^{2+}-binding sites of two distinct classes (24). Sites 1 and 2 appear to bind Ca^{2+} specifically with a binding constant of $3.2 \times 10^5\ M^{-1}$. These sites are termed the Ca^{2+}-specific sites. Sites 3 and 4, however, bind both Ca^{2+} and Mg^{2+} with binding constants of $2.1 \times 10^7\ M^{-1}$ and $5 \times 10^3\ M^{-1}$, respectively. This class of sites is referred to as the Ca^{2+}-Mg^{2+} sites. It should be noted that the Ca^{2+}-Mg^{2+} sites have a higher affinity for Ca^{2+} than the Ca^{2+}-specific sites. Under physiological conditions, however, the Ca^{2+}-Mg^{2+} sites may be occupied by Mg^{2+} since the intracellular concentration of Mg^{2+} is about 1 mM. Studies on the Ca^{2+} dependency of actomyosin ATPase activity also indicate that the Ca^{2+}-specific sites may be the regulatory sites of skeletal muscle contraction (24).

Although cardiac and skeletal muscle troponin C have similar functions, they are not identical proteins and have distinct biochemical properties (15, 27). Based on the sequence homology of these two proteins, one of the Ca^{2+}-specific sites may be defective and no longer bind Ca^{2+}. Thus the Ca^{2+}-Mg^{2+} sites are present in cardiac troponin (15, 17), but there may be only one Ca^{2+}-specific site (17). Physical studies on Ca^{2+}-binding properties indicate that the properties of the Ca^{2+}-specific regulatory sites of skeletal and cardiac muscle troponin C are similar. Thus in the troponin complex in skeletal muscle, the interaction of actin and myosin may be regulated by Ca^{2+} binding to two specific Ca^{2+}-binding sites, whereas in cardiac muscle Ca^{2+} binds to one Ca^{2+}-specific site. A physiological role for the Ca^{2+}-Mg^{2+} sites has not been identified, although it seems likely that these two divalent binding sites play some role in the regulation of the contractile process in cardiac and skeletal muscles.

Pharmacological agents that may inhibit Ca^{2+} binding to troponin and the activation of actin-myosin interaction have been described for the cardiac system. Halothane, enflurane, and other general anesthetics depress myocardial contractility. These agents inhibit the maximum activated tension in myocardial skinned fibers and alter troponin regulation by increasing the concentration of Ca^{2+} required to activate the muscle fibers (29, 30). The degree of depression of contractile function in these preparations is not as great as that seen in living muscles, however, so that the primary pharmacological effects are probably related to alterations in membranes responsible for Ca^{2+} delivery and removal from the contractile elements. Thus the effect of these agents on contractile proteins at concentrations required for anesthesia is probably minimal.

Calmodulin and Thick Filament Regulation

Myosin from mammalian muscles contains a class of light-chain subunits that have many similar biochemical properties. The molecular weights of these myosin light chains range from 18,500 to 20,000, and they can bind divalent cations. Because these light chains from vertebrate tissues have the unique property of being phosphorylated by a specific type of a Ca^{2+}-dependent protein kinase (myosin light-chain kinase), they have been referred to as phosphorylatable light chains or P light chains (1, 10, 26, 31).

The P light chains of myosin from cardiac, skeletal, and smooth muscles are phosphorylated at a specific serine residue by myosin light-chain kinase (Fig. 3). The dephosphorylation of these light chains is catalyzed by another enzyme, myosin light-chain phosphatase (Fig. 3). The dependence of myosin light-chain kinase activity on Ca^{2+} was first described by Pires (23), and the Ca^{2+} concentrations required for activation were similar to those required to activate actomyosin ATPase activity. The enzyme was later identified in smooth and cardiac muscles as well as in several nonmuscle tissues (1, 10, 26). More recently it was shown that the Ca^{2+} sensitivity of the enzyme was mediated by a low-molecular-weight protein identified as calmodulin. Calmodulin has been shown to confer Ca^{2+} sensitivity on a variety of enzymatic reactions involving Ca^{2+}-dependent cyclic nucleotide phosphodiesterase, brain adenylate cyclase, erythrocyte membrane ATPase, and other enzymes (for recent review see 19). The general mechanism of activation for most of the calmodulin-dependent enzymes, including myosin light-chain kinase, involves a sequential process. Calcium first binds to calmodulin to form a Ca^{2+}-calmodulin complex that subsequently binds to the inactive catalytic subunit, resulting in a holoenzyme complex with enzymatic activity (Fig. 4).

Although myosin phosphorylation was first discovered in skeletal muscle, a biochemical function in skeletal muscle is less clear than it is in smooth muscle (26). Phosphorylation of the P light chain of skeletal muscle myosin is not essential for actin activation of myosin ATPase activity. Furthermore tension generated by threads made from purified actin and myosin is not affected by P light-chain phosphorylation.

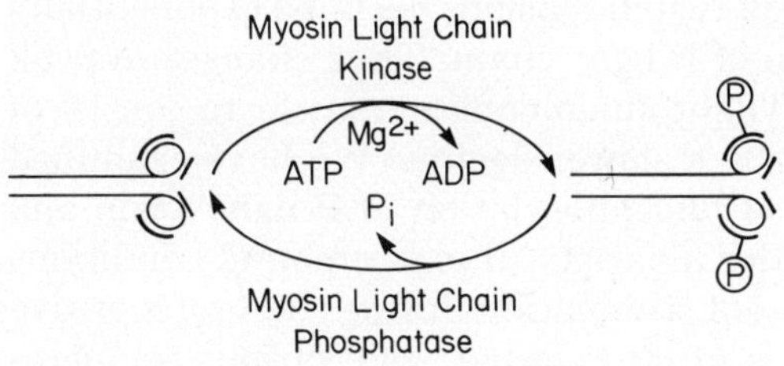

FIG. 3. Phosphorylation and dephosphorylation of myosin. A single type of light-chain subunit (P light chain) is phosphorylated by myosin light-chain kinase in presence of Ca^{2+}. Dephosphorylation is catalyzed in a Ca^{2+}-independent manner by myosin light-chain phosphatase.

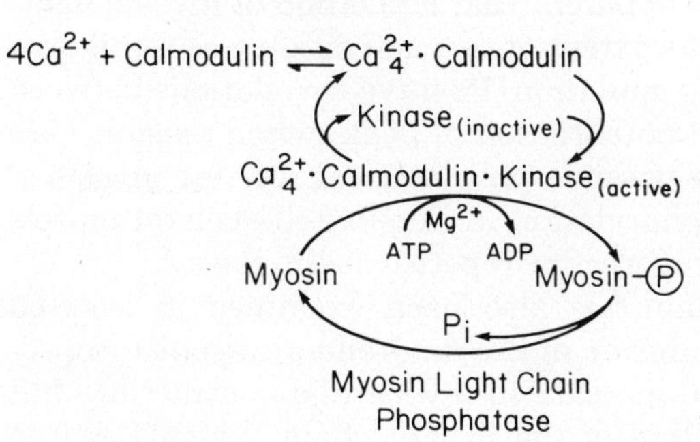

FIG. 4. Regulation of myosin light-chain kinase by Ca^{2+} and calmodulin.

To fully understand the mechanisms involved in the regulation of myosin light-chain kinase activity, we studied the activation properties of the purified enzyme (4) and compared them with the apparent activation properties observed in intact muscle. This biochemical approach allows control over particular experimental conditions such as ion concentration, pH, and temperature. Activation of purified myosin light-chain kinase activity follows a sequential, fully reversible process in which four Ca^{2+} bind to the four divalent metal binding sites on calmodulin. The Ca_4^{2+}-calmodulin then forms a complex with the inactive catalytic subunit of myosin light-chain kinase, which results in activation of enzymatic activity. The activation constant of the Ca_4^{2+}-calmodulin was 0.86 nM, indicating a very high affinity of calmodulin for the enzyme. The activation constant for Ca^{2+} was about 2 μM at a free-Mg^{2+} concentration of 2 mM. This value is in the range of Ca^{2+} concentrations that are thought to exist in muscles during a contractile process. Thus events leading to contraction via increased concentrations of cytoplasmic Ca^{2+} would be expected to initiate myosin phosphorylation.

We have reported changes in myosin light-chain phosphate content occurring in response to tetanic stimulation of fast-twitch skeletal muscle and have also proposed a possible mechanism for myosin light-chain phosphorylation in regulating contractile performance (20). Upon tetanic stimulation for 1 s, P light-chain phosphate increased in a time-dependent manner from a pretetanic resting value of 0.10 mol phosphate/mol P light chain to a maximum value of 0.70 mol phosphate/mol P light chain at 20 s and thereafter declined more slowly in a monoexponential fashion. During the 1-s tetanic contraction, the phosphate content of the P light chain did not differ significantly from pretetanic controls. A transient increase in peak twitch tension was observed after a 1-s tetanic stimulation. The peak twitch tension increased to maximal value at about 20 s, and subsequent peak twitch tensions declined exponentially at a rate similar to dephosphorylation of P light chain. Thus changes in peak twitch tension similar to changes in P light-chain content appear to consist of a moderately rapid increase followed by a slower decline. We have examined this relationship between the extent of phosphorylation of P light chain and potentiation of isometric twitches under a variety of experimental conditions (28). These include various frequencies of stimulation during the 1-s repetitive stimulation period, different durations of stimulation, and various temperatures. From these experiments it was apparent that activation of myosin light-chain kinase was extremely rapid. The extent of this activation was significant within 50 ms of the onset of tetanic stimulation. Positive correlations between P light-chain phosphate content and potentiation of peak twitch tension were established in these studies. Thus a possible physiological role for myosin P light-chain phosphorylation-dephosphorylation in fast-twitch skeletal muscle may be related to potentiation of tension after repetitive stimulation.

Phosphorylation of P light chain has also been examined in isolated working rabbit and rat hearts (14). Infusion of the β-adrenergic agonist isoproterenol for 30 s resulted in a marked increase in myocardial contractility but had no effect on the phosphate content of the P light chain, which was 0.44

mol phosphate/mol P light chain. Infusion of a high-Ca^{2+} concentration for as long as 18 min caused marked increases in left ventricular +dP/d*t*, but produced no significant effect on the phosphate content of the P light chain. Perfusing the isolated working rabbit heart with 0-mM Ca^{2+} buffer caused left ventricular +dP/d*t* to decrease to 0 within 30 s. At 30 s there was no significant decrease in the phosphate content of the P light chain. Continued perfusion with the 0-mM Ca^{2+} buffer, however, resulted in a slow rate of dephosphorylation of the P light chain so that after 30 min the phosphate content was 50% of the control value. Reperfusion with control buffer did not result in restoration of contractile activity, indicating myocardial viability had been severely compromised. Reversible arrest of contractile activity could be accomplished by increasing the perfusate K^+ concentration to 22 mM. This intervention caused complete cessation of contractile activity within 15 s. The P light-chain phosphate content was significantly decreased after 30 min. When the inactive hearts were reperfused with control buffer, left ventricular dP/d*t* was restored to control values within 10 min. Thus the hearts remained viable even though contraction was not occurring in the presence of 22 mM K^+. During the reperfusion with control buffer for 10 min, however, there was no increase in the phosphate content of the P light-chain control values. Thus a slow rate of dephosphorylation of the P light chain was found under two experimental conditions in which contractile activity was abolished. These results are analogous to the slow rate of dephosphorylation of the P light chain in skeletal muscle. They also indicate that there is no apparent correlation between acute changes in inotropic state and the extent of phosphorylation of the P light chain in cardiac muscle.

A slow rate of myosin light-chain kinase inactivation may explain the relatively high phosphate content of P light chain from cardiac muscle. In control hearts the phosphate content of the P light chain was 0.44 mol phosphate/mol light chain, which represents 44% of the sites being phosphorylated. The rate of contraction in the control perfused hearts was about 2 beats/s. Hence the period of relaxation during which the sarcoplasmic Ca^{2+} concentration remains low is very short. Assuming that the rate of inactivation of cardiac myosin light-chain kinase is comparable to that of the skeletal muscle enzyme ($t_{\frac{1}{2}}$ = 1.3 s; 28), the continuous contractile activity in cardiac muscle results in continuous activation of the kinase. Thus the net phosphate content of the P light chain would reflect the steady-state equilibrium between the rates of phosphorylation and dephosphorylation catalyzed by myosin light-chain kinase and myosin light-chain phosphatase, respectively. The fact that the P light chain in cardiac muscle is not fully phosphorylated under physiological conditions is consistent with the findings that cardiac muscle contains considerably less myosin light-chain kinase activity relative to myosin light-chain phosphatase activity than fast-twitch skeletal muscle. Further implications of the results from studies on phosphorylation of P light chain in cardiac muscle need to be explored. At this time, however, the physiological importance of P light-chain phosphorylation in cardiac muscle is not clear. It seems likely that if it does help modulate myocardial contraction, it is probably under a chronic long-term regulation associated with particular adaptations rather than acute changes in myocardial contractility.

SMOOTH MUSCLE CONTRACTILE ELEMENTS

Leiotonin and Thin Filament Regulation

Calcium regulation of contractile elements from smooth muscles is different from the type of regulation seen in cardiac and skeletal muscle (1, 10, 22). In smooth muscle systems, the presence of Ca^{2+} results in activation of an inactive state, whereas in cardiac and skeletal muscle the presence of Ca^{2+} results in the reversal of an inhibition of an active state. Ebashi and colleagues (see 22 for recent review) proposed that this regulatory scheme in smooth muscle is located in thin filaments. This system is composed of two distinct proteins, leiotonin A and leiotonin C, which are distinct from the troponin subunits found in cardiac and skeletal muscles. Actin activation of smooth muscle myosin ATPase activity did not occur unless leiotonin A, leiotonin C, and tropomyosin were added in the presence of Ca^{2+}. Leiotonin A has a molecular weight of 80,000, whereas leiotonin C has a molecular weight of 18,000 and is the Ca^{2+}-binding component of the leiotonin system. Leiotonin C has a number of biochemical properties that distinguish it from calmodulin and troponin C. Leiotonin exerts its maximal effect at a low ratio of 1:10 with tropomyosin, which is in contrast to the 1:1 molar ratio of troponin:tropomyosin required for regulation in skeletal and cardiac muscles. Leiotonin does not appear to phosphorylate myosin; the identification of the mechanism by which it exerts its effects is not known.

Calmodulin and Thick Filament Regulation

Evidence obtained from numerous studies (1, 10, 26) on actomyosin ATPase activity, tension of skinned fibers, and correlations linking myosin phosphorylation to tension generation in intact muscle suggests that myosin phosphorylation plays a key role in the contraction of various types of smooth muscles (1, 10, 26). The critical element that distinguishes this system from that described for cardiac and skeletal muscle is the effect of phosphorylation on the actin-myosin interaction. Phosphorylation of the P light chain appears to be required for actin activation of myosin ATPase activity. Many of the initial studies were done with partially purified preparations of actomyosin that also contained myosin light-chain kinase and calmodulin. The addition of Ca^{2+} resulted in an activation of myosin ATPase activity that paralleled light-chain phosphorylation, and the Ca^{2+} concentration required for activation of the ATPase activity was similar to that required for P light-chain phosphorylation. The extent of actin activation of the myosin ATPase activity depended directly on the extent of myosin phosphorylation. The biochemical properties of the regulation of the myosin phosphorylation were similar to those described for striated muscles (6). Calcium first binds to calmodulin, and the Ca^{2+}-calmodulin complex then activates the inactive catalytic subunit of myosin light-chain kinase with a subsequent phosphorylation of the P light chain of myosin. This sequence of biochemical reactions leads to a form of myosin that can interact with actin. A decrease in the Ca^{2+} concentration results in a reversal of this sequential activation process. The myosin light-chain kinase becomes inactive due to dissociation of the Ca^{2+}-calmodulin complex from the catalytic subunit, and myosin light-chain phosphatase dephosphorylates the P light chain of myosin, resulting in inhibition of actin activation of myosin ATPase activity. From this general biochemical scheme, one could then propose that myosin phosphorylation leads to contraction, whereas dephosphorylation results in relaxation of smooth muscles.

In skinned smooth muscle fiber preparations in which the sarcolemma has been functionally removed, P light-chain phosphorylation and generation of tension were equally sensitive to Ca^{2+} and both were reversed when Ca^{2+} concentrations were lowered to 0.01 μM (18). Calcium-dependent thiophosphorylation of the P light chains resulted in irreversible activation of tension generation that was no longer Ca^{2+} dependent. The thiophosphorylated P light chain is resistant to hydrolysis by myosin light-chain phosphatase and is irreversibly phosphorylated. These results suggest that Ca^{2+} regulation occurs at the level of myosin light-chain kinase and that once myosin is phosphorylated, the actin-myosin interaction is not regulated by Ca^{2+}. Also phosphorylation of smooth muscle myosin light chain appears to be essential or obligatory for actin-myosin interactions in skinned fibers and is consistent with data obtained on actin activation of the ATPase activity of phosphorylated myosin. Although the evidence suggests that phosphorylation is a primary mechanism for Ca^{2+} regulation of contractile elements in smooth muscle, Ca^{2+} control mechanisms involving leiotonin might also contribute to regulation of smooth muscle actin-myosin interactions.

Recent studies on the phosphorylation of myosin in intact smooth muscle cells are consistent with the general hypothesis that phosphorylation of smooth muscle myosin may play an important role in contraction and that the reaction may be regulated intracellularly by Ca^{2+}. Investigators have used swine carotid arteries (2, 8), rat uterine smooth muscle (16), and canine tracheal smooth muscle (7). The P light chain of myosin was phosphorylated in these different types of smooth muscles after the initiation of contraction by either high KCl concentrations or pharmacological agonists. Relaxation was correlated to dephosphorylation of the P light chain. When Ca^{2+} was removed from the physiological salt solutions, phosphorylation of the P light chain was reversibly inhibited.

Although these results generally support the conclusion that myosin phosphorylation is important in smooth muscle contraction, the mechanisms by which protein phosphorylation leads to shortening or tension development may be rather complex. An important study by Murphy and colleagues (8), with strips prepared from media of swine carotid arteries, has shown that with high levels of K^+, the extent of P light-chain phosphorylation increased during the first 30 s to a maximum and then decreased over the next 5 min to about one-half the maximum value. The isometric tension, however, continued to increase while the phosphate content in the P light chain was decreasing. The tension then maintained a steady high value. The velocities of isotonic contractions were recorded for the various tensions, and the force-velocity relationship was constructed at different times after stimulation with K^+. The maximum contraction velocity increased in the early phase of the contraction and then decreased in parallel with the level of myosin phosphorylation. Thus there was a good correlation between the velocity of muscle shortening and the extent of P light-chain phosphorylation. These results suggest that with a subsequent decline in velocity there would be a decrease in energy utilization with a maintenance of tension. The biochemical mechanism associated with a decrease in P light-chain phosphorylation is not clear, although it must reflect a decrease in the activity of myosin light-chain kinase activity compared with myosin light-chain phosphatase activity. Clearly more information is needed on the properties of myosin phosphorylation in isolated smooth muscle to determine if there are distinct properties associated with different kinds of smooth muscle and to develop mechanistic models by which myosin phosphorylation leads to tension and shortening in smooth muscle.

CALMODULIN AND CALCIUM ANTAGONISTS

The phenothiazine antipsychotic drugs are potent inhibitors of the calmodulin-activated form of cyclic nucleotide phosphodiesterase (32). The mechanism of inhibition of cyclic nucleotide phosphodiesterase by antipsychotic drugs seems to be the result of the drug binding to Ca^{2+}-calmodulin, thereby inhibiting activation. Since the primary effect of antipsychotic drugs on this enzyme system is apparently an inactivation of the Ca^{2+}-calmodulin complex, these compounds and similar drugs might also be expected to inhibit various other enzyme systems activated by Ca^{2+}-calmodulin (e.g., myosin light-chain kinase). It was shown in skinned smooth muscle fibers that inhibition of tension development correlated with the potency of the antipsychotic drugs for binding to Ca^{2+}-calmodulin and that this inhibition of tension correlated with inhibition of myosin phosphorylation (5). There was also a positive correlation between the extent of inhibition of chicken gizzard smooth muscle actomyosin ATPase activity and the extent of inhibition of P light-chain phosphorylation (11). Pharmacological agents such as *N*-(6-aminohexyl)-5-chloro-1-naphthalenesulfonamide (W 7), chlorpromazine, prenylamine, and N^2-dansyl-L-arginine-4-*tert*-butylpiperidine amide (No. 233) interact with calmodulin and inhibit in a dose-dependent manner Ca^{2+} activation of chicken gizzard actomyosin and myosin phosphorylation. These agents bind to calmodulin (12) and cause relaxation of vascular smooth muscle (13). These particular compounds do not inhibit the ATPase activity of gizzard smooth muscle actomyosin when myosin is in the phosphorylated form. Thus the inhibitory effects of these compounds are probably related to the inhibition of myosin light-chain kinase activity via interaction with Ca^{2+}-calmodulin. Although these agents inhibit the activation of smooth muscle myosin light-chain kinase by Ca^{2+}-calmodulin, it is not clear whether the antipsychotic phenothiazines would cause relaxation of smooth muscle via this mechanism under therapeutic situations.

CONCLUSION

Calcium plays a ubiquitous role in excitation-contraction coupling. This chapter has emphasized the differences in Ca^{2+} modulation of contraction in skeletal, cardiac, and smooth muscles. Contractile elements of skeletal and cardiac muscles share some common structural and physiological properties. In both, Ca^{2+} binds to a specific contractile protein regulatory subunit, troponin, to remove inhibition of actin-myosin interaction. Calcium also activates a Ca^{2+}-dependent protein kinase that in turn phosphorylates myosin light chains. Skeletal and cardiac muscles appear to differ, however, in their physiological responses to light-chain phosphorylation. In skeletal muscle, light-chain phosphorylation may alter tension development in a dynamic manner. In cardiac muscle, light-chain phosphorylation does not appear to play a significant role in the phasic regulation of actin-myosin interaction. The potential tonic roles of Ca^{2+}-mediated cardiac light-chain phosphorylation remain to be demonstrated. The mechanisms of Ca^{2+} modulation of contraction appear quite different in smooth muscle. Although there is some controversy about whether there are regulatory proteins in the thin filaments, the biochemical properties of Ca^{2+} regulation of actin-myosin interaction are distinct from those described for cardiac and skeletal muscles. In addition Ca^{2+} stimulation of myosin light-chain phosphorylation probably plays a pivotal role in activating contractile element interaction. Details of these activation processes remain to be elucidated.

The unique aspects of Ca^{2+} regulation of the interaction between contractile elements of these basic muscle types may help explain differences in the responses to drugs that alter Ca^{2+} availability. These responses are discussed in the following chapters.

REFERENCES

1. ADELSTEIN, R. S., AND E. EISENBERG. Regulation and kinetics of the actin-myosin-ATP interaction. *Annu. Rev. Biochem.* 49: 921–956, 1980.
2. BARRON, J. T., M. BÁRÁNY, K. BÁRÁNY, AND R. V. STORTI. Reversible phosphorylation and dephosphorylation of the 20,000 dalton light chain of myosin during contraction-relaxation-contraction cycle of arterial smooth muscle. *J. Biol. Chem.* 255: 6238–6244, 1980.
3. BLINKS, J. R., R. RÜDEL, AND S. R. TAYLOR. Calcium transients in isolated amphibian skeletal muscle fibers: detection with aequorin. *J. Physiol. London* 277: 291–323, 1978.
4. BLUMENTHAL, D. K., AND J. T. STULL. Activation of skeletal muscle myosin light chain kinase by calcium (2+) and calmodulin. *Biochemistry* 19: 5608–5614, 1980.
5. CASSIDY, P., P. E. HOAR, AND W. G. L. KERRICK. Inhibition of Ca^{2+}-activated tension and myosin light chain phosphorylation in skinned smooth muscle strips by the phenothiazines. *Pfluegers Arch.* 387: 115–120, 1980.
6. DABROWSKA, R., J. M. F. SHERRY, D. K. AROMATORIO, AND D. J. HARTSHORNE. Modulator protein as a component of the myosin light chain kinase from chicken gizzard. *Biochemistry* 17: 253–258, 1978.
7. DELANEROLLE, P., AND J. T. STULL. Myosin phosphorylation during contraction and relaxation of tracheal smooth muscle. *J. Biol. Chem.* 255: 9993–10000, 1980.

8. DILLON, P. F., M. O. AKSOY, S. P. DRISKA, AND R. A. MURPHY. Myosin phosphorylation and the cross-bridge cycle in arterial smooth muscles. *Science* 211: 495–497, 1981.
9. FAY, F. S., H. H. SHLEVIN, W. C. GRANGER, JR., AND S. R. TAYLOR. Aequorin luminescence during activation of single isolated smooth muscle cells. *Nature London* 280: 506–508, 1979.
10. HARTSHORNE, D. J., AND A. GORECKA. Biochemistry of the contractile proteins of smooth muscle. In: *Handbook of Physiology. The Cardiovascular System*, edited by D. F. Bohr, A. P. Somlyo, and H. V. Sparks, Jr. Bethesda, MD: Am. Physiol. Soc., 1980, sect. 2, vol. II, chapt. 4, p. 93–120.
11. HIDAKA, H., M. NAKA, AND T. YAMAKI. Effect of novel specific myosin light chain kinase inhibitors on Ca^{2+}-activated Mg^{2+}-ATPase of chicken gizzard actomyosin. *Biochem. Biophys. Res. Commun.* 90: 694–699, 1979.
12. HIDAKA, H., T. YAMAKI, M. NAKA, T. TANAKA, H. HAYASHI, AND R. KOBAYASHI. Calcium-regulated modulator protein interacting agents inhibit smooth muscle calcium-stimulated protein kinase and ATPase. *Mol. Pharmacol.* 17: 66–72, 1980.
13. HIDAKA, H., T. YAMAKI, T. TOTSUKA, AND M. ASANO. Selective inhibitors of Ca^{2+}-binding modulator of phosphodiesterase produce vascular relaxation and inhibit actin-myosin interaction. *Mol. Pharmacol.* 15: 49–59, 1979.
14. HIGH, C. W., AND J. T. STULL. Phosphorylation of myosin in perfused rabbit and rat hearts. *Am. J. Physiol.* 239 (*Heart Circ. Physiol.* 8): H756–H764, 1980.
15. HOLROYDE, M. J., S. P. ROBERTSON, J. D. JOHNSON, R. J. SOLARO, AND J. D. POTTER. The calcium and magnesium binding sites on cardiac troponin and their role in the regulation of myofibrillar adenosine triphosphatase. *J. Biol. Chem.* 255: 11688–11693, 1980.
16. JANIS, R. A., B. M. MOATS-STAATS, AND R. T. GUALTIERI. Protein phosphorylation during spontaneous contraction of smooth muscle. *Biochem. Biophys. Res. Commun.* 96: 265–270, 1980.
17. JOHNSON, J. D., J. H. COLLINS, S. P. ROBERTSON, AND J. D. POTTER. A fluorescent probe study of Ca^{2+} binding to the Ca^{2+} specific sites of cardiac troponin and troponin C. *J. Biol. Chem.* 255: 9635–9640, 1980.
18. KERRICK, W. G. L., P. E. HOAR, AND P. S. CASSIDY. Calcium activated tension: the role of myosin light chain phosphorylation. *Federation Proc.* 39: 1558–1563, 1980.
19. KLEE, C. B., T. H. CROUCH, AND P. G. RICHMAN. Calmodulin. *Annu. Rev. Biochem.* 49: 489–515, 1980.
20. MANNING, D. R., AND J. T. STULL. Myosin light chain phosphorylation and phosphorylase *a* activity in rat extensor digitorum longus muscle. *Biochem. Biophys. Res. Commun.* 90: 164–170, 1979.
21. MARBAN, E., T. J. RINK, R. W. TSIEN, AND R. Y. TSIEN. Free calcium in heart muscle at rest and during contraction measured with Ca^{2+}-sensitive microelectrodes. *Nature London* 286: 845–850, 1980.
22. NONOMURA, Y., AND S. EBASHI. Calcium regulatory mechanism in vertebrate smooth muscle. *Biomed. Res.* 1: 1–14, 1980.
23. PIRES, E., S. V. PERRY, AND M. A. W. THOMAS. Myosin light chain kinase, a new enzyme from striated muscle. *FEBS Lett.* 41: 292–296, 1974.
24. POTTER, J. D., AND J. GERGELY. The calcium and magnesium binding sites on troponin and their role in the regulation of myofibrillar adenosine triphosphatase. *J. Biol. Chem.* 250: 4628–4633, 1975.
25. RAY, K. P., AND P. J. ENGLAND. Phosphorylation of the inhibitory subunit of troponin and its effect on the calcium dependency of cardiac myofibril adenosine triphosphatase. *FEBS Lett.* 70: 11–16, 1976.
26. STULL, J. T. Phosphorylation of contractile proteins in relation to muscle function. *Adv. Cyclic Nucleotide Res.* 13: 39–93, 1980.
27. STULL, J. T., AND J. E. BUSS. Calcium binding properties of beef cardiac troponin. *J. Biol. Chem.* 253: 5932–5938, 1978.
28. STULL, J. T., C. F. SANFORD, D. R. MANNING, D. K. BLUMENTHAL, AND C. W. HIGH. Phosphorylation of myofibrillar proteins in striated muscles. *Cold Spring Harbor Conferences on Cell Proliferation: Protein Phosphorylation*, edited by E.G. Krebs and O.K. Rosen. Cold Spring Harbor, N.Y.: Cold Spring Harbor Lab., in press.
29. SU, J. Y., AND W. G. L. KERRICK. Effects of halothane on Ca^{2+}-activated tension development in mechanically disrupted rabbit myocardial fibers. *Pfluegers Arch.* 375: 111–117, 1978.
30. SU, J. Y., AND W. G. L. KERRICK. Effects of enflurane on functionally skinned myocardial fibers from rabbits. *Anesthesiology* 52: 385–389, 1980.
31. TAYLOR, E. W. Mechanism of actomyosin ATPase and the problem of muscle contraction. *CRC Crit. Rev. Biochem.* 6: 103–164, 1980.
32. WEISS, B., AND R. M. LEVIN. Mechanism for selectively inhibiting the activation of cyclic nucleotide phosphodiesterase and adenylate cyclase by antipsychotic agents. *Adv. Cyclic Nucleotide Res.* 9: 285–303, 1978.
33. WILKINSON, J. M., AND R. J. A. GRAND. Comparison of amino acid sequence of troponin I from different striated muscles. *Nature London* 277: 31–35, 1978.

4

Effects of Calcium Antagonists on Sarcoplasmic Reticular and Sarcolemmal Vesicles From Canine Heart

HENRY R. BESCH, JR., LARRY R. JONES, AND KURT A. MADDOCK

Departments of Pharmacology and Toxicology and of Medicine and the Krannert Institute of Cardiology, Indiana University School of Medicine, Indianapolis, Indiana

THE CONCEPT THAT ORGANIC COMPOUNDS can competitively interfere with the actions of the inorganic cation Ca^{2+} developed from the early studies of Lindner with congeners of verapamil (23). Originally thought to be β-adrenergic receptor antagonists, verapamil and its methoxy derivative D 600 were soon recognized to be blockers of the slow inward current of the cardiac action potential, which is carried principally by Ca^{2+} (8). The importance of this action led not only to systematic efforts to synthesize other Ca^{2+}-channel blockers, but also to new efforts to interpret the actions of a variety of other drugs in terms of potential Ca^{2+} antagonistic effects.

Recent studies have shown that Ca^{2+} antagonism by organic compounds is not limited to just one of the multiple roles of Ca^{2+} in cellular homeostasis. In the broadest sense, Ca^{2+} antagonists are a class of chemically heterogeneous compounds that share only the ability to interfere with recognized cellular effects of Ca^{2+}. Thus verapamil and its congeners may be subclassified as Ca^{2+} entry antagonists, whereas some other chemical classes of compounds (Fig. 1), such as the methylenedioxyindenes (29) and ryanodine (13, 33), appear to antagonize the effect of Ca^{2+} at specific intracellular sites.

MEMBRANE REGULATION OF CELLULAR CALCIUM HOMEOSTASIS

Acute Ca^{2+} homeostasis in cardiac tissue depends principally on the operation of components of two main membrane systems: the sarcolemma (Sl)

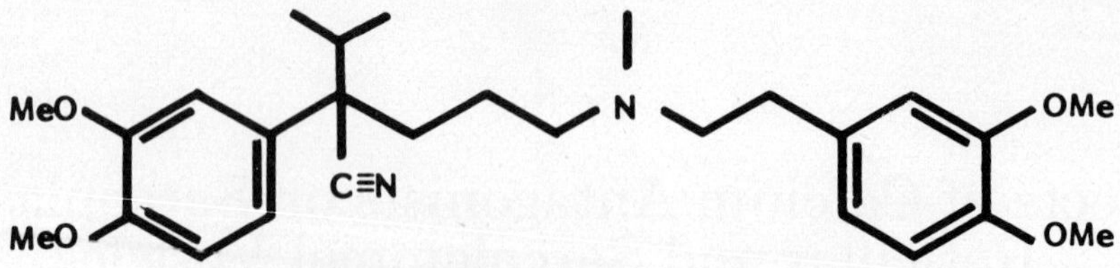

Verapamil

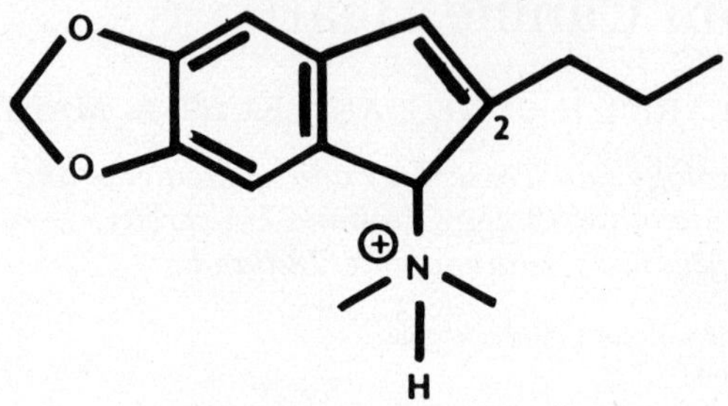

2 - Propyl - MDI

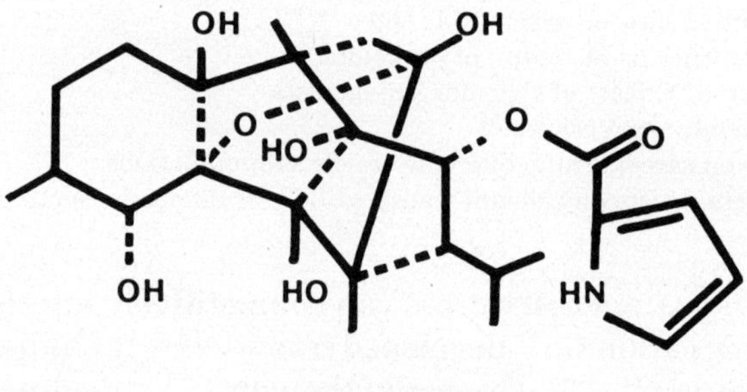

Ryanodine

FIG. 1. Structures of verapamil, 2-propyl-methylenedioxyindene (2-propyl-MDI), and ryanodine. Although less potent, 2-propyl-MDI had effects similar to those of verapamil and D 600 (not shown).

and the sarcoplasmic reticulum (SR). The Sl contains at least three functional components subserving Ca^{2+} handling: *1)* the slow Ca^{2+} channel that provides the major pathway for Ca^{2+} entry into the cell interior; *2)* the Na^{+}-Ca^{2+} exchange diffusion carrier, which may operate to move Ca^{2+} in either direction across the Sl; and *3)* an adenosine triphosphate (ATP)-driven Ca^{2+} pump that can be functionally distinguished from the SR Ca^{2+} pump by use of Na^{+} (4, 34). Cardiac SR contains not only a powerful Ca^{2+} pump to remove Ca^{2+} from the cytosol, but also a Ca^{2+} channel(s) that permits Ca^{2+} passage from the SR interior to the cytosol. Under some conditions Ca^{2+} efflux from the SR can occur via the Ca^{2+} pump itself, operating in a "backward" mode. However, much evidence suggests that distinct channels are also present (6, 18, 26). The recent direct demonstration of unit conductances of monovalent-cation channels in SR vesicles melded with black lipid membranes (27) supports the notion that monovalent cations participate in regulation of the SR Ca^{2+} pump (14, 15).

Cardiac Microsomal Preparations

Cardiac microsomes form a convenient in vitro model of the operation of the SR. These microsomes [or membrane vesicles (MV)] consist largely (60–70%) of vesicles of SR origin that can accumulate and release Ca^{2+}. Cardiac microsomes also contain (about 15–30%) vesicles of Sl origin that are relatively tightly sealed (2) and that can be used to demonstrate Ca^{2+} accumulation occurring via Na^+-Ca^{2+} exchange diffusion (28, 30). Cardiac microsomes cannot reliably be used to examine the active, ATP-driven Sl Ca^{2+} pumps present in this mixture of vesicles because of the large proportion of highly active SR Ca^{2+} pumps. Selective analyses of the effects of Ca^{2+} antagonists therefore require studies with more highly purified Sl and SR vesicle subpopulations.

Functional Subfractions of Cardiac Membrane Vesicles

Recently it has become possible to study functional subfractions of SR vesicles. Skeletal muscle SR fractions have been subdivided into heavy and light fractions presumably representing terminal cisternae enriched in Ca^{2+}-K^+-ATPase and longitudinal tubular vesicles, respectively (24). Using a different methodology, we have prepared functional subfractions of cardiac SR, which we term heavy and light particulate fractions (10). We adopted this terminology for convenience, even though our subfractions are not comparable to the skeletal SR fractions. We have found that our heavy SR fraction is more enriched in Ca^{2+}-K^+-ATPase activity than the light fraction (16). It appears likely that further functional subfractionation (e.g., into transverse tubular and peripheral vesicles) of cardiac Sl will soon be experimentally available, following the leads established with skeletal muscle membranes (3, 21, 31).

EFFECTS OF CALCIUM ANTAGONISTS ON CALCIUM ACCUMULATION BY CARDIAC MEMBRANE VESICLES

Calcium antagonists at high concentrations have been shown to decrease nonspecifically active Ca^{2+} accumulation by MVs (36); it is well known that they also produce blockade of the slow Ca^{2+} channel (8). This chapter focuses on two means by which Ca^{2+} accumulation by cardiac SR vesicles is enhanced by Ca^{2+} antagonists. First, we show that the magnitude and even the net direction (inhibitory or stimulatory) of the effect of the Ca^{2+} entry antagonist D 600 depend on the ionic conditions of the incubation. Second, we present data showing that the intracellular Ca^{2+} antagonist ryanodine may well act specifically on a Ca^{2+}-efflux channel of cardiac SR. (The actions of ryanodine on intact skeletal muscle and on isolated skeletal muscle microsomes differ from those on cardiac tissue. For further discussion of this point, see 13, 33.) With relatively purified subfractions of cardiac MVs, the effects of ryanodine could be localized to the light SR fraction, whereas the effects of D 600 were not limited to light SR. Neither ryanodine nor verapamil could be shown to affect Ca^{2+} transport by Sl vesicles.

Effects of Monovalent Cations on Action of D 600

Cardiac MVs were prepared and assayed as previously described (2, 13–15; for recent reviews see 9, 12). Calcium accumulation by cardiac MVs represents in large part the action of the SR vesicular component. The rates of the Ca^{2+} accumulation and the associated ATP hydrolysis are modulated in parallel by monovalent cations (Fig. 2). Potassium is the most effective of the monovalent cations (18). Since no synergism occurs between Na^+ and K^+, the monovalent-cation effect on Ca^{2+}-ATPase activity is clearly distinguishable from the monovalent-cation effect on the Na^+ pump of the Sl (11). The rate of Ca^{2+} uptake, but not the vesicular Ca^{2+} capacity, is diminished in the absence of K^+.

The magnitude of effect of drugs on Ca^{2+} accumulation by SR vesicles is also influenced by monovalent cations. This is shown for 10^{-4} M D 600 in Figure 3. Concentration-effect curves (Fig. 4) show that D 600 stimulates Ca^{2+} accumulation about 1.5-fold at lower drug concentrations but completely inhibits Ca^{2+} transport at higher concentrations. In the presence of K^+, all the responses to D 600 are attenuated. Parallel experiments on the Ca^{2+}-ATPase activity show that the stimulatory effects of D 600 occur with no change in the rate of ATP hydrolysis, whereas the inhibitory effects at higher D 600 concentrations occur on Ca^{2+}-ATPase as well as on Ca^{2+} accumulation (data not shown).

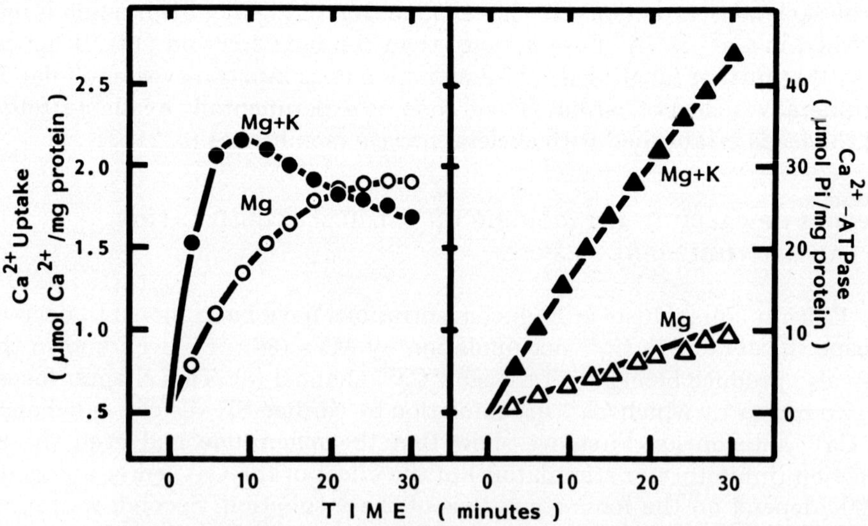

FIG. 2. Stimulation by K^+ of Ca^{2+} uptake (*left panel*) and Ca^{2+}-ATPase activity (*right panel*) of cardiac membrane vesicles. For Ca^{2+} uptake vesicles were incubated at 37°C for times indicated on abscissa in a medium containing 50 mM histidine (pH 6.8), 3 mM $MgCl_2$, 50 μM $^{45}CaCl_2$, 10–20 μg/ml protein, and 3 mM tris(hydroxymethyl)aminomethane (Tris)-oxalate, with (●, ▲) or without (○, △) 100 mM KCl. Reaction was started with 3 mM Tris-ATP and terminated by millipore filtration. ATPase assays were carried out in parallel under identical conditions, except that duplicate tubes were run with and without 1 mM EGTA. Values plotted represent EGTA-inhibitable activity.

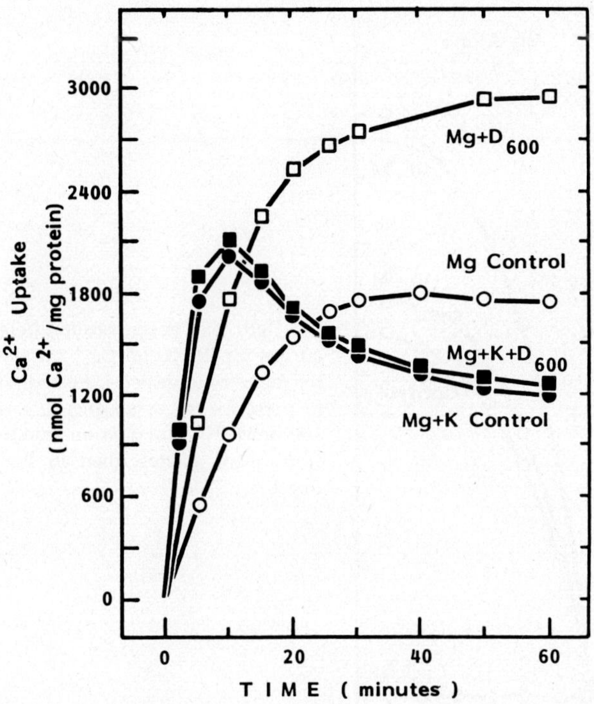

FIG. 3. Alteration by K^+ of effect of 10^{-4} M D 600 on Ca^{2+} uptake by cardiac membrane vesicles. Uptake conditions were identical to those described in Figure 2, except for presence of D 600.

The varying effectiveness of D 600, which depends on the presence of K^+, is not based simply on the effects of ionic strength. It also occurs when choline is substituted for K^+, even though choline itself does not stimulate Ca^{2+} transport (Fig. 5).

Effects of Monovalent Cations on Action of Ryanodine

The intracellular Ca^{2+} antagonist ryanodine augments both the rate and the extent of Ca^{2+} uptake by cardiac microsomes, and the degree of this stimulation varies with the ionic conditions (Fig. 6). Calcium uptake measured in the absence of monovalent cations exhibits only a low sensitivity to ryanodine, and Ca^{2+} uptake is not enhanced by addition of choline ions. In contrast, ryanodine augments Ca^{2+} uptake three- to fivefold when K^+ is present at optimal concentrations. Oxalate-supported Ca^{2+} uptake by cardiac MVs can reach 10–12 μmol/mg protein in the presence of ryanodine when the extravesicular Ca^{2+} concentration is maintained at about 10 μM by a calcium ethylene glycol-bis(β-aminoethylether)-*N,N'*-tetraacetic acid (EGTA) buffer system of adequate buffer capacity. These results demonstrate that the Ca^{2+} uptake capacity of cardiac muscle microsomes approaches that of fast skeletal muscle microsomes. The effect of ryanodine occurs without a measurable increase in Ca^{2+}-dependent ATPase activity, suggesting that the Ca^{2+} pump turnover rate

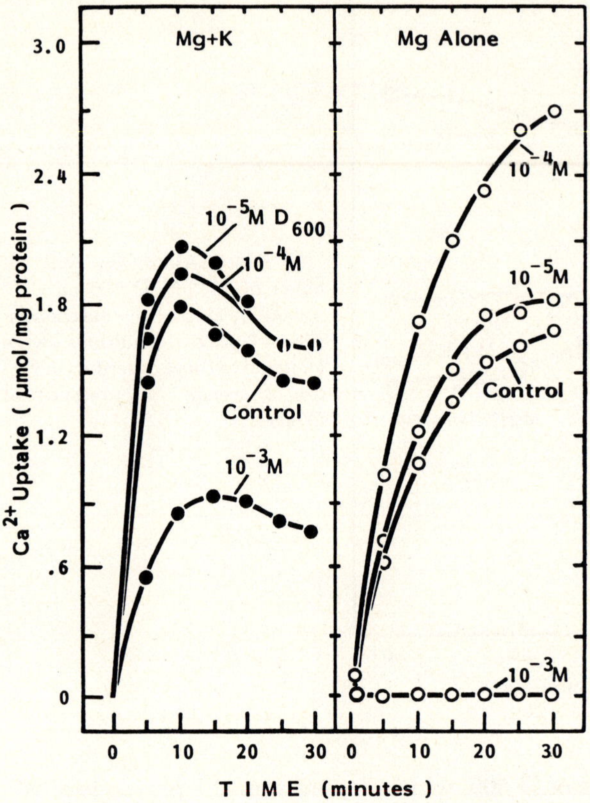

FIG. 4. Concentration-effect curves for D 600 on Ca^{2+} uptake by membrane vesicles, measured in presence (●) or absence (○) of 100 mM K^{+}. Incubation conditions were as described in Figure 2.

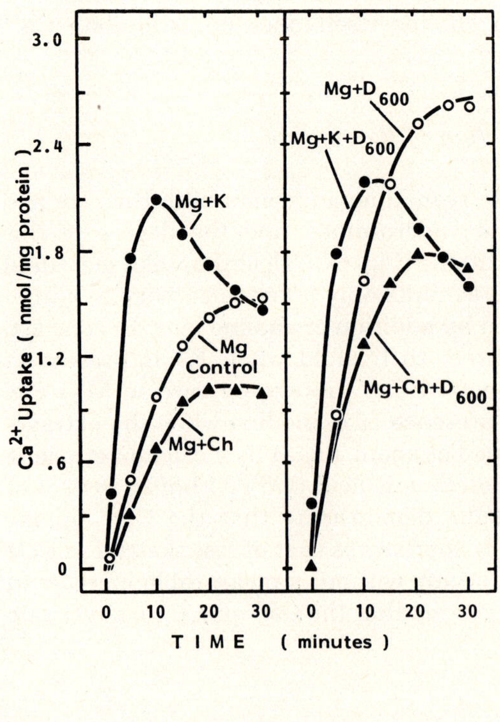

FIG. 5. Comparison of effects of no monovalent cation (○), K^{+} (●), and choline (Ch; ▲) on Ca^{2+} uptake by cardiac membrane vesicles, measured in absence (*left*) and presence (*right*) of 10 μM D 600. Incubation conditions were as described in Figure 2.

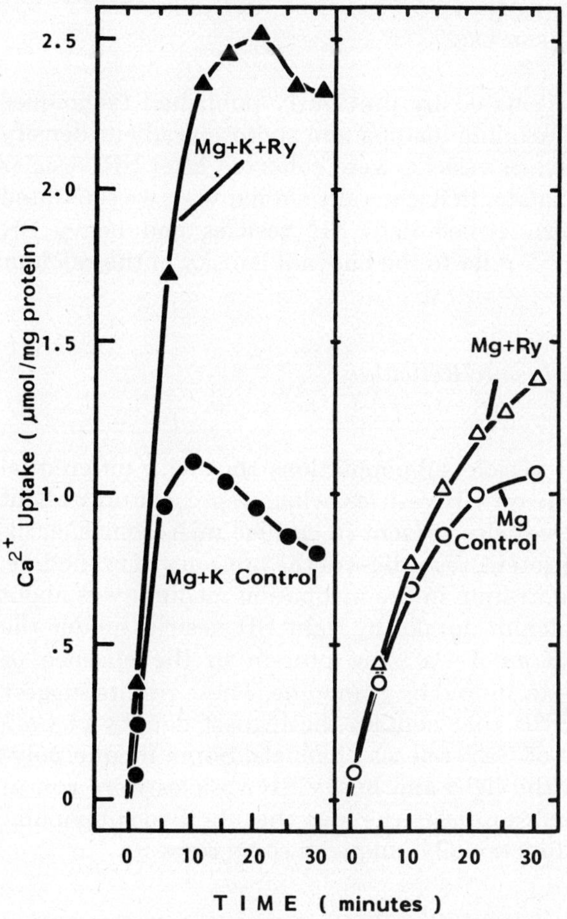

FIG. 6. Ryanodine (Ry)-induced stimulation of Ca^{2+} uptake by cardiac membrane vesicles in presence (*left*) or absence (*right*) of 100 mM K^+. Ryanodine concentration was 100 μM. Incubation conditions were as described in Figure 2.

is not increased. It thus seems reasonable to suggest that the augmented rate of net Ca^{2+} accumulation in the presence of ryanodine results from a decrease in the rate of Ca^{2+} efflux from the vesicles. Thus the levels of Ca^{2+} accumulation by cardiac microsomes usually reported seem limited by a concomitant Ca^{2+} release that opposes maximal Ca^{2+} accumulation within the vesicles. It is equally plausible that ryanodine directly increases the coupling efficiency of the cardiac SR Ca^{2+} pump, since coupling ratios considerably less than 2 are routinely found for cardiac microsomes. Our very recent results indeed show that Ca^{2+} turnover rates (occurring after net accumulation has reached a steady state) are higher in the presence of ryanodine than in its absence (16). The rates of Ca^{2+} turnover under steady-state conditions have been suggested (19, 20) to represent simply unidirectional Ca^{2+} influx. The observation that Ca^{2+}-influx rates (and necessarily efflux rates) apparently are increased when a higher steady-state level of Ca^{2+} has been achieved within the vesicles does not establish a molecular mechanism of action. At this point we are not able to demonstrate unequivocally whether ryanodine's primary effect involves faster Ca^{2+}-influx rates or slower (relative) efflux rates.

SUBCELLULAR LOCALIZATION OF EFFECTS OF CALCIUM ANTAGONISTS: SUBFRACTIONS OF MEMBRANE VESICLES

Cardiac MVs were subfractionated by previously published techniques based partly on limited calcium oxalate loading and sucrose gradient density centrifugation (5, 11, 17). Briefly, Sl vesicles were collected after SR vesicles had been loaded with calcium oxalate. In a generally similar way, we separated SR vesicles into subpopulations termed light SR vesicles and heavy SR vesicles, where "light" and "heavy" refer to the buoyant density of the calcium oxalate–loaded vesicles.

Effects of Ryanodine on Sarcoplasmic Reticular Vesicle Subpopulations

Calcium uptake by the SR vesicle subpopulations showed a differential response to ryanodine (Fig. 7). Heavy SR vesicles, which represent only about 15% of the total number of SR vesicles present in cardiac microsomes, accumulate almost 9 μmol Ca^{2+}/mg protein, regardless of the presence of ryanodine. Note that the added Ca^{2+} concentration in the incubation medium was about 100 μM in this experiment. Calcium uptake by light SR vesicles under the same conditions was about 1.8 μmol Ca^{2+}/mg protein in the absence of ryanodine and was stimulated 5- to 10-fold by ryanodine. These results suggest that those areas of the cardiac SR that contain the highest density of Ca^{2+} pumps have the lowest density of Ca^{2+}-release channels. Some unique polypeptide bands were seen when the light and heavy SR vesicles were run in polyacrylamide gels (16). This substantiates the idea that the two subpopulations are both structurally and functionally unique in some respects.

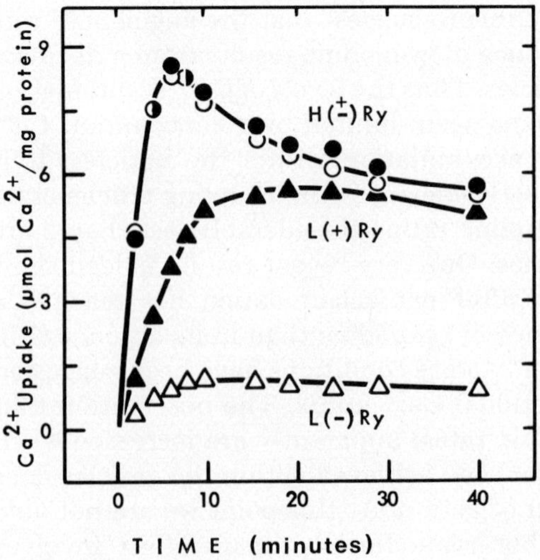

FIG. 7. Effect of ryanodine (Ry) on Ca^{2+} uptake by subpopulations of cardiac SR membrane vesicles. L, light SR vesicles; H, heavy SR vesicles. Ryanodine concentration was 100 μM. Incubation conditions were as described in Figure 2.

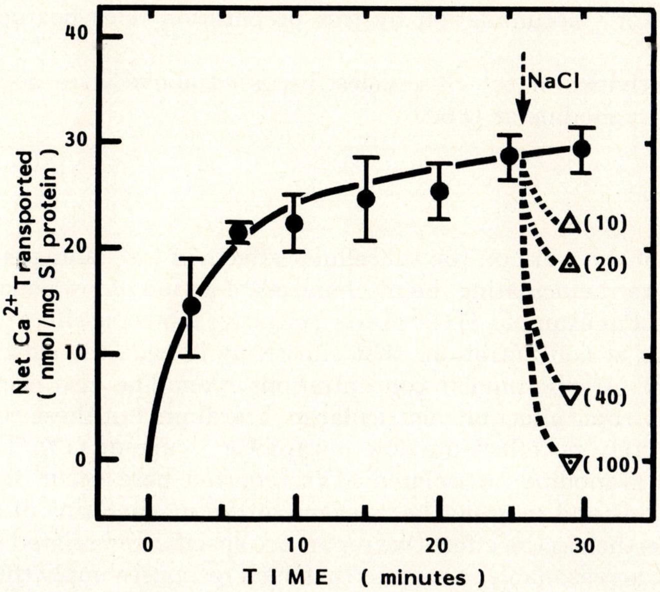

FIG. 8. The ATP-dependent Ca^{2+} accumulation by Sl membrane vesicles. Reaction medium contained 20 mM morpholinopropane sulfonic acid (MOPS)-Tris (pH 7.4), 3 mM $MgCl_2$, 50 μM $^{45}CaCl_2$, 100 mM KCl, and 3 mM Tris-ATP. Vesicles were diluted 30-fold at time indicated by arrow into medium containing NaCl in the millimolar concentrations shown in parentheses. Reactions were stopped by filtration through GF-C filters. Values are means ± SE for 4 measurements on a single preparation but are typical of those in 8 other preparations.

Effects of Calcium Antagonists on Calcium Transport by Sarcolemmal Vesicles

Isolation procedures for obtaining cardiac Sl vesicles have recently been improved to increase yield and purity (17). The preparations used in the studies presented here generally yield full Na^+-K^+-ATPase activities of 100–175 μmol $P_i \cdot$ mg protein$^{-1} \cdot h^{-1}$, representing an enrichment of up to 80-fold over the homogenate activities. Fully activated adenylate cyclase activity of these Sl vesicles is 10–20 nmol cyclic adenosine 3′,5′-monophosphate$\cdot$mg$^{-1}\cdot$ min^{-1}. At most about 15% of the native Sl vesicles are leaky. Sarcoplasmic reticular contamination (estimated from Ca^{2+}-K^+-ATPase activity) represents less than 5% of the total ATPase activity of the preparation. These Sl vesicles thus appear suitable for functional and enzymatic studies, the specific activities of the positive enzyme markers being 3–7 times higher than those of other recently reported Sl preparations (1, 22, 32).

About 15% of these Sl vesicles are oriented inside-out and can accumulate Ca^{2+} by ATP-dependent active transport. Virtually all the accumulated Ca^{2+} can be released by addition of Na^+ in sufficient concentrations (Fig. 8). These data, the fact that oxalate has no effect on this Ca^{2+} accumulation (data not shown), and several other distinguishing criteria (34) demonstrate that ATP-dependent Ca^{2+} accumulation by this preparation does not occur into SR vesicles.

The activities of the Sl vesicles discussed above were not consistently affected by ryanodine or D 600.

CONCLUSION

Much of the work on the subcellular effects of Ca^{2+} antagonists has been directed toward elucidating the mechanisms of action of low concentrations of these drugs. For example, D 600 exerts a negative inotropic effect on ventricular myocardium at concentrations that effectively impede the slow inward Ca^{2+} current (35). At micromolar concentrations ryanodine also exerts an initial negative inotropic effect on ventricular myocardium, but these concentrations have absolutely no effect on slow inward Ca^{2+} current (13). The effects of D 600 and ryanodine on isolated MVs reported here occur at higher drug concentrations and may not be relevant to the mechanisms of action of the drugs. Nevertheless the effects we report are specific and related to concentration (13). It is reasonable to suggest that such responses imply the existence of specific ion-flux pathways across the SR membranes.

Both D 600 and ryanodine increase the net amount of Ca^{2+} accumulation by cardiac SR under certain conditions. The action of D 600 on Ca^{2+} accumulation by cardiac SR vesicles was antagonized by the presence of the regulatory monovalent cation K^+, whereas the effect of ryanodine was augmented by K^+. Since the SR Ca^{2+} pump seems to operate electrogenically (7, 25, 37), the differential effectiveness of the drugs that depends on the K^+ concentration may somehow reflect interactions between the molecular substrates of these drugs and the SR vesicle membrane potential.

REFERENCES

1. ALSTYNE, E., D. K. BARTSCHAT, N. V. WELLSMITH, S. L. POE, W. P. SCHILLING, AND G. E. LINDENMAYER. Isolation of a highly enriched sarcolemma membrane fraction from canine heart. *Biochim. Biophys. Acta* 553: 388–395, 1979.
2. BESCH, H. R., JR., L. R. JONES, AND A. M. WATANABE. Intact vesicles of canine cardiac sarcolemma: evidence from vectorial properties of Na^+, K^+-ATPase. *Circ. Res.* 39: 586–595, 1976.
3. CAMPBELL, K. P., C. FRANZINI-ARMSTRONG, AND A. E. SHAMOO. Further characterization of light and heavy sarcoplasmic reticulum vesicles. Identification of the "sarcoplasmic reticulum feet" associated with heavy sarcoplasmic reticulum vesicles. *Biochim. Biophys. Acta* 602: 97–116, 1980.
4. CARONI, P., AND E. CARAFOLI. An ATP-dependent Ca^{2+}-pumping system in dog heart sarcolemma. *Nature London* 283: 765–767, 1980.
5. CARSTEN, M. E., AND M. K. REEDY. Cardiac sarcoplasmic reticulum: chemical and electron microscope studies of calcium accumulation. *J. Ultrastruct. Res.* 35: 554–574, 1971.
6. CARVALHO, A. P., AND B. LEO. Effects of ATP on the interaction of Ca^{++}, Mg^{++}, and K^+ with fragmented sarcoplasmic reticulum isolated from rabbit skeletal muscle. *J. Gen. Physiol.* 50: 1327–1352, 1967.
7. FARMEN, R. H., L. R. JONES, AND H. R. BESCH, JR. Transmembrane potential and Ca^{2+} release from cardiac sarcoplasmic reticulum vesicles: modulation by selective cation ionophores. *Circulation* 59: 2–223, 1979.
8. FLECKENSTEIN, A., H. KAMMERMEIER, H. J. DÖRING, AND H. J. FREUND. Zum Wirkungsmechanismus neuartiger Koronadilatatoren mit gleichzeitig Sauerstoffeinsparenden Myokard-Effekten, Prenylamin und Iproveratril. *Z. Kreislaufforsch.* 56: 716–744, 1967.
9. JONES, L. R., AND H. R. BESCH, JR. Calcium handling by cardiac sarcoplasmic reticulum. *Tex. Rep. Biol. Med.* 39: 19–36, 1979.
10. JONES, L. R., AND H. R. BESCH, JR. Isolation of functionally distinct subpopulations of cardiac sarcoplasmic reticulum. *Federation Proc.* 38: 744, 1979.
11. JONES, L. R., H. R. BESCH, JR., J. W. FLEMING, M. M. McCONNAUGHEY, AND A. M. WATANABE. Separation of vesicles of cardiac sarcolemma from vesicles of cardiac sarcoplasmic reticulum. Comparative biochemical analysis of component activities. *J. Biol. Chem.* 254: 530–539, 1979.

12. JONES, L. R., H. R. BESCH, JR., AND K. MADDOCK. Isolation of canine cardiac sarcolemmal vesicles. In: *Methods in Pharmacology. Myocardial Biology*, edited by A. Schwartz. New York: Plenum, vol. 5, in press.
13. JONES, L. R., H. R. BESCH, JR., J. L. SUTKO, AND J. T. WILLERSON. Ryanodine-induced stimulation of net Ca^{2+}-uptake by cardiac sarcoplasmic reticulum vesicles. *J. Pharmacol. Exp. Ther.* 209: 48–55, 1979.
14. JONES, L. R., H. R. BESCH, JR., AND A. M. WATANABE. Monovalent cation stimulation of Ca^{2+}-uptake by cardiac membrane vesicles: correlation with stimulation of Ca^{2+}-ATPase activity. *J. Biol. Chem.* 252: 3315–3323, 1977.
15. JONES, L. R., H. R. BESCH, JR., AND A. M. WATANABE. Regulation of the calcium pump of the cardiac sarcoplasmic reticulum: interactive roles of potassium and ATP on the phosphoprotein intermediate of the K^+, Ca^{2+}-ATPase. *J. Biol. Chem.* 253: 1643–1653, 1978.
16. JONES, L. R., AND S. E. CALA. Biochemical evidence for functional heterogeneity of cardiac sarcoplasmic reticulum vesicles. *J. Biol. Chem.* in press.
17. JONES, L. R., S. W. MADDOCK, AND H. R. BESCH, JR. Unmasking effect of alamethicin on the (Na^+, K^+)-ATPase, beta-adrenergic receptor coupled adenylate cyclase, and cAMP-dependent protein kinase activities of cardiac sarcolemmal vesicles. *J. Biol. Chem.* 255: 9971–9980, 1980.
18. KASAI, M., AND T. KOMETANI. Inhibition of anion permeability of sarcoplasmic reticulum vesicles by 4-acetoamido-4′-isothiocyano-stilbene-2,2′-disulfonate. *Biochem. Biophys. Acta* 557: 243–247, 1979.
19. KATZ, A. M., C. F. LOUIS, D. I. REPKE, G. FUDYMA, P. A. KUPSAWY, AND M. SHIGEKAWA. Time-dependent changes of calcium influx and efflux in rabbit skeletal muscle sarcoplasmic reticulum. *Biochim. Biophys. Acta* 596: 94–107, 1980.
20. KIRCHBERGER, M. A., AND D. WONG. Calcium efflux from isolated cardiac sarcoplasma reticulum. *J. Biol. Chem.* 253: 6941–6945, 1978.
21. LAU, Y. H., A. H. CASWELL, AND J. P. BRUNSCHWIG. Isolation of transverse tubules by fractionation of triad junctions of skeletal muscle. *J. Biol. Chem.* 252: 5565–5574, 1977.
22. LEVITSKY, D. O., M. K. ALIEV, A. V. KUZMIN, T. S. LEVCHENKO, V. M. SMIRNOV, AND E. I. CHAZOV. Isolation of calcium pump system and purification of calcium ion-dependent ATPase from heart muscle. *Biochim. Biophys. Acta* 443: 468–484, 1976.
23. LINDER, E. Phenyl-propyl-diphenyl-propyl-amin, eine neue Substanz mit coronargefasserweiternder Wirkung. *Arzneim. Forsch.* 10: 569–573, 1960.
24. MEISSNER, G. Isolation and characterization of two types of sarcoplasmic reticulum vesicles. *Biochem. Biophys. Acta* 389: 51–68, 1975.
25. MEISSNER, G. Calcium transport and monovalent cation and proton fluxes in sarcoplasmic reticulum vesicles. *J. Biol. Chem.* 256: 636–643, 1981.
26. McKINLEY, D., AND G. MEISSNER. Evidence for a K^+, Na^+ permeable channel in sarcoplasmic reticulum. *J. Membr. Biol.* 44: 159–186, 1978.
27. MILLER, C. Voltage-gated cation conductance channel from fragmented sarcoplasmic reticulum: steady-state electrical properties. *J. Membr. Biol.* 40: 1–23, 1978.
28. PITTS, B. J. R. Stoichiometry of sodium-calcium exchange in cardiac sarcolemmal vesicles. *J. Biol. Chem.* 254: 6232–6235, 1979.
29. RAHWAN, R. G., M. M. FAUST, AND D. T. WITIAK. Pharmacologic evaluation of new calcium antagonists: 2-substituted 3-methylamine-5, 6-methlene-dioxyindenes. *J. Pharmacol. Exp. Ther.* 201: 126–137, 1977.
30. REEVES, J. P., AND J. L. SUTKO. Sodium-calcium exchange in cardiac membrane vesicles. *Proc. Natl. Acad. Sci. USA* 76: 590–594, 1979.
31. SCALES, D., AND R. SABBADINI. Microsomal T system: a stereological analysis of purified microsomes derived from normal and dystrophic skeletal muscle. *J. Cell Biol.* 83: 33–46, 1979.
32. SULAKHE, P. V., AND P. J. ST. LOUIS. Passive and active fluxes across plasma membranes. *Prog. Biophys. Mol. Biol.* 35: 135–195, 1980.
33. SUTKO, J. L., J. T. WILLERSON, H. R. BESCH, JR., AND L. R. JONES. Ryanodine: alterations of papillary muscle contractile state and responsiveness to inotropic interventions. *J. Pharmacol. Exp. Ther.* 209: 37–47, 1979.
34. TRUMBLE, W. R., J. L. SUTKO, AND J. P. REEVES. ATP-dependent calcium transport in cardiac sarcolemmal membrane vesicles. *Life Sci.* 27: 207–214, 1980.
35. WATANABE, A. M., AND H. R. BESCH, JR. Cyclic AMP modulation of slow calcium influx channels in guinea pig hearts. *Circ. Res.* 35: 316–324, 1974.
36. WATANABE, A. M., AND H. R. BESCH, JR. Subcellular myocardial effects of verapamil and D 600: comparisons with propranolol. *J. Pharmacol. Exp. Ther.* 191: 241–251, 1974.
37. ZIMMIAK, P., AND E. RACKER. Electrogenicity of Ca^{2+} transport catalyzed by the Ca^{2+}-ATPase from sarcoplasmic reticulum. *J. Biol. Chem.* 253: 4631–4637, 1978.

5

Fundamental Actions of Calcium Antagonists on Myocardial and Cardiac Pacemaker Cell Membranes

A. FLECKENSTEIN

Physiological Institute, University of Freiburg, Freiburg, West Germany

EFFECTS OF CALCIUM ANTAGONISTS ON CARDIOVASCULAR FUNCTIONS

The discovery of Ca^{2+} antagonism as a new principle of action of coronary drugs has given a new orientation to cardiovascular research and therapy. The roots of this development reach back to 1964, when it was reported by Fleckenstein that two new coronary dilators, later given the names verapamil (iproveratril) and prenylamine, mimic the cardiac effects of simple Ca^{2+} withdrawal (4). Both substances, like Ca^{2+} deficiency, diminish Ca^{2+}-dependent contractile force without a major change in action potential. Both verapamil and prenylamine reduce Ca^{2+}-dependent high-energy phosphate utilization of the beating heart and lower extra-oxygen consumption connected with contractile activity. But these inhibitory effects can promptly be neutralized with elevated Ca^{2+}, β-adrenergic catecholamines, or cardiac glycosides—measures that restore the Ca^{2+} supply to the contractile system. In an extensive search for other Ca^{2+} antagonists of this type, a considerable number of substances that met these criteria were identified in our laboratory during the following years (see Table 1). In fact, Ca^{2+} antagonism turned out to represent a new pharmacodynamic approach to cardiac inhibition that could be clearly distinguished from β-receptor blockade (10, 15). Therefore in 1969 we introduced the term *competitive Ca^{2+} antagonist* as a novel drug designation (17). At low extracellular Ca^{2+} concentrations, the effects of Ca^{2+} antagonists were greatly increased, whereas they promptly disappeared when the Ca^{2+} concentration was elevated above normal. Moreover we further demonstrated *1)* that specific Ca^{2+} antagonists can also block excitation-contraction coupling of vascular smooth muscle and, in this manner, lower Ca^{2+}-dependent coronary vascular tone and neutralize all types of experimental coronary spasms (13, 18, 21, 22);

TABLE 1. *Specific Ca^{2+}-antagonistic inhibitors of excitation-contraction coupling in myocardium and vascular smooth muscle*

Group A

Verapamil, Mol wt 454.59	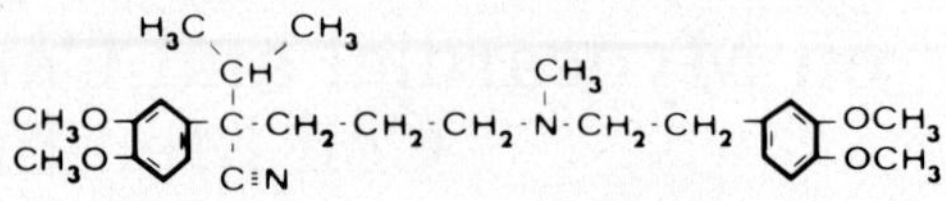
Compound D 600, Mol wt 485.59	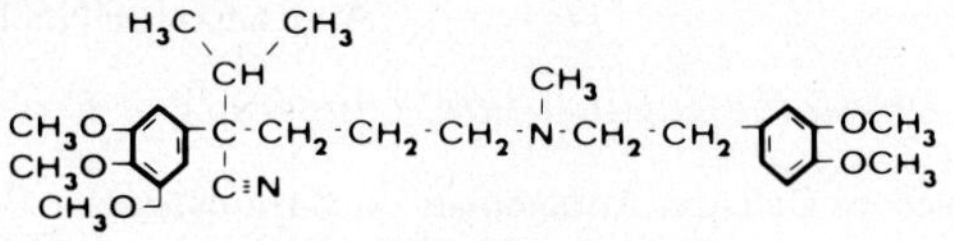
Nifedipine, Mol wt 346.34	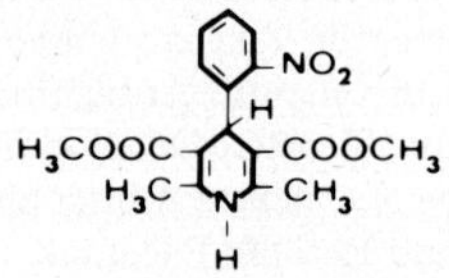
Niludipine, Mol wt 490.55	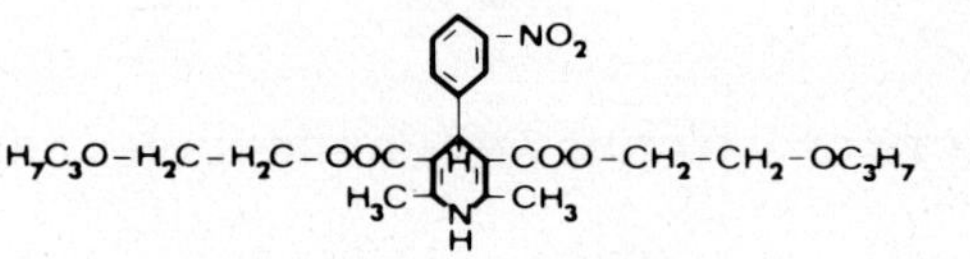
Nimodipine, Mol wt 418.45	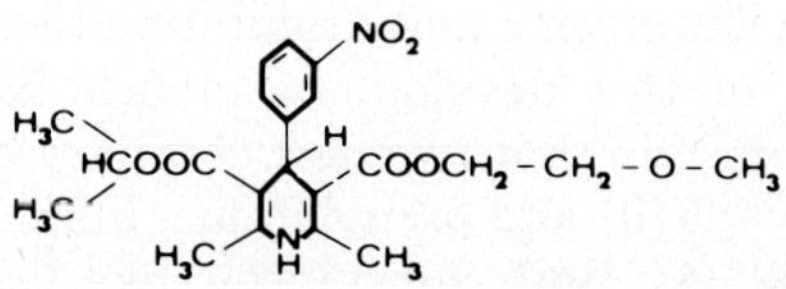
Diltiazem, Mol wt 414.52	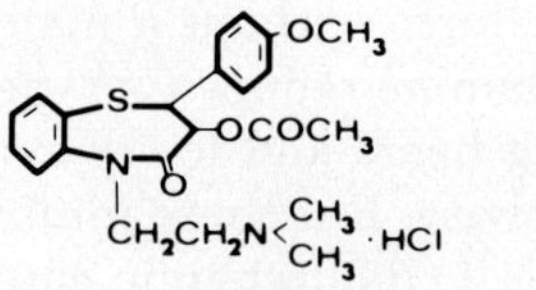

2) that the Ca^{2+} antagonists interfere with the uptake of labeled ^{45}Ca into the myocardium and prevent myocardial necrotization arising from deleterious intracellular Ca^{2+} overload (5, 7, 9, 14); and *3*) that according to voltage-clamp studies with verapamil, D 600, and nifedipine, Ca^{2+} antagonists act as specific inhibitors of the slow transsarcolemmal Ca^{2+} influx but do not (or only slightly) affect the fast Na^{+} current (25, 27). In addition we and other research groups simultaneously found that the inhibition by Ca^{2+} antagonists of cardiac pacemaker activity and atrioventricular conduction is also due to an interference with transmembrane Ca^{2+} influx because Ca^{2+} necessarily serves as an electrical charge carrier in nodal excitation (2, 23, 26, 37, 43, 45, 46).

Verapamil, D 600 (a methoxy derivative of verapamil), nifedipine and its

TABLE 1.—*Continued*

Group B

Prenylamine,
Mol wt 329.46

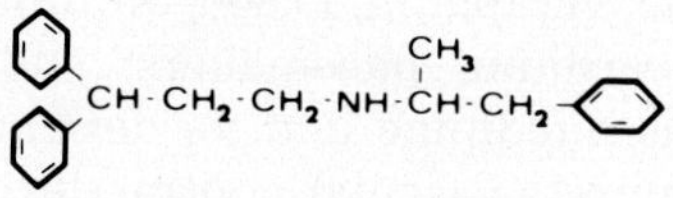

Fendiline,
Mol wt 315.46

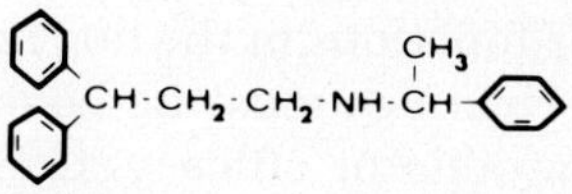

Terodiline,
Mol wt 281.0

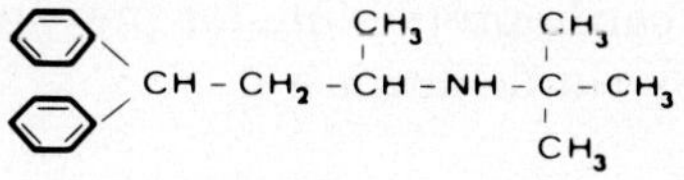

Perhexiline,
Mol wt 277.50

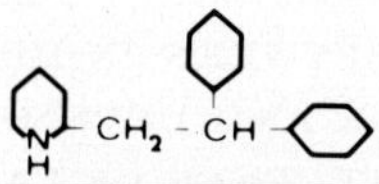

From studies of Fleckinstein and colleagues (6–9, 11–13, 15–19, 21, 25, 27). The Ca^{2+}-antagonistic features of diltiazem were determined by Nakajima et al. (32).

derivatives, nimodipine, and niludipine are the most specific and powerful Ca^{2+} antagonists (12, 15, 16, 17). In 1975 Japanese pharmacologists (32) contributed diltiazem to the Ca^{2+}-antagonist family. We have subdivided the substances listed in Table 1 into two groups based on our comparative studies. Group A comprises the Ca^{2+} antagonists of outstanding efficacy such as verapamil, D 600, nifedipine, niludipine, nimodipine, and diltiazem. These substances can inhibit Ca^{2+}-dependent excitation-contraction coupling of the mammalian ventricular myocardium by 90% or more before the fast Na^{+} influx during the rising phase of the action potential is also affected. Moreover they do not interfere with transmembrane Mg^{2+} conductivity. Group B, on the other hand, includes prenylamine (15), fendiline (11), terodiline, and perhexiline (19). These drugs are somewhat less potent and specific than those in Group A. This means that they induce a concomitant inhibition of the Na^{+}-dependent excitatory process when Ca^{2+}-dependent contractile tension development of isolated papillary muscles has been reduced by 50–70%. Such strong contractile depression can never be observed in vivo, however, as long as the drug concentrations remain in the therapeutic range. In other words, the significance of the influence of the group B Ca^{2+} antagonists on the fast transmembrane Na^{+} influx is more academic than practical. Interestingly, the substances of group B also reduce the Mg^{2+} influx through the excited myocardial fiber membranes.

Thus there are certain peculiarities that indicate that the Ca^{2+} antagonists are not a homogeneous group. Nevertheless ample evidence exists to indicate that all Ca^{2+} antagonists listed in Table 1 belong to the same pharmacological group. This group will undoubtedly grow. However, new compounds must be properly verified by a series of measurements on myocardium and vascular smooth muscle with electrophysiological, biochemical, and isotopic techniques before they can be identified as specific Ca^{2+} antagonists. It should also be noted in this context that the relative effects on myocardium, cardiac pacemakers, and vascular smooth muscle differ for each type of Ca^{2+} antagonist used. For example, verapamil, D 600, or diltiazem exert their Ca^{2+}-antagonistic actions on myocardium, pacemakers, and vasculature with almost equal strength, whereas nifedipine and its derivatives (niludipine and nimodipine) preferentially suppress vascular contractility and have only a moderate effect on cardiac pacemaker functions in the dog and human. Due to this divergence, nifedipine and its derivatives must be considered the Ca^{2+} antagonists with the most powerful vasodilator efficacy. Conversely, verapamil is particularly useful for direct cardioprotection, for prevention of myocardial Ca^{2+} overload, and for antidysrhythmic purposes.

CORRELATION BETWEEN INHIBITION OF SLOW CHANNEL–MEDIATED CALCIUM INFLUX AND DEVELOPMENT OF CONTRACTILE TENSION UNDER THE INFLUENCE OF CALCIUM ANTAGONISTS

Figure 1 illustrates our experiments (6) on an isolated guinea pig papillary muscle treated with a high concentration (10^{-5} M) of verapamil. Voltage-clamp measurements showed that this concentration of verapamil almost completely blocked the slow Ca^{2+} inward current (I_{si}). Consequently, in this papillary muscle, contractility was also rapidly abolished. However, resting potential and action potential parameters such as upstroke velocity and height of the

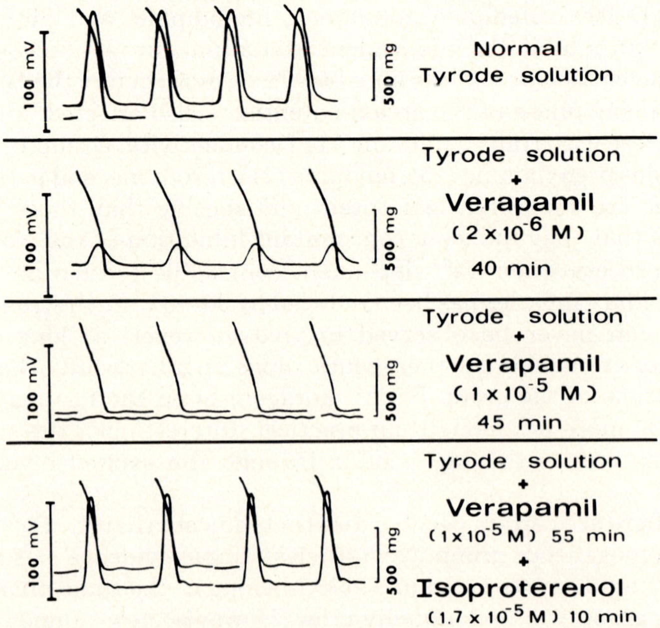

FIG. 1. Selective loss of cardiac contractility under influence of a high overdose of verapamil and the reversal of this effect by isoproterenol. Experiments on an electrically driven (2 shocks/min) guinea pig isolated papillary muscle in Tyrode's solution with 2 mM Ca^{2+}. [From Fleckenstein (6).]

overshoot remained practically unchanged. Only the plateau of action potential appeared to be slightly shortened by verapamil. This is because Ca^{2+} can no longer contribute to the maintenance of depolarization during the late phase of the plateau when the transmembrane Ca^{2+} conductivity has been blocked. The I_{si} is affected in an opposite manner by Ca^{2+} antagonists and sympathetic transmitter substances. Calcium antagonists restrict the transmembrane Ca^{2+} influx, whereas β-receptor stimulation promotes the Ca^{2+} inward current even in the presence of Ca^{2+} antagonists. Accordingly, as shown at the bottom of Figure 1, β-receptor stimulation with isoproterenol totally reversed the verapamil-induced suppression of myocardial contractility.

These investigations, however, were not sufficiently precise for deciding whether the measurable decrease of transsarcolemmal Ca^{2+} influx by Ca^{2+} antagonists really determines the degree of contractile inhibition in a strictly quantitative manner or whether excitation-contraction uncoupling by Ca^{2+} antagonists is so predominantly due to this specific action on the slow Ca^{2+} channel that any additional interference with intracellular Ca^{2+} movements can be disregarded. A number of investigators have attempted to answer these questions. It was shown in our laboratory that control of contractile force by Ca^{2+} antagonists is lost in skinned myocardial fibers that are prepared by glycerin-water extraction or according to Winegrad's (42) technique. In such skinned fibers, Ca^{2+} antagonists were not inhibitory when contractile responses were directly induced by addition of adenosine triphosphate (ATP) and Ca^{2+} (8). This indicated that the structural integrity of the cardiac sarcolemma

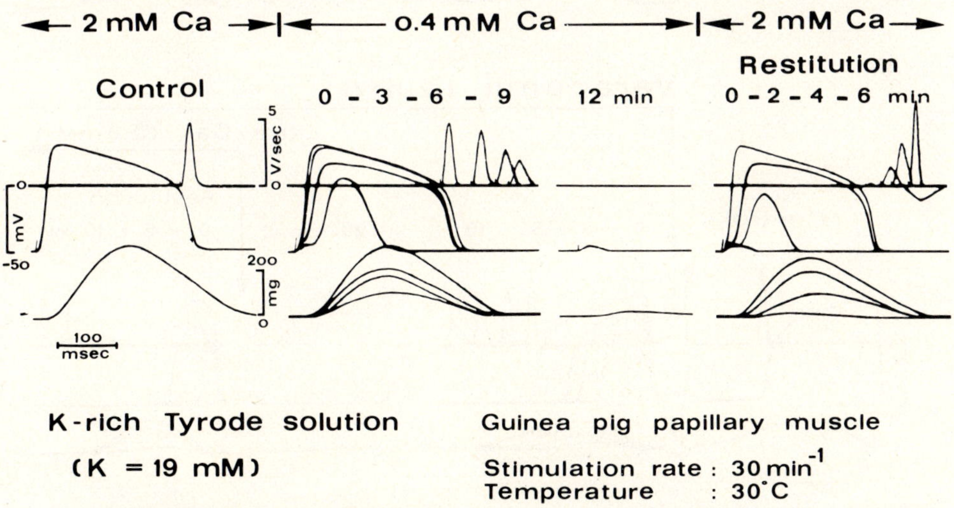

FIG. 2. Parallel suppression of Ca^{2+}-dependent electrical excitability and contractility of a partially depolarized guinea pig papillary muscle in K^{+}-rich (19 mM) Tyrode's solution by lowering the Ca^{2+} content from 2 to 0.4 mM. The dV/dt_{max} of the single-fiber action potentials is directly registered in the top set of graphs. Superimposed traces below of the action potentials and the corresponding mechanograms show the progressive decay of the bioelectrical and contractile functions at time intervals of 3, 6, 9, and 12 min after muscle was transferred to solution with less Ca^{2+}. After readmission of 2 mM Ca^{2+} at the end of the experiment, there was full recovery of all bioelectrical and mechanical parameters to their original heights within 6 min. Bioelectrical activity was recorded during 70 min with an inserted microelectrode on 1 single myocardial fiber.

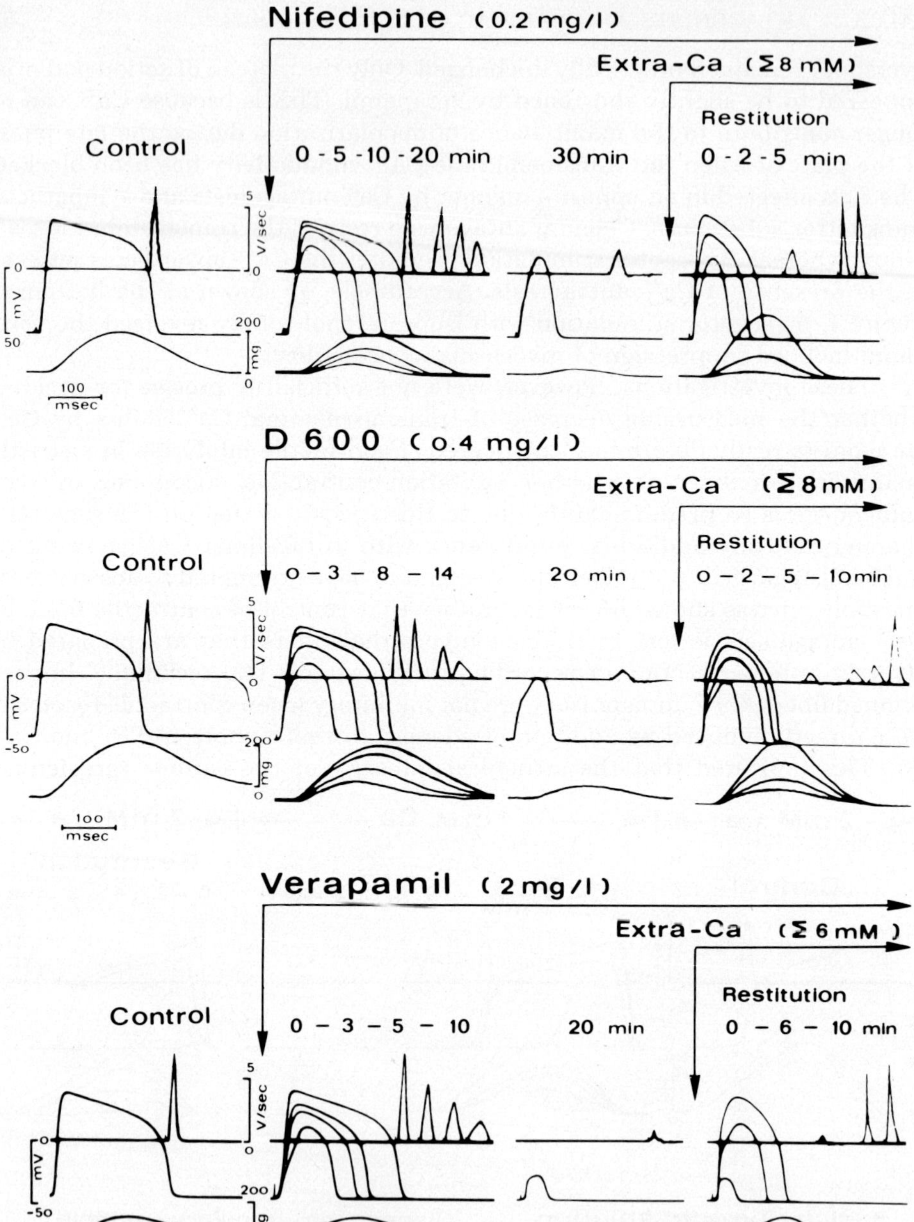

FIG. 3. Parallel suppression by Ca^{2+} antagonists nifedipine, D 600, and verapamil of the Ca^{2+}-dependent bioelectrical and mechanical activities of partially depolarized guinea pig papillary muscles in K^+-rich (19 mM) Tyrode's solution with normal Ca^{2+} content (2 mM). There were 3 identical experiments with 3 different Ca^{2+} antagonists. In each experiment the *upper curve* shows the gradual reduction of dV/dt_{max} during an observation period of 20–40 min after addition of

membrane is an indispensable prerequisite for the action of Ca^{2+} antagonists. Obviously these agents lose their effectiveness if Ca^{2+} has free access to the myofibrils where ATP is split. Hence there is certainly no direct effect of Ca^{2+} antagonists on the Ca^{2+}-dependent activation of myofibrillar ATPase and on energy transformation in the myofilaments. Similarly, in fractionized sarcoplasmic reticulum, several attempts to show an inhibitory effect of negative inotropic doses of verapamil on binding, accumulation, or exchange of Ca^{2+} have failed (3, 34, 38). Moreover in isolated cardiac mitochondria, only an excessive concentration of verapamil (more than 1,000 times greater than that required to produce contractile failure) suppressed Ca^{2+} uptake (20). Thus evidence of a significant action of Ca^{2+} antagonists on intracellular sites is lacking.

Therefore it seemed more promising to concentrate on the cardiac sarcolemma membrane and to analyze, under the influence of Ca^{2+} antagonists, the quantitative correlation between inhibition of Ca^{2+} influx and the decrease in contractile force. For this study isolated cat and guinea pig papillary muscles, in which the fast Na^+ transport system had previously been blocked, were used. For instance, if the cardiac fiber membrane potential is reduced to −45 mV, the fast Na^+ inward current becomes inactivated, whereas the slow Ca^{2+} channels continue to respond to electrical stimulation. The simplest method to produce this critical degree of depolarization is to add excess K^+ to the experimental solution (31). Such predepolarized myocardial fibers can still conduct propagated impulses. In this situation, however, both contractile performance and transmembrane bioelectrical activity depend on the slow inward Ca^{2+} current. Clearly such Ca^{2+}-mediated action potentials in partially depolarized myocardial fibers are highly susceptible to changes in the extracellular Ca^{2+} concentration. Thus upstroke velocity, as well as the height and duration of Ca^{2+}-carried action potentials, rapidly decreases in a Ca^{2+}-deficient medium. In the experiment illustrated in Figure 2, bioelectrical activity and contractile function totally disappeared within 12 min after the Ca^{2+} concentration had been reduced from 2 mM to 0.4 mM. The Ca^{2+}-dependent action potentials and contractile force, however, recovered within 6 min after return to the original medium containing 2 mM Ca^{2+}.

The decisive result, however, was that in partially depolarized myocardium, the Ca^{2+} antagonists perfectly imitate the effects of simple Ca^{2+} deficiency. Figures 3 and 4 illustrate results obtained with nifedipine (0.2 mg/liter), D 600 (0.4 mg/liter), verapamil (2 mg/liter), diltiazem (4 mg/liter), perhexiline maleate (20 mg/liter), and fendiline (30 mg/liter). In all these experiments, the time course of inhibition of the Ca^{2+}-mediated action potentials coincides with the development of contractile blockade. At the end of the experiments represented in Figures 3 and 4, the effects of the Ca^{2+} antagonists were overcome by increasing the Ca^{2+} concentration to 6 or 8 mM. The bioelectrical

drugs. *Superimposed records* underneath show corresponding decrease of height and duration of Ca^{2+}-dependent action potentials and of isometric mechanograms. Finally, the cardiodepressant effects of the Ca^{2+} antagonists were overcome by raising Ca^{2+} content of K^+-rich Tyrode's solutions from 2 to 6 or 8 mM. Bioelectrical records were derived in each experiment from microelectrode insertions on the same myocardial fiber. [From Späh and Fleckenstein (36).]

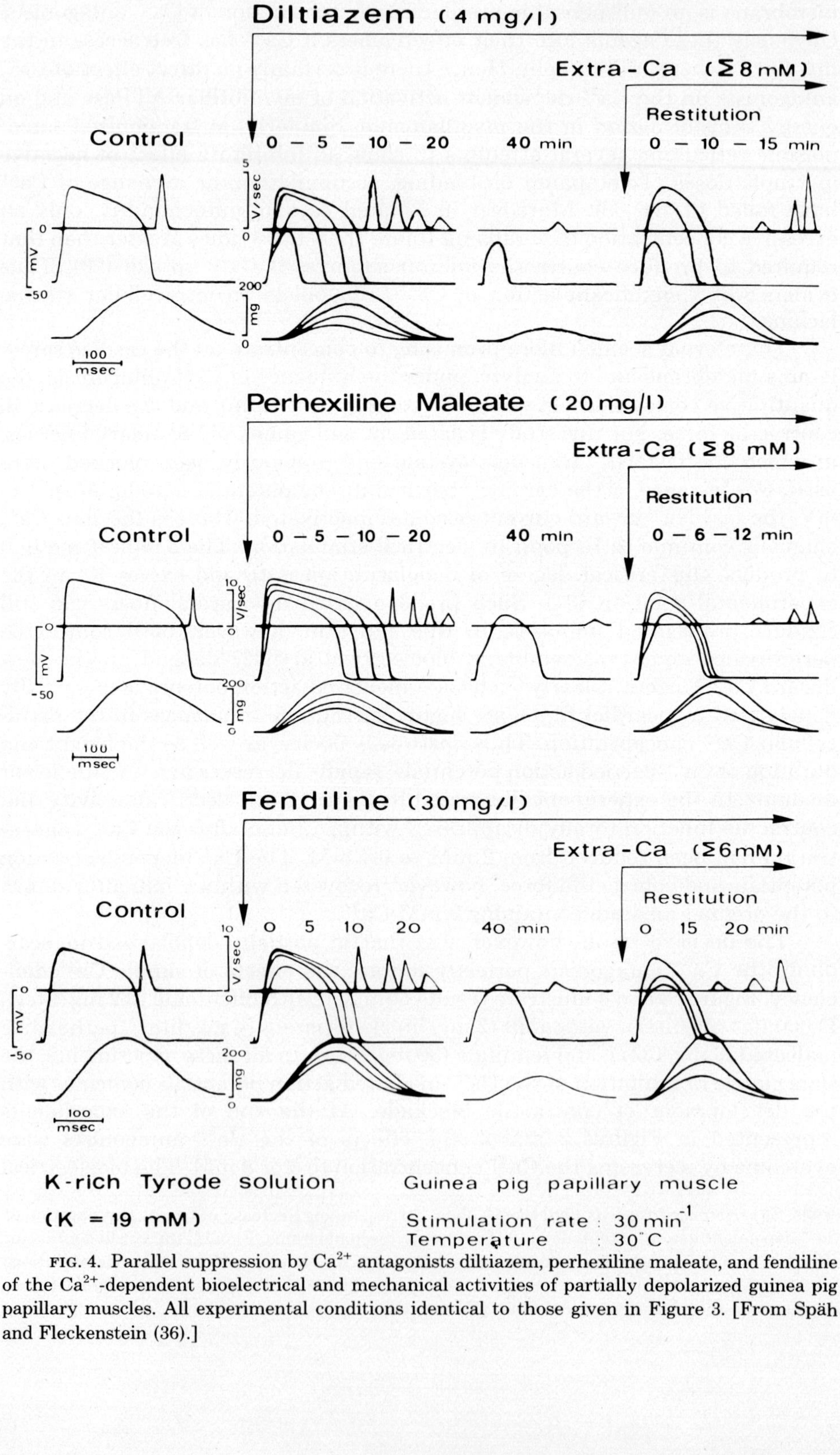

FIG. 4. Parallel suppression by Ca^{2+} antagonists diltiazem, perhexiline maleate, and fendiline of the Ca^{2+}-dependent bioelectrical and mechanical activities of partially depolarized guinea pig papillary muscles. All experimental conditions identical to those given in Figure 3. [From Späh and Fleckenstein (36).]

membrane activity of the partially depolarized myocardium recovered more or less in parallel with the mechanical performance.

Partially depolarized ventricular myocardium offers unique possibilities for testing the quantitative correlation between transmembrane Ca^{2+} influx (as indicated by the bioelectrical parameters of the Ca^{2+}-mediated action potentials) and mechanical responses. For instance, the maximum upstroke velocity (dV/dt_{max}) of Ca^{2+}-mediated action potentials indicates the intensity of transsarcolemmal Ca^{2+} influx through the slow channel. It is evident from Figure 5 that isometric peak tension always declines by approximately the same amount that each of the different Ca^{2+} antagonists inhibits the dV/dt_{max} of the Ca^{2+}-mediated action potentials (regardless of which Ca^{2+} antagonist is administered). Obviously contractile force declines in parallel with the decrease in Ca^{2+} influx. In this type of experiment another bioelectrical parameter that can easily be compared with the magnitude of development of contractile tension is the duration of the Ca^{2+}-mediated action potentials. This reflects the length of time of Ca^{2+} entry, and again there is a close linear relationship between this bioelectrical parameter and the contractile force developed as shown in Figure 6. Finally, in Figure 7, the areas of the Ca^{2+}-dependent isometric mechanograms are plotted against the areas of the Ca^{2+}-mediated action potential curves. Again there is a perfect linearity between the magnitude of the Ca^{2+}-dependent bioelectrical and mechanical responses. Thus the

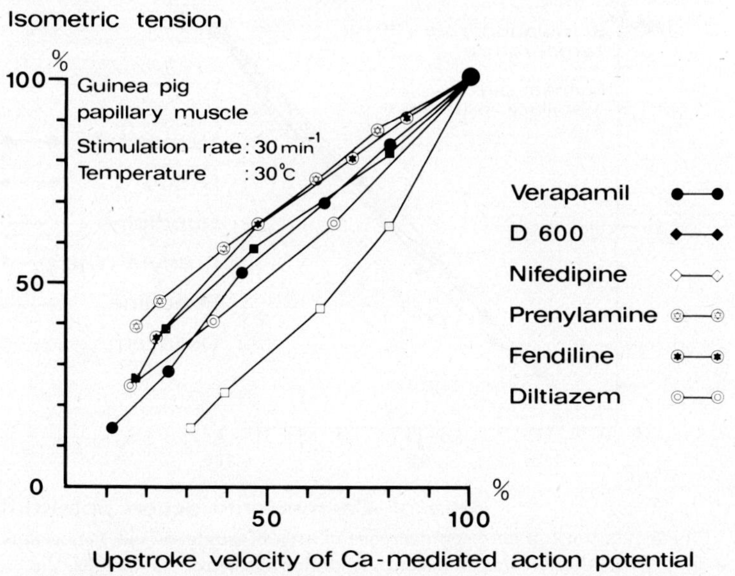

FIG. 5. Linear relationship between the decrease of isometric peak tension and reduction of dV/dt_{max} of Ca^{2+}-mediated action potentials under influence of various Ca^{2+} antagonists. Control data were obtained on partially depolarized guinea pig papillary muscles after incubation in a K^+-rich (19 mM), Ca^{2+}-containing (2 mM) Tyrode's solution for 30 min. Then each Ca^{2+} antagonist was applied in different effective concentrations for 20–40 min until definite changes were measured on the Ca^{2+}-dependent action potentials and mechanograms. Data are given as percentages of control value. [From Späh and Fleckenstein (36).]

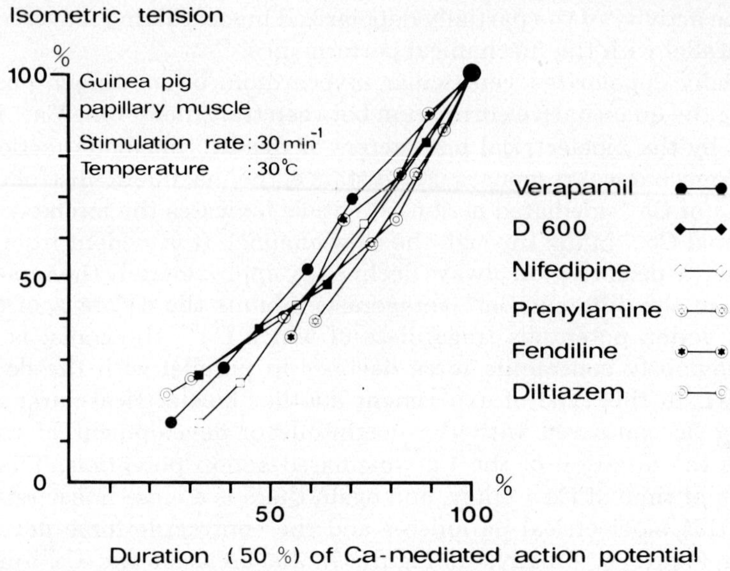

FIG. 6. Linear relationship between decrease of isometric peak tension and reduction of plateau phase (determined at 50% repolarization) of Ca^{2+}-mediated action potentials under influence of various Ca^{2+} antagonists. Muscles used were same as those in experiments of Figure 5. [From Späh and Fleckenstein (36).]

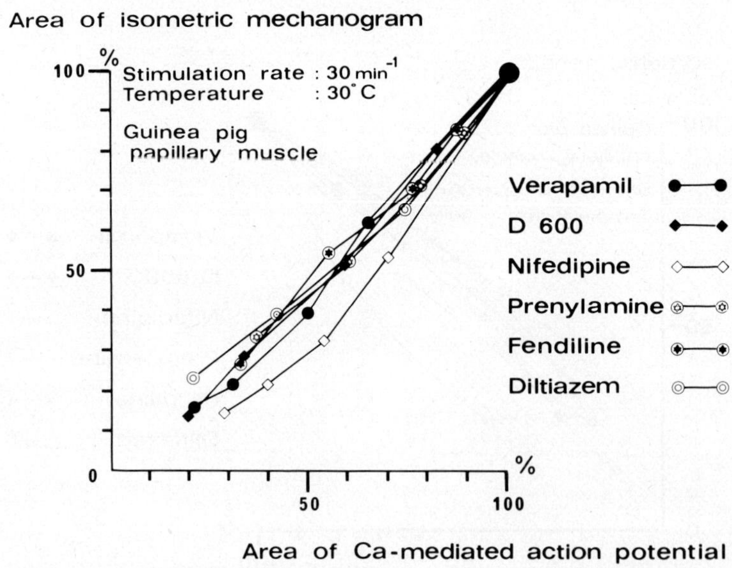

FIG. 7. Linear relationship between decrease of area of isometric mechanograms and reduction of area of Ca^{2+}-mediated action potentials under influence of various Ca^{2+} antagonists. Experimental procedure same as in Figures 5 and 6. [From Späh and Fleckenstein (36).]

conclusion is justified that the various Ca^{2+} antagonists, tested in this study on partially depolarized myocardium, act uniformly to reduce contractile force as a direct function of their actual inhibitory effects on the electrogenic I_{si}, irrespective of the drug, concentration, or time of exposure chosen.

OPPOSITE ACTIONS OF CALCIUM ANTAGONISTS AND CALCIUM PROMOTERS ON TRANSSARCOLEMMAL CALCIUM INFLUX, TENSION DEVELOPMENT, AND OXYGEN CONSUMPTION

The inhibitory effects of simple Ca^{2+} deficiency as well as those of Ca^{2+} antagonists are not only overcome with additional Ca^{2+} but also with the Ca^{2+} promoters [β-adrenergic catecholamines, histamine, and dibutyryl cyclic adenosine monophosphate (DBcAMP)]. Figure 8 shows two typical experiments on partially depolarized guinea pig papillary muscles in which the Ca^{2+}-dependent action potentials and the contractile responses were first suppressed by exposure to a medium with low Ca^{2+} for 10 min. Then isoproterenol or DBcAMP was added to the Ca^{2+}-deficient solution. Interestingly, in both completely paralyzed papillary muscles, the Ca^{2+}-dependent action potentials and contractility fully recovered. Obviously isoproterenol and DBcAMP restore a sufficient transmembrane Ca^{2+} supply even if the extracellular Ca^{2+} concentration is far below normal. However, the same reactivation of Ca^{2+} influx also occurs in partially depolarized guinea pig papillary muscles previously paralyzed by administration of Ca^{2+} antagonists. Figure 9 represents two experiments with nifedipine and verapamil in which isoproterenol reestablished the I_{si} within 5 min so that upstroke velocity, height, and duration of the Ca^{2+}-mediated action potentials as well as contractile force recovered to a considerable extent in the presence of the Ca^{2+} antagonists. Dibutyryl cyclic AMP produced the same recovery as the β-adrenergic drugs.

To explain these observations, it is reasonable to assume that the sarcolemma membrane not only regulates the inward diffusion of Ca^{2+} into the myoplasm, but also plays an important role as a Ca^{2+}-accumulating system that recruits Ca^{2+} actively from the environment. As illustrated by Figure 10, a functionally important Ca^{2+} pool seems to be located in the superficial layers of the sarcolemma membrane. Calcium ions (represented by balls) are taken up from the extracellular space and accumulated in this sarcolemmal Ca^{2+} store. Presumably the highest Ca^{2+} concentration is attained in those parts of the sarcolemma membrane that are invaginated in the form of T tubules. From here, Ca^{2+} enters the interior of the myocardial fibers through the slow channels (marked by arrows) as soon as the excitation wave depolarizes these membrane structures. Although the basement membrane contains plenty of negatively charged free or protein-bound mucopolysaccharide material with specific affinity for Ca^{2+} (28, 29), the decisive reaction that leads to an accumulation of Ca^{2+} in the sarcolemma membrane probably involves the production of cyclic AMP (39, 44). According to this concept, β-adrenergic catecholamines, through the formation of cyclic AMP, activate a membrane-bound protein kinase that in turn induces phosphorylation of membrane proteins in the vicinity of the slow channels or even within the slow channels. By this reaction the number of fixed negative phosphate groups available for accumulation of Ca^{2+} is increased. In this way β-adrenergic catecholamines probably fill the superficial Ca^{2+} pool with more loosely bound Ca^{2+} that is instantaneously available to the slow channels as soon as depolarization has occurred. In other words, β-adrenergic catecholamines seem to increase the

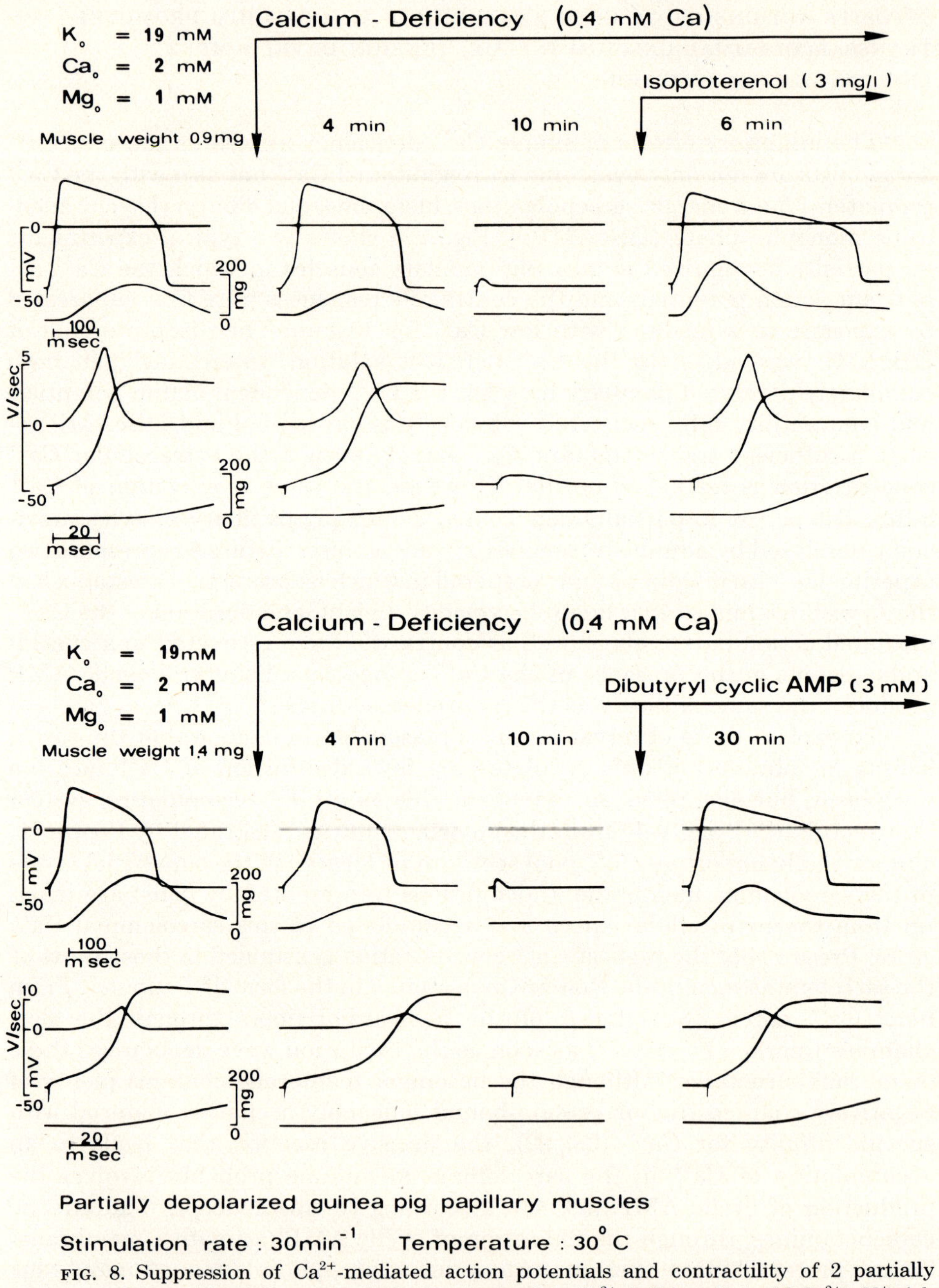

FIG. 8. Suppression of Ca^{2+}-mediated action potentials and contractility of 2 partially depolarized guinea pig papillary muscles by immersion in a Ca^{2+}-deficient (0.4 mM Ca^{2+}), K^+-rich (19 mM) Tyrode's solution for 10 min. After cessation of bioelectrical and mechanical activities, a more or less complete recovery was obtained by adding isoproterenol or DBcAMP to the Ca^{2+}-deficient media. Isoproterenol was fully effective within 6 min, whereas restitution with DBcAMP required 25–30 min. This is probably due to slow rate of transformation of DBcAMP into cyclic AMP. Microelectrode insertion on same myocardial fiber in each experiment. (From A. Fleckenstein and F. Späh, unpublished data.)

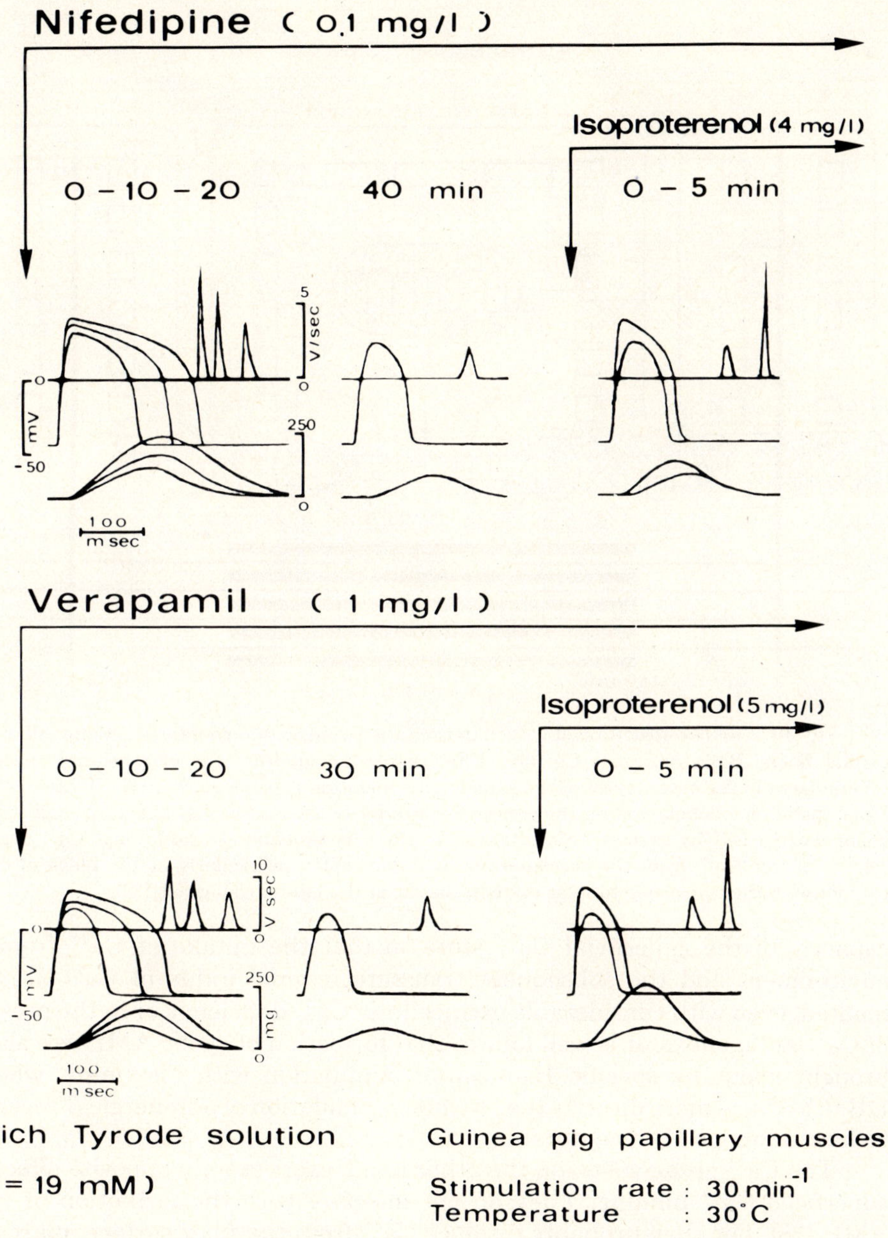

FIG. 9. Depression by nifedipine and verapamil of Ca^{2+}-mediated action potentials and mechanical responses of 2 partially depolarized guinea pig papillary muscles in a K^+-rich (19 mM) Tyrode's solution with normal Ca^{2+} content (2 mM). After muscles had been exposed to media containing nifedipine or verapamil for 30–40 min, the β-adrenergic stimulant isoproterenol was added to restore bioelectrical and mechanical activity. In fact, as in the case of simple Ca^{2+} deficiency, isoproterenol and other Ca^{2+} promoters such as histamine and DBcAMP can induce recovery of Ca^{2+}-mediated action potentials and contractile responses in the presence of Ca^{2+} antagonists. (From A. Fleckenstein and F. Späh, unpublished data.)

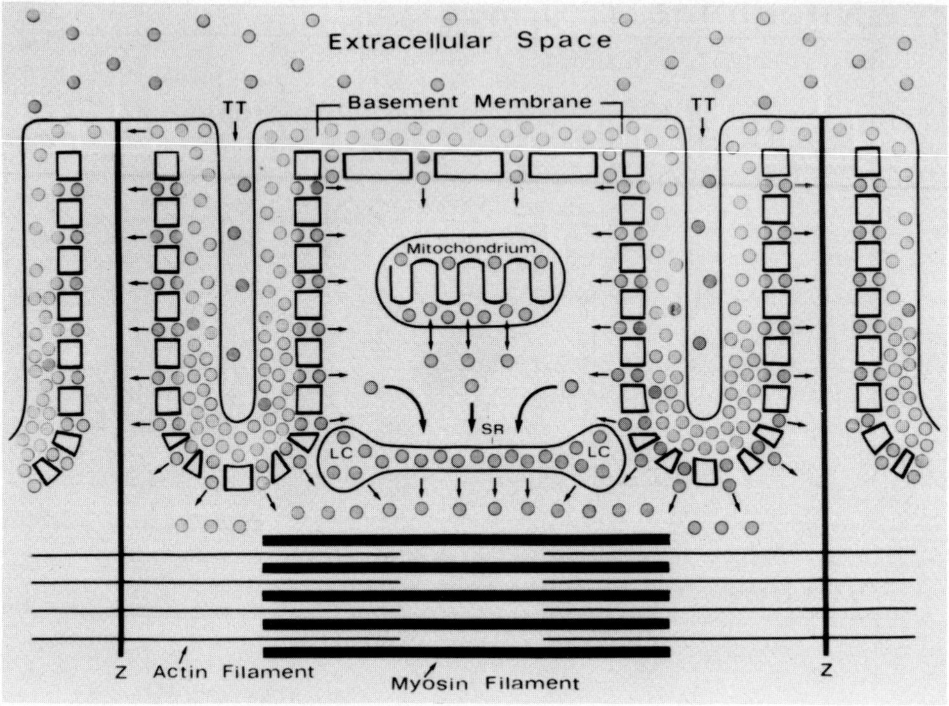

FIG. 10. Cellular structures involved in excitation-contraction coupling of mammalian myocardial fibers. Balls represent Ca ions. They are taken up from the extracellular space and accumulated in the outer layers of the sarcolemma membrane, particularly in the T tubules (TT). Upon membrane depolarization they enter the interior of the myocardial fibers through the slow channels (marked by arrows). Simultaneously there is probably an additional Ca^{2+}-triggered release of free Ca^{2+} from the sarcoplasmic reticulum (SR), particularly at the place of closest contact with the transverse tubular system, that is, at the lateral cisternae (LC).

capacity of the superficial Ca^{2+} store so that the uptake of Ca^{2+} from the environment and the subsequent transsarcolemmal influx of Ca^{2+} are augmented, even with considerable extracellular Ca^{2+} deficiency or in the presence of Ca^{2+} antagonists. It is well known that formation of cyclic AMP can also be brought about by specific H_2-receptor stimulation with histamine, whereas DBcAMP acts more directly (i.e., without stimulation of adrenergic β-receptors or histaminergic H_2 receptors).

The Ca^{2+} antagonists, on the other hand, exert exactly opposite effects on superficial Ca^{2+} binding. They do not interfere with the formation of cyclic AMP (38), but they probably displace Ca^{2+} from the fiber surface apart from any direct interference with the slow channel–mediated Ca^{2+} inward current. Accordingly, isolated cardiac sarcolemma membranes could be depleted of bound Ca^{2+} by verapamil (34). There are two different types of Ca^{2+} storage in cardiac sarcolemma membranes; one is at low-affinity binding sites, whereas the other is at high-affinity binding sites (41). Verapamil specifically inhibits low-affinity Ca^{2+} binding. The effective verapamil concentration that blocks low-affinity Ca^{2+} binding in isolated cardiac sarcolemma membranes may be as low as 1.4 μM (44). Thus verapamil and other Ca^{2+} antagonists apparently impair primarily the Ca^{2+}-binding capacity or the refilling of the superficial Ca^{2+} depots in the sarcolemma membrane so that the availability of Ca^{2+} to the slow channels is critically diminished.

The cardiac effects of β-adrenergic substances and Ca^{2+} antagonists counteract each other in almost every respect. Calcium binding in the superficial membrane pool, transmembrane Ca^{2+} influx through the slow channels, splitting of ATP by the Ca^{2+}-activated myofibrillar ATPase, contractile tension development, and the oxygen requirement of the beating heart are all increased by the β-adrenergic catecholamines. Conversely, the Ca^{2+} antagonists decrease all these Ca^{2+}-dependent parameters. It is commonly known that the intensity of oxidative cardiac metabolism depends on the rate of ATP consumption. Therefore the Ca^{2+} antagonists, in a manner similar to simple Ca^{2+} withdrawal, lower the oxygen requirement as they reduce Ca^{2+}-dependent splitting of ATP and isometric tension development (7, 9). In this respect, the Ca^{2+} antagonists may exert a beneficial influence on patients with a hyperkinetic heart function or angina pectoris just like the adrenergic β-receptor blocking agents do, despite the fact that the mode of action is different. Calcium antagonists inhibit the transmembrane Ca^{2+} supply directly by acting on the slow channels. The β-receptor blocking agents, on the other hand, diminish transmembrane Ca^{2+} influx indirectly by neutralizing the promoter effects of β-adrenergic catecholamines. Nevertheless both Ca^{2+} antagonists and β-blockers reduce the availability of Ca^{2+} to the contractile system so that eventually the splitting of ATP, the contractile energy expenditure, and the cardiac oxygen requirement are lowered.

We initially demonstrated that the well-known cardiotoxic effects of β-adrenergic overstimulation by overdoses of sympathetic catecholamines are simply due to an excessive transmembrane Ca^{2+} uptake (5, 7, 9, 14). In fact, Ca^{2+} overload of the cardiac cytoplasm has two toxic consequences involving *1)* excessive consumption of ATP due to supernormal activation of Ca^{2+}-dependent ATPases of myofibrils, sarcoplasmic reticulum, and mitochondria and *2)* impairment of ATP synthesis due to functional and structural damage to the mitochondria. Both effects summate in a deleterious exhaustion of the high-energy phosphate stores and finally in myocardial cell necrosis. Thus in 1968 we emphasized the significance of myocardial Ca^{2+} overload as a new principle in cardiac pathophysiology and simultaneously demonstrated that necrotization could be fully prevented with the help of Ca^{2+} antagonists that counteract excessive myocardial Ca^{2+} accumulation. In addition to sympathetic overstimulation, overdoses of vitamin D or dihydrotachysterol, as well as alimentary K^+ and Mg^{2+} deficiency or genetic defects in hereditary cardiomyopathic hamsters (30), can produce structural decay of myocardial tissue by this pathogenic mechanism. In all these cases, Ca^{2+} antagonists minimize excessive Ca^{2+} uptake and high-energy phosphate breakdown and, in this manner, protect structural integrity. As recently shown by other groups, Ca^{2+} overload is also responsible to a significant extent for impairment of the mitochondria in anoxic or ischemic cardiac tissue. Accordingly, Ca^{2+} antagonists such as verapamil (33), nifedipine (24), or diltiazem (40) can even protect hypoxic and ischemic hearts against the additional Ca^{2+}-induced myocardial fiber damage that precipitates the inevitable structural decay.

CONTRARY INFLUENCES OF CALCIUM ANTAGONISTS AND CALCIUM PROMOTERS ON CALCIUM-DEPENDENT NOMOTOPIC AND ECTOPIC PACEMAKER ACTIVITY

Calcium antagonists and β-adrenergic catecholamines also have opposite effects on cardiac pacemakers (as schematically shown in Fig. 11). In fact, not only the initiation of myocardial contraction, but also the fundamental processes of sinoatrial (SA) node and atrioventricular (AV) node automaticity and AV conduction are necessarily linked with a transmembrane influx of Ca^{2+}. Moreover the Ca^{2+}-transport systems that operate in the nodal cell membranes closely resemble the slow membrane channels of ordinary myocardial fibers. Accordingly, in both nodal cells and ordinary myocardial fibers, these channels respond to virtually the same activators and inhibitors of the Ca^{2+} inward current (2, 43, 45, 46). Thus β-adrenergic agents and DBcAMP, by promoting the transfer of Ca^{2+} through the slow channels, not only cause an increase in

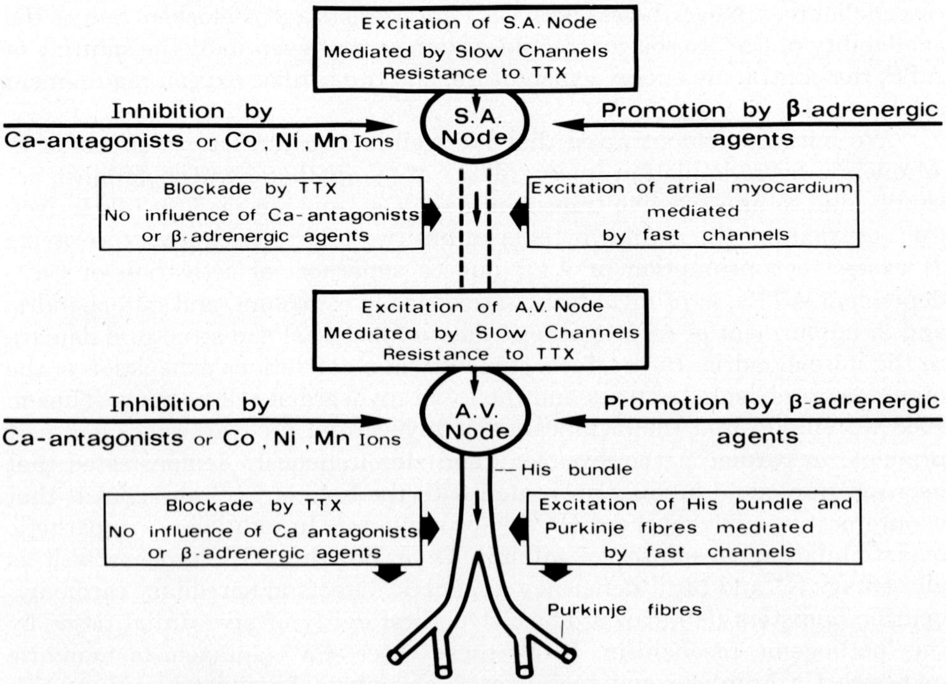

FIG. 11. Opposite effects of Ca^{2+} antagonists and Ca^{2+} promoters (β-adrenergic agents) on slow channel-mediated, TTX-insensitive SA node automaticity and AV node conductivity. Refractoriness to Ca^{2+} antagonists and β-adrenergic catecholamines of normal TTX-sensitive impulse propagation in atrial, His' bundle, Purkinje, and ventricular fibers.

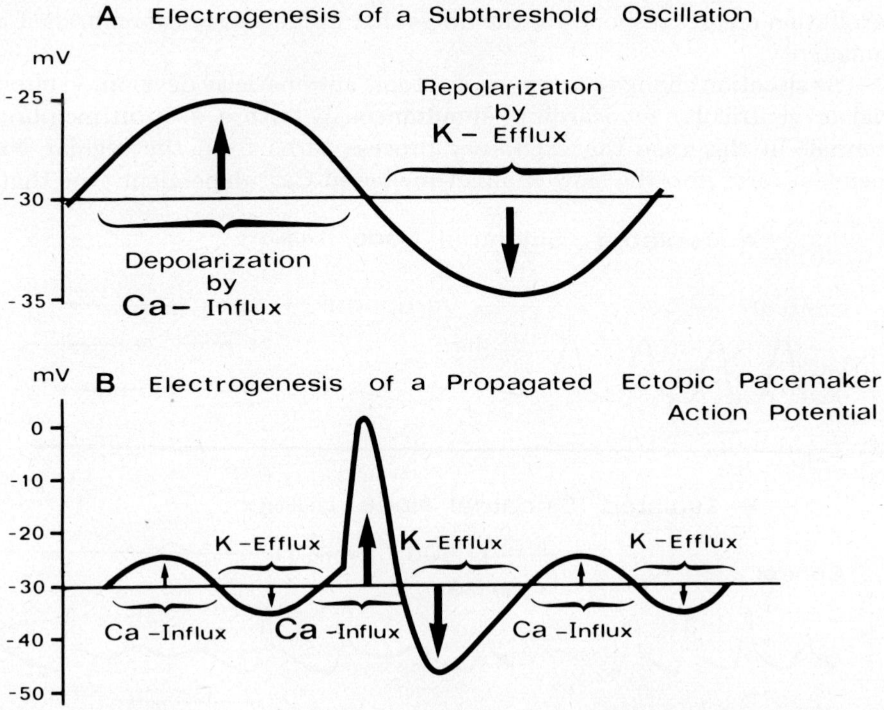

FIG. 12. Scheme of coupled Ca^{2+}-K^{+} exchange that underlies electrogenesis of ectopic subthreshold oscillations (*A*) and propagated ectopic pacemaker action potentials (*B*) in a low range of membrane potential (below −50 mV). In both *A* and *B* the depolarizing slow channel-mediated inward current is carried by Ca^{2+}, whereas the subsequent repolarization is brought about by the efflux of an equivalent amount of K^{+} that drives the membrane potential back to the diastolic values. The greater influx of Ca^{2+}, in case of a propagated ectopic pacemaker action potential, is apparently compensated by an enhancement of the K^{+} efflux so that an overshooting repolarization is produced.

contractile force, but also produce an analogous rise in heart rate and AV conduction velocity. Conversely, organic Ca^{2+} antagonists exert negative inotropic, chronotropic, and dromotropic influences. The Ca^{2+}-antagonistic divalent Co, Ni, and Mn ions act similarly in that by inhibiting Ca^{2+} inflow across the slow membrane channels, they also suppress contractility and pacemaker excitation. The same functional impairment results from simple Ca^{2+} withdrawal. On the other hand, tetrodotoxin (TTX), known to block the fast Na^{+} channels, does not inhibit SA node and AV node automaticity or AV conduction. Thus there are two contrasting types of cardiac excitation that can be easily distinguished by their different sensitivities to drugs and by other criteria. The first is the fast Na^{+}-dependent excitatory events in atrial myocardium, His' bundle, Purkinje fibers, and ventricular myocardium (dV/dt_{max} = 170–400 V/s; propagation rate = 60–200 cm/s). This type of excitation is rather refractory to Ca^{2+} antagonists and Ca^{2+} promoters. The second is the slow Ca^{2+}-dependent impulse generation and impulse conduction in the SA and AV nodes (dV/dt_{max} = 2–7 V/s; propagation rate = 2–6 cm/s). This type of excitation readily responds to the drugs that influence transmembrane Ca^{2+} conductivity.

The situation changes, however, if ectopic automaticity develops in altered atrial or ventricular myocardium simultaneously with a drop in membrane potential. In this case the excitatory process turns from the regular Na^{+}-dependent form into the slow channel–mediated Ca^{2+}-dependent type that is

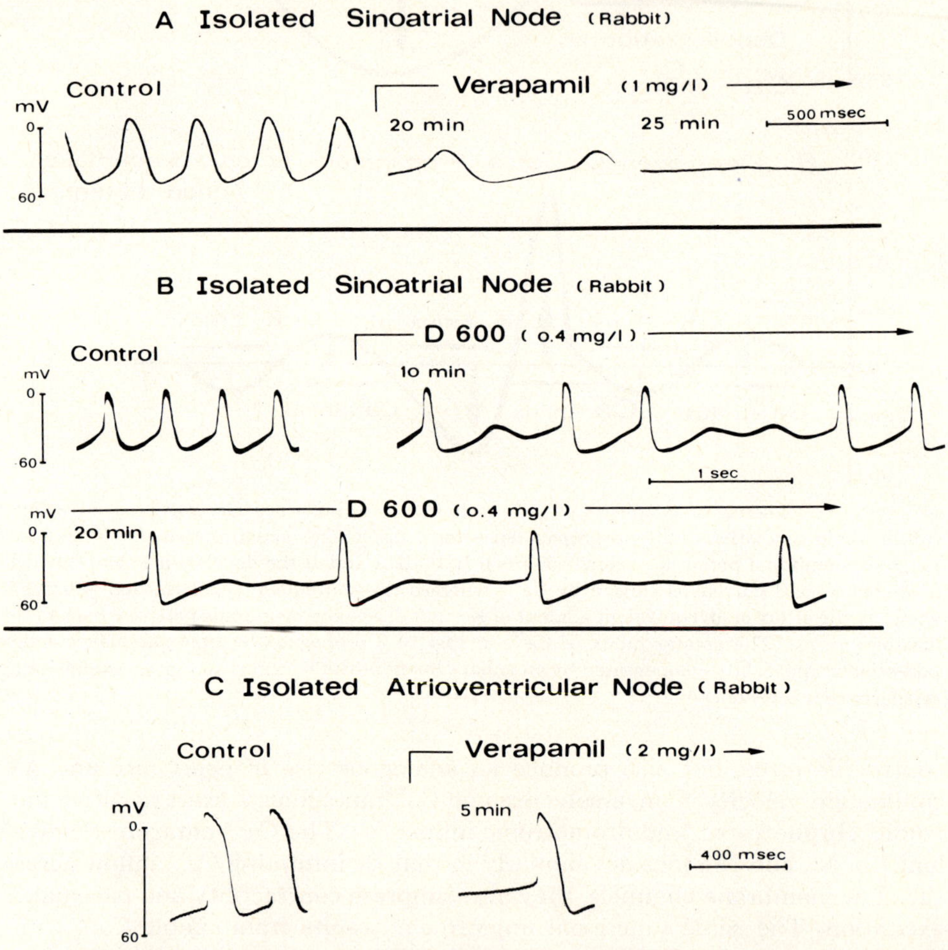

FIG. 13. Inhibition by verapamil and D 600 of pacemaker function of isolated SA and AV nodes from rabbits. Verapamil and D 600 reduce both steepness of slow diastolic depolarization, thereby producing a negative chronotropic effect, and upstroke velocity of propagated nodal action potentials. Height of overshoot also decreases upon prolonged exposure to the drugs, whereas there is a growing tendency toward hyperpolarization during diastole. As shown in *B*, a nodal cell, treated with the Ca^{2+} antagonist D 600, increasingly fails to elicit propagated pacemaker action potentials. This is due both to a reduction of amplitude of the intrinsic oscillatory potential changes and to a shift of threshold potential toward zero. Thus an increasing number of spontaneous local depolarization waves remain unpropagated. Each record in experiments *A*, *B*, and *C* was taken by steady impalement of 1 single cell. [*A* from Kohlhardt et al. (26); *B* from H. Tritthart, M. Kohlhardt, and A. Fleckenstein, unpublished data; and *C* from Tritthart et al. (37).]

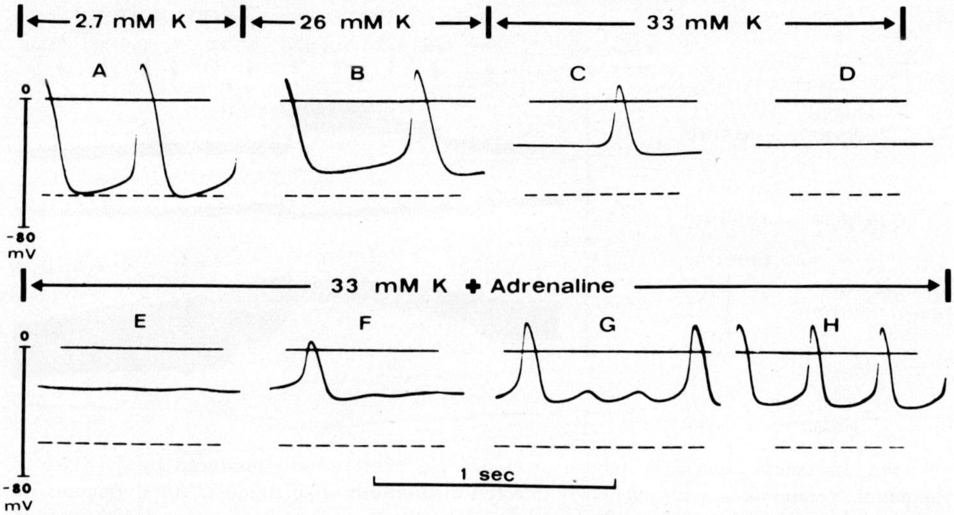

FIG. 14. Resuscitation by adrenaline of automaticity of a K^+-depolarized pacemaker cell of isolated SA node from a rhesus monkey. *A*: regular pacemaker action potential in Tyrode's solution with slightly reduced K^+ content (2.7 mM). *B*: depolarization and slowing of pacemaker rate 2 min after exposure to K^+-rich Tyrode's solution (26 mM K^+). *C*: last spontaneous action potential 14 min after increasing extracellular K^+ concentration to 33 mM. *D*: blockade of automaticity. *E*: commencement of slow oscillations of membrane potential 45 s after addition of adrenaline (5 μg/ml) to K^+-rich (33 mM) Tyrode's solution. *F*: first propagated action potential 1 min after administration of noradrenaline. *G*: occurrence of local subthreshold oscillations and propagated action potentials in ratio of 2:1 after time lapse of 1.5 min following addition of adrenaline. *H*: restoration of quasi-regular pacemaker function in K^+-rich Tyrode's solution 4 min after administration of adrenaline although maximal diastolic potential remains reduced by 25 mV compared with its original height in *A*. Intracellular recordings from same nodal cell throughout experiment at 36°C. [From Antoni et al. (1).]

characteristic of partially depolarized myocardium. Thus Ca^{2+} antagonists [the "fourth class" of antidysrhythmic drugs according to Singh and Vaughan Williams (35)] can abolish the uncontrolled firing of such Ca^{2+}-dependent atrial or ventricular ectopic foci and also block Ca^{2+}-dependent slow channel–mediated reentry pathways. Figure 12 illustrates the electrogenesis of ectopic automaticity in a low range of membrane potentials (−30 to −50 mV). In this range, automatic impulses appear in two typical forms, either as spontaneous local subthreshold oscillations of membrane potential or as propagated pacemaker action potentials if the depolarizing phase of these local oscillations reaches the threshold. The scheme further indicates that Ca^{2+} influx is the primary step responsible for spontaneous depolarization. This is true of both the depolarizing phase of the local oscillations and the Ca^{2+}-dependent upstroke of the propagated pacemaker action potentials. In a second step, after the entry of Ca^{2+}, a delayed K^+ outward current is elicited that drives the membrane potential back to the higher diastolic values. The Ca^{2+} antagonists, by inhibiting this Ca^{2+}- K^+ exchange cycle, interrupt the fundamental process of ectopic automaticity, whereas β-adrenergic agents, by stimulating the Ca^{2+} inward current, promote spontaneous impulse discharge. With respect to these

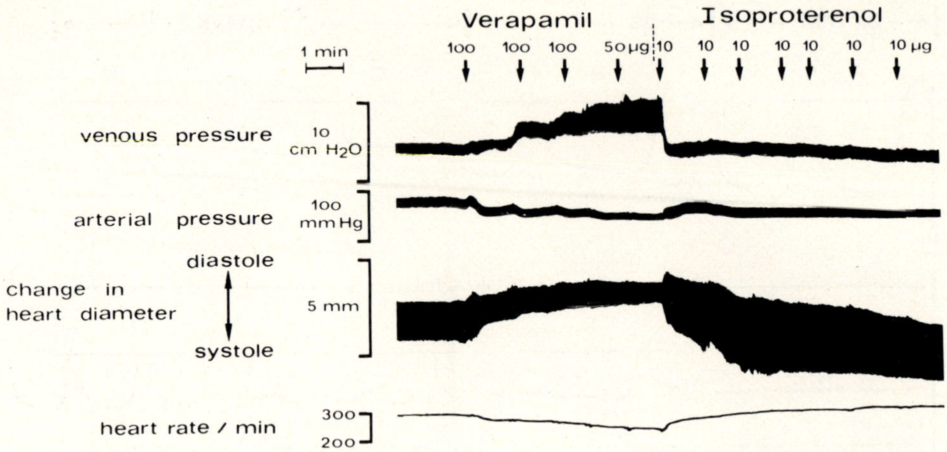

FIG. 15. Acute contractile failure of guinea pig heart in situ produced by overdose of verapamil. Verapamil was intravenously injected in the form of single doses (total amount: 1.2 mg/kg body wt) until cardiac incompetence developed. As indicated by upward deflection of record of heart diameter, verapamil produced large ventricular dilation in that it rapidly increased both the end-diastolic and, particularly, the end-systolic cross diameters. Moreover the contraction amplitude, as visualized by the difference between diastolic and systolic diameter, became progressively smaller. For cardiac contractile failure, central venous pressure, measured in left jugular vein with the Statham element P23BB, rose, whereas there was a simultaneous drop in arterial blood pressure, measured with the Statham element P23Db in the right carotid artery. Heart rate also decreased. However, the first therapeutic dose of 10 μg isoproterenol (which corresponds to 30–35 μg/kg body wt) restored the contractile function of the verapamil-treated heart almost completely within 20–30 s. Experiment was done on an open-chest preparation under Nembutal-ether anesthesia; artificial respiration was given after the pericardium was removed. Systolic and diastolic heart diameters were continuously registered with the method of H. Kammermeier and H. J. Döring (*Pfluegers Arch. Gesamte Physiol. Menschen Tiere* 273: 311, 1961). Recordings of heart rate were done with an electronic device as indicated by H. Krause (*Z. Kreislaufforsch.* 52: 128, 1963). [From Fleckenstein (4).]

typical drug effects, nomotopic and ectopic pacemaker cells behave, at least in principle, very similarly. For instance, Figure 13 shows the inhibition of nomotopic SA node and AV node automaticity by the Ca^{2+} antagonists verapamil and D 600. Conversely, in Figure 14, the adrenaline-induced resuscitation of spontaneous impulse discharge in a previously paralyzed SA node of a rhesus monkey is demonstrated.

This report reflects the long road from our first description in 1964 of the contrary actions of verapamil and β-adrenergic agents on ventricular contractility and SA node frequency of guinea pig hearts in situ (Fig. 15) to the definite clarification of this antagonism on a cellular level, that is, in terms of an opposite influence on fundamental Ca^{2+} conductivity of myocardial and pacemaker cells.

SUMMARY

During the past decade, evidence has been obtained for the existence of a dual electrogenic membrane carrier system in atrial and ventricular myocardium. This system consists of a fast channel for Na^+, which initiates the cardiac fiber action potentials, and a separate slow channel for Ca^{2+}, which contributes to the late plateau phase of the action potential and induces Ca^{2+}-dependent splitting of ATP in the contractile machinery. Furthermore transmembrane Ca^{2+} transport through the slow channel is also involved in nomotopic and ectopic pacemaker activity as well as in AV impulse conduction. Therefore with the discovery of the new group of highly potent organic Ca^{2+} antagonists that specifically block Ca^{2+} transport through the slow channels, it became possible to limit cardiac energy expenditure and oxygen requirements as well as automatic impulse discharge from pacemaker cells both in vitro and in vivo to any extent. In this respect, Ca^{2+} antagonists operate in a manner opposite that of β-adrenergic catecholamines, since the latter agents exert their positive inotropic, chronotropic, and dromotropic effects by acting as promoters of the slow channel–mediated transmembrane Ca^{2+} influx. This also explains why overdoses of sympathetic catecholamines produce Ca^{2+} overload of cardiac muscle with two toxic consequences, namely *1)* exhaustive consumption of ATP due to supernormal activation of Ca^{2+}-dependent myofibrillar, sarcoplasmic, and mitochondrial ATPases, and *2)* disintegration of mitochondria leading to a blockade of ATP synthesis and finally to myocardial necrotization. Excessive Ca^{2+} uptake, high-energy phosphate breakdown, and structural decay of myocardial fibers can be prevented by Ca^{2+} antagonists.

REFERENCES

1. ANTONI, H., . HERKEL, AND A. FLECKENSTEIN. Die Restitution der automatischen Erregungsbildung in Kalium-gelähmten Schrittmacher-Geweben durch Adrenalin. Elektrophysiologische Studien am isolierten Sinusknoten (Meerschweinchen, Rhesusaffe) sowie am Purkinje-Faden (Rhesusaffe). *Pfluegers Arch. Gesamte Physiol. Menschen Tiere* 277: 633–649, 1963.
2. CRANEFIELD, P. F. *The Conduction of the Cardiac Impulse—the Slow Response and Cardiac Arrhythmias.* New York: Futura, 1975.
3. ENTMAN, M. L., J. C. ALLEN, E. P. BORNET, P. C. GILLETTE, E. T. WALLICK, AND A. SCHWARTZ. Mechanisms of calcium accumulation and transport in cardiac relaxing system (sarcoplasmic reticulum membranes): effects of verapamil, D 600, × 537 A and A 23187. *J. Mol. Cell. Cardiol.* 4: 681–687, 1972.
4. FLECKENSTEIN, A. Die Bedeutung der energiereichen Phosphate für Kontraktilität und Tonus des Myokards. *Verh. Dtsch. Ges. Inn. Med.* 70: 81–99, 1964.
5. FLECKENSTEIN, A. Myokardstoffwechsel und Nekrose. In: *Herzinfarkt und Schock*, edited by L. Heilmeyer and H.-J. Holtmeier. Stuttgart: Thieme, 1968, p. 94–109.
6. FLECKENSTEIN, A. Experimental heart failure due to disturbances in high-energy phosphate metabolism. In: *Proc. Vth Eur. Congr. Cardiol., Athens*, 1968, p. 255–269. (Also *Verh. Dtsch. Ges. Kreislaufforsch.* 34: 15–34 1968.)
7. FLECKENSTEIN, A. Specific inhibitors and promoters of calcium action in the excitation-contraction coupling of heart muscle and their role in the prevention or production of myocardial lesions. In: *Calcium and the Heart*, edited by P. Harris and L. Opie. London: Academic, 1971, p. 135–188.
8. FLECKENSTEIN, A. Specific pharmacology of calcium in myocardium, cardiac pacemakers, and vascular smooth muscle. *Annu. Rev. Pharmacol. Toxicol.* 17: 149–166, 1977.
9. FLECKENSTEIN, A., H. J. DÖRING, J. JANKE, AND Y. K. BYON. Basic actions of ions and drugs on myocardial high-energy phosphate metabolism and contractility. In: *Handbook of Experimental Pharmacology. Vol. XVI*, edited by J. Schmier and O. Eichler. Berlin: Springer-Verlag, 1975, p. 345–405.
10. FLECKENSTEIN, A., H. J. DÖRING, AND H. KAMMERMEIER. Experimental heart failure due to inhibition of utilisation of high-energy phosphates. In: *Int. Symp. Coronary Circ. Energ. Myocardium, Milan, 1966*, edited by G. Marchetti and B. Taccardi. Basel: Karger, 1967, p. 220–236.
11. FLECKENSTEIN, A., G. FLECKENSTEIN-GRÜN, AND Y. K. BYON. Fundamentale Herz- und Gefaesswirkungen des Ca^{++}-antagonistischen Koronartherapeutikums Fendilin. *Arzneim. Forsch.* 27: 562–571, 1977.
12. FLECKENSTEIN, A., G. FLECKENSTEIN-GRÜN, Y. K. BYON, H. P. HAASTERT, AND F. SPÄH. Vergleichende Untersuchungen über die Ca^{++}-antagonistischen Grundwirkungen von Niludipin (Bay a 7168) und Nifedipin (Bay a 1040) auf Myokard, Myometrium und glatte Gefaessmuskulatur. *Arzneim. Forsch.* 29: 230–246, 1979.
13. FLECKENSTEIN, A., G. GRÜN, Y. K. BYON, H. J. DÖRING, AND H. TRITTHART. The basic Ca-antagonistic actions of nifedipine on cardiac energy metabolism and vascular smooth muscle tone. In: *New Therapy of Ischemic Heart Disease*, edited by K. Hashimoto, E. Kimura, and T. Kobayashi. Tokyo: Univ. of Tokyo Press, 1975, p. 31–44. (1st Int. Adalat Symp., Tokyo, 1973.)
14. FLECKENSTEIN, A., J. JANKE, H. J. DÖRING, AND O. PACHINGER. Ca overload as the determinant factor in the production of catecholamine-induced myocardial lesions. In: *Cardiomyopathies, Recent Advances in Studies on Cardiac Structure and Metabolism*, edited by E. Bajusz

and G. Rona. Baltimore: University Park, 1973, vol. 2, p. 455–466.

15. FLECKENSTEIN, A., H. KAMMERMEIER, H. J. DÖRING, AND H. J. FREUND. Zum Wirkungsmechanismus neuartiger Koronardilatatoren mit gleichzeitig Sauerstoff-einsparenden Myokard-Effekten, Prenylamin und Iproveratril. *Z. Kreislaufforsch.* 56: 716–744; 839–853, 1967.
16. FLECKENSTEIN, A., H. TRITTHART, H. J. DÖRING, AND Y. K. BYON. Nifedipine (Bay a 1040)-ein hochaktiver Ca^{++}-antagonistischer Inhibitor der elektro-mechanischen Koppelungsprozesse im Warmblüter-Myokard. *Arzneim. Forsch.* 22: 22–33, 1972.
17. FLECKENSTEIN, A., H. TRITTHART, B. FLECKENSTEIN, A. HERBST, AND G. GRÜN. A new group of competitive Ca-antagonists (Iproveratril, D 600, Prenylamine) with highly potent inhibitory effects on excitation-contraction coupling in mammalian myocardium (abstr.). *Pfluegers Arch.* 307: R25, 1969.
18. FLECKENSTEIN-GRÜN, G., AND A. FLECKENSTEIN. Ca-dependent changes in coronary smooth muscle tone and the action of Ca-antagonistic compounds with special reference to adalat (nifedipine). In: *New Therapy of Ischemic Heart Disease*, edited by W. Lochner, W. Braasch, and G. Kroneberg. Berlin: Springer-Verlag, 1975, p. 66–75. (2nd Int. Adalat Symp., Amsterdam, 1974.)
19. FLECKENSTEIN-GRÜN, G., A. FLECKENSTEIN, Y. K. BYON, AND K. W. KIM. Mechanism of action of Ca^{2+}-antagonists in the treatment of coronary disease with special reference to perhexiline maleate. In: *Proc. Symp. Perhexiline Maleate, Strasbourg, 1976.* Amsterdam: Excerpta Med., 1978, p. 1–22.
20. FREY, M., AND J. JANKE. The effect of organic Ca-antagonists (verapamil, prenylamine) on the calcium transport system in isolated mitochondria of rat cardiac muscle (abstr.). *Pfluegers Arch.* 359, Suppl.: R26, 1975.
21. GRÜN, G., AND A. FLECKENSTEIN. Die elektromechanische Entkoppelung der glatten Gefaessmuskulatur als Grundprinzip der Coronardilatation durch 4-(2′-Nitrophenyl)-2,6-dimethyl-1,4-dihydropyridin-3,5-dicarbonsäure-dimethylester (Bay a 1040, Nifedipine). *Arzneim. Forsch.* 22: 334–344, 1972.
22. GRÜN, G., A. FLECKENSTEIN, AND Y. K. BYON. Ca-antagonism, a new principle of vasodilation. In: *Proc. Int. Union Physiol. Sci., XXV Int. Congr., Munich,* 1971, vol. IX, p. 221.
23. HAASTERT, H. H. Besonderheiten der chronotropen, dromotropen und inotropen Wirkung von Ca^{++}-Antagonisten (Verapamil, D 600, Nifedipin) am Herzen in situ und an isolierten Vorhöfen sowie deren Beeinflussung durch Extra-Calcium und Isoproterenol. *Naunyn-Schmiedebergs Arch. Pharmakol.* 227, Suppl.: R26, 1973.
24. HENRY, P. D., R. SHUCHLEIB, L. J. BORDA, R. ROBERTS, J. R. WILLIAMSON, AND B. E. SOBEL. Effects of nifedipine on myocardial perfusion and ischemic injury in dogs. *Circ. Res.* 43: 372–380, 1978.
25. KOHLHARDT, M., B. BAUER, H. KRAUSE, AND A. FLECKENSTEIN. Differentiation of the transmembrane Na and Ca channel in mammalian cardiac fibres by the use of specific inhibitors. *Pfluegers Arch.* 335: 309–322, 1972.
26. KOHLHARDT, M., H. R. FIGULLA, AND O. TRIPATHI. The slow membrane channel as the predominant mediator of the excitation process of the sinoatrial pacemaker cell. *Basic Res. Cardiol.* 71: 17–26, 1976.
27. KOHLHARDT, M., AND A. FLECKENSTEIN. Inhibition of the slow inward current by nifedipine in mammalian ventricular myocardium. *Naunyn-Schmiedebergs Arch. Pharmakol.* 298: 267–272, 1977.
28. LANGER, G. A. Excitation-contraction coupling. *Annu. Rev. Physiol.* 35: 55–86, 1973.
29. LANGER, G. A., J. S. FRANK, L. M. NUDD, AND K. SERAYDARIAN. Sialic acid: effect of removal on calcium exchangeability of cultured heart cells. *Science* 193: 1013–1015, 1976.
30. LOSSNITZER, K., J. JANKE, B. HEIN, M. STAUCH, AND A. FLECKENSTEIN. Disturbed myocardial calcium metabolism: a possible pathogenetic factor in the hereditary cardiomyopathy of the Syrian hamster. In: *Recent Advances in Cardiac Structure and Metabolism*, edited by A. Fleckenstein and G. Rona. Baltimore: University Park, 1975, vol. 6, p. 207–217.
31. MASCHER, D. Electrical and mechanical responses from ventricular muscle fibres after inactivation of the sodium carrying system. *Pfluegers Arch.* 317: 359–372, 1970.
32. NAKAJIMA, H., M. HOSHIYAMA, K. YAMASHITA, AND A. KIYOMOTO. Effect of diltiazem on electrical and mechanical activity of isolated cardiac ventricular muscle of guinea pig. *Jpn. J. Pharmacol.* 25: 383–392, 1975.
33. NAYLER, W. G., A. GRAU, AND A. SLADE. A protective effect of verapamil on hypoxic heart muscle. *Cardiovasc. Res.* 10: 650–662, 1976.
34. NAYLER, W. G., AND J. SZETO. Effect of verapamil on contractility, oxygen utilization, and calcium exchangeability in mammalian heart muscle. *Cardiovasc. Res.* 6: 120–128, 1972.
35. SINGH, B. N., AND E. M. VAUGHAN WILLIAMS. A fourth class of antidysrhythmic action? Effect of verapamil on ouabain toxicity, on atrial and ventricular intracellular potentials and on other features of cardiac function. *Cardiovasc. Res.* 6: 109–119, 1972.
36. SPÄH, F., AND A. FLECKENSTEIN. Nachweis einer strengen quantitativen Korrelation zwischen Hemmung des transmembranären Calcium-Influx und der mechanischen Spannungsentwicklung des Myocards unter dem Einfluss verschiedener Calcium-Antagonisten. In: *Calcium-Antagonism*, edited by A. Fleckenstein and H. Roskamm. Berlin: Springer-Verlag, 1980, p. 29–41.
37. TRITTHART, H., B. FLECKENSTEIN, AND A. FLECKENSTEIN. Some fundamental actions of antiarrhythmic drugs on the excitability and the contractility of single myocardial fibres. *Naunyn-Schmiedebergs Arch. Pharmakol.* 269: 212–219, 1971. (Proc. Joint Meeting Ital. Ger. Pharmacol. Soc., Heidelberg, 1970.)
38. WATANABE, A. M., AND H. R. BESCH, JR. Subcellular myocardial effects of verapamil and D 600: comparison with propranolol. *J. Pharmacol. Exp. Ther.* 191: 241–251, 1974.
39. WATANABE, A. M., AND H. R. BESCH, JR. The relationship between adenosine 3′,5′-monophosphate levels and systolic transmembrane calcium flux. In: *Basic Functions of Cations in Myocardial Activity*, edited by A. Fleckenstein and N. S. Dhalla. Baltimore: University Park, 1975, vol. 5, p. 95–102. (Recent Adv. Stud. Card. Struct. Metab. Ser.)
40. WEISHAAR, R., K. ASHIKAWA, AND R. J. BING. Effect of diltiazem, a calcium antagonist, on myocardial ischemia. *Am. J. Cardiol.* 43: 1136–1143, 1979.
41. WILLIAMSON, J. R., M. L. WOODROW, AND A. SCARPA. Calcium binding to cardiac sarcolemma. In: *Basic Function of Cations in Myocardial Activity*, edited by A. Fleckenstein and N. S. Dhalla. Baltimore: University Park, 1975, vol. 5, p. 61–71. (Recent Adv. Stud. Card. Struct. Metab. Ser.)
42. WINEGRAD, S. Studies of cardiac muscle with a high permeability to calcium produced by treatment with ethylenediamine-tetraacetic acid. *J. Gen. Physiol.* 58: 71–93,

1971.

43. WIT, A. L., AND P. F. CRANEFIELD. Effect of verapamil on the sinoatrial and atrioventricular nodes of the rabbit and the mechanism by which it arrests reentrant atrioventricular nodal tachycardia. *Circ. Res.* 35: 413–425, 1974.

44. WOLLENBERGER, A. The role of cyclic AMP in the adrenergic control of the heart. In: *Contraction and Relaxation in the Myocardium*, edited by W. G. Nayler. New York: Academic, 1975, p. 113–190.

45. ZIPES, D. P., AND J. C. FISCHER. Effects of agents which inhibit the slow channel on sinus node automaticity and atrioventricular conduction in the dog. *Circ. Res.* 34: 184–192, 1974.

46. ZIPES, D. P., AND C. MENDEZ. Action of manganese ions and tetrodotoxin on atrioventricular nodal transmembrane potentials in isolated rabbit hearts. *Circ. Res.* 32: 447–454, 1973.

6

Sites of Action of Calcium Antagonists in Vascular Smooth Muscle

GEORGE B. WEISS

Department of Pharmacology, University of Texas Health Science Center, Dallas, Texas

THE ADVENT OF AGENTS that appear to block vascular smooth muscle Ca^{2+} entry channels with varying degrees of specificity has led to a number of questions about the precise sites of action of various vasodilators. As defined previously (42), the actions of inhibitory agents not acting at receptors for stimulatory agents can be described in terms of different mechanisms that directly or indirectly decrease the level of free intracellular Ca^{2+}. This essential final step results from changes in uptake, binding, or release of Ca^{2+} that either prevent increases in intracellular free-Ca^{2+} concentrations elicited with stimulatory agents or directly alter patterns of uptake and distribution of Ca^{2+} in vascular muscle in the absence of stimulatory agents. It is important to consider whether a given inhibitor acts in a manner that results in relaxation of vascular smooth muscle or acts less directly by disrupting the Ca^{2+}-dependent excitation-contraction coupling sequence when defining more precise sites and mechanisms of action for each type of inhibitory agent. These sites of action include different types of binding sites for Ca^{2+} as well as varied mechanisms by which Ca^{2+} entry into the intracellular compartment might occur.

Detection of quantitatively small changes in Ca^{2+} uptake, binding, or release in the presence of relatively large nonspecific Ca^{2+} binding has been a problem for workers in this field of research. In the past few years, however, several techniques that facilitate the delineation and separation of qualitatively distinct Ca^{2+} components have been developed. Combinations of these newer approaches can be used to measure small and specific changes in Ca^{2+} sequestration and mobilization and this, in turn, can help investigators define more precisely the differing actions of stimulatory agents and the corresponding sites of action of agents that decrease contractile tone or responsiveness. Finally, the cellular basis of action of Ca^{2+}-channel blocking agents can be contrasted with the effects of other types of smooth muscle vasodilators in a vascular smooth muscle model that contains several different types of Ca^{2+} channels and sites.

MEASUREMENT OF DIFFERENT CALCIUM FRACTIONS

Use of La^{3+} to separate different sites at which Ca^{2+} acts was originally suggested by Lettvin et al. (33) on the basis that La^{3+} binds at superficially located Ca^{2+} sites in a less reversible manner than Ca^{2+}. The blockade of Ca^{2+} uptake and of superficial binding of Ca^{2+} was subsequently demonstrated in a variety of tissues, including intestinal (45), uterine (10), and vascular (11) smooth muscle. In an attempt to dissociate physiologically significant Ca^{2+} uptake from quantitatively large and nonspecific Ca^{2+} binding in smooth muscle, Van Breemen and co-workers (40, 41) proposed an approach in which tissues could be exposed to a variety of stimulatory agents or conditions in the presence of ^{45}Ca and subsequently placed into washout solutions containing a concentration of La^{3+} high enough to replace all superficially bound ^{45}Ca and to prevent any further uptake or efflux of cellular ^{45}Ca. The assumption that a sufficiently high concentration of extracellular La^{3+} displaces extracellular Ca^{2+} and blocks Ca^{2+} uptake appears to be clearly justified in intestinal (5) and vascular (12) smooth muscle. The assumption that ^{45}Ca efflux is readily blocked by La^{3+} proved incorrect (5, 7, 12, 26, 34, 36), however, and it was necessary to modify the procedure by increasing the La^{3+} concentration (8) or by altering other parameters such as temperature (6) to decrease the rate of La^{3+}-resistant ^{45}Ca efflux during the washout period in La^{3+}-containing solution. To minimize ^{45}Ca efflux in the presence of La^{3+}, a solution was used in which 80.8 mM La^{3+} was substituted for all other cations and temperature was lowered to 0.5°C (26). Under these conditions, loss of slow-component ^{45}Ca in canine coronary artery (Fig. 1) was minimized ($t_{\frac{1}{2}}$ was greater than 462 min) and the rate coefficient of ^{45}Ca efflux decreased to approximately 0.15%/min (17). Similar results have been reported for rabbit aorta (26) and canine renal arteries (25) and veins (15). Thus under these conditions, the retention of an La^{3+}-resistant component of ^{45}Ca uptake can be optimized (and the removal of an La^{3+}-sensitive component can be accomplished) during 60 min of washout.

A second approach used to identify and characterize discrete components of ^{45}Ca uptake is a Scatchard coordinate plot. This type of plot was originally devised by Scatchard (37) to describe the binding of small molecules to macromolecules and has been used to estimate the relative affinity of sites and number of binding sites for such biological parameters as the binding of Tb^{3+} to ribosomes and ribonucleic acid (2) and the effects of neuraminidase and phospholipase C on binding of Ca^{2+} in the guinea pig taenia coli cell membrane (24). The simplifying assumptions that most of the ^{45}Ca taken up by the cells is present in bound form and that the extracellular Ca^{2+} concentration is equivalent to the free-Ca^{2+} level in the solution are used to obtain two-component Scatchard coordinate plots for canine (Fig. 2; 32), rabbit (29, 43, 44), and rat (48) aortic smooth muscle, bovine basilar artery (48), canine coronary arteries (17), and canine renal arteries (25) and veins (15) as well as for a variety of other smooth muscle preparations (including intestinal and tracheal) (G. B. Weiss, unpublished data). Thus this type of coordinate plot can be used to define the conditions useful for measurement of ^{45}Ca uptake

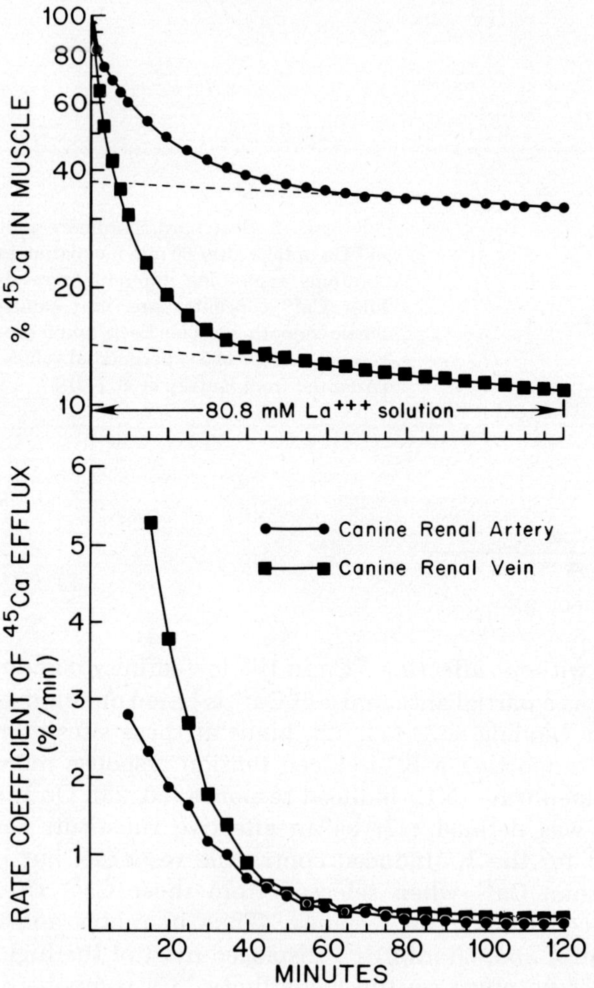

FIG. 1. Effect of La^{3+}-substituted (80.8 mM) solution at 0.5°C on ^{45}Ca efflux from canine coronary arteries. Desaturation (*top*) and corresponding rate coefficient (*bottom*) curves were obtained from average washouts of the same 12 muscles. Dashed line (*top*) indicates extrapolation of the slow component of the desaturation curve. All muscles were incubated in normal (1.5-mM Ca^{2+}) solution plus ^{45}Ca for 60 min before beginning washout. [From Hester et al. (17) by permission of the American Heart Association, Inc.]

that is predominantly either high- or low-affinity ^{45}Ca. If ^{45}Ca uptake is measured at sufficiently low extracellular concentrations of Ca^{2+} (e.g., 0.03 mM Ca^{2+}), then cells are Ca^{2+} deficient and ^{45}Ca is present at only high-affinity Ca^{2+} sites. Conversely, if high extracellular Ca^{2+} concentrations (e.g., 1.5 or 5.0 mM Ca^{2+}) are used, then most of the Ca^{2+} taken up is bound at low-affinity Ca^{2+} sites (or sequestered in cellular depots) even though the smaller number of high-affinity sites present would also be saturated. The uptake of ^{45}Ca at the kinetically distinct high- and low-affinity sites differed; ^{45}Ca uptake was slow ($t_{\frac{1}{2}}$ was approximately 30 min) at high-affinity sites and much more rapid ($t_{\frac{1}{2}}$ was not more than 5 min) at low-affinity sites (27). Furthermore approximately 20% of the ^{45}Ca at high-affinity sites and 10% of the ^{45}Ca at the more numerous low-affinity sites are not removed by a 60-min washout in isosmotic La^{3+} solution at 0.5°C (26).

Addition of an appropriate concentration of Sr^{2+} can be used to remove

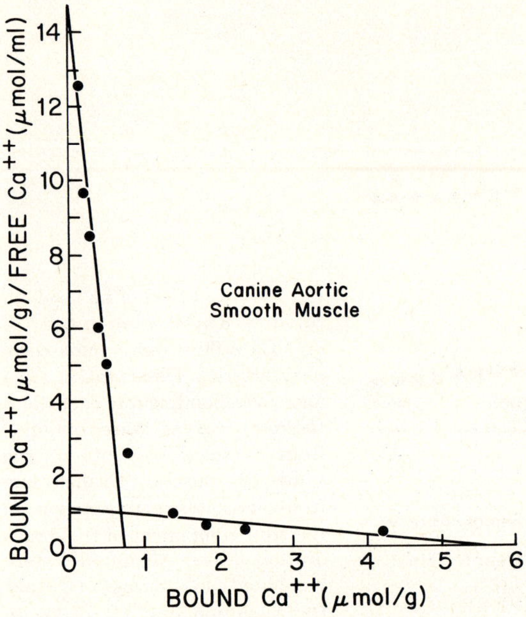

FIG. 2. Scatchard-coordinate plot of ^{45}Ca uptake after 60 min incubation in solutions containing different extracellular Ca^{2+} concentrations into canine aortic smooth muscle. Each point represents average of 5 experimental values. [Adapted from Kutsky et al. (32).]

^{45}Ca from high-affinity sites without affecting ^{45}Ca in the low-affinity component (28, 44). The use of Sr^{2+} as a partial antagonist of Ca^{2+} is based on findings that Sr^{2+} displaces ^{45}Ca from binding sites (14, 22), binds at these sites (46), and substitutes for Ca^{2+} in supporting a K^+-induced tension response more than in maintaining a norepinephrine (NE)-induced response (20, 23). On the basis of these effects, Sr^{2+} was defined (42) as an effective substitute for extracellular Ca^{2+} important for the K^+-induced contractile response, but it was less readily available than Ca^{2+} when released from those Ca^{2+} sites important for the NE-induced contractile response. Thus in rabbit aortic smooth muscle (28, 44), 1.5 mM and 5.0 mM Sr^{2+} displaced most of the high-affinity ^{45}Ca and had little or no effect on the low-affinity ^{45}Ca component. Note that Sr^{2+} may well have some activity at the low-affinity Ca^{2+} sites and channels, but at higher extracellular Ca^{2+} concentrations, an effect of Sr^{2+} on ^{45}Ca mobilization is not observed. What the above procedure represents is a specific variation of Sr^{2+} and Ca^{2+} concentrations so that experimental high- and low-affinity ^{45}Ca components can be affected differentially by Sr^{2+}.

ACTIONS OF AGONISTS ON SPECIFIC CALCIUM-45 COMPONENTS

The concept that different stimulatory agents can elicit contractile responses by using different sources of Ca^{2+} was originally based on the findings that in vascular smooth muscle, depletion of Ca^{2+} inhibited K^+-induced responses more rapidly than NE-induced responses (18, 19, 21). The K^+-induced responses depended on Ca^{2+} originating from extracellular or relatively superficial membrane Ca^{2+} stores, whereas catecholamine-induced contractility initially depended on release of a sequestered Ca^{2+} fraction (3, 39). Identification of specific Ca^{2+} binding and uptake components affected by either high K^+ or NE can now be more readily accomplished with the three approaches defined in the preceding section.

When La^{3+}-resistant ^{45}Ca uptake (Ca^{2+} retention) was measured under conditions favoring either high- or low-affinity Ca^{2+} binding (0.03 mM Ca^{2+} and 1.5 mM Ca^{2+}, respectively), qualitatively different effects were obtained with K^+ and NE (Fig. 3). The uptake of low-affinity Ca^{2+} was approximately doubled by K^+, whereas NE had no effect in control (polarized) muscle and further increased Ca^{2+} retention in K^+-depolarized preparations. The concentration of NE used in these studies elicited only minimal changes in polarization (35). When higher (depolarizing) concentrations of NE (up to 10^{-4} M) were used in isolated arteries (15, 25, 26), then an increased retention of low-affinity Ca^{2+} was also elicited with NE. Thus an uptake of approximately 300 nmol low-affinity Ca^{2+} occurs in resting (control) muscles, and this is increased by depolarizing stimuli. Conversely, nondepolarizing concentrations of NE decrease high-affinity Ca^{2+} retention and either do not alter low-affinity Ca^{2+} retention (control muscles) or further increase low-affinity Ca^{2+} retention (K^+-depolarized muscles). The effect of 5.0 mM Sr^{2+} on high- and low-affinity Ca^{2+}-retention changes elicited by K^+ and NE confirms this dissociation of components (28). The increases in low-affinity Ca^{2+} retention obtained with high K^+ and with high K^+ plus NE are not affected by Sr^{2+}, but Sr^{2+} does remove the major portion of high-affinity retained Ca^{2+}, including that portion otherwise displaced by NE.

Additionally the rate at which low-affinity retained Ca^{2+} was increased by

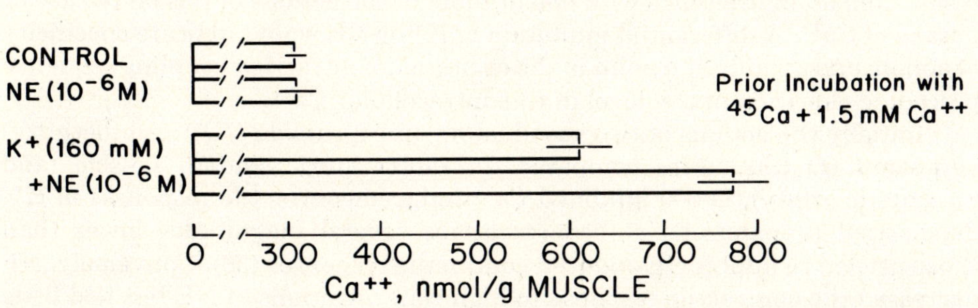

FIG. 3. Effect of NE and K^+ on Ca^{2+} retention after prior incubation with ^{45}Ca in solutions favoring binding at high-affinity Ca^{2+} sites (0.03 mM extracellular Ca^{2+}) or low-affinity Ca^{2+} sites (1.5 mM extracellular Ca^{2+}); n = 6–18. [Adapted from Karaki and Weiss (26).]

high K^+ and by NE (plus high K^+) and the rate at which high-affinity retained Ca^{2+} was decreased by NE were examined (27) by adding K^+ or NE after equilibration of muscles with ^{45}Ca and then removing muscles after short additional incubation periods. In either case changes in retained Ca^{2+} equal to those previously seen after 60 min of ^{45}Ca incubation were observed 1–2 min after exposure to NE or K^+ and were maintained. Thus regardless of duration of loading of high- and low-affinity Ca^{2+} sites in resting muscle, agonist-induced changes in uptake or release of Ca^{2+} occur rapidly (and are maintained for the duration of exposure to the stimulatory agent). This indicates that drug-induced mobilization of either high- or low-affinity Ca^{2+} is sufficiently fast to correlate with the effective time course for development of tension and that one way for an inhibitory agent to prevent NE-induced or K^+-induced increases in tension or contractility is to alter at least one of these defined components of Ca^{2+} release or uptake.

ACTIONS OF INHIBITORS ON SPECIFIC CALCIUM-45 COMPONENTS

Inhibitory agents not acting directly at receptors for agonists have generally been considered as nonspecific blocking agents or, in the case of vascular smooth muscle, as nonspecific vasodilators. Within this context *nonspecific* could mean either a lack of precise mechanism of action or a variety of specific but undefined mechanisms by which blocking of either agonist-induced increases in tension or altered resting tone occurs. In the latter situation, one obvious point of action would be a direct or indirect antagonism of one or more Ca^{2+}-associated effects. If K^+ and NE alter high- and/or low-affinity Ca^{2+} in specific ways that can be described in terms of sensitivity (or lack of sensitivity) to La^{3+} displacement and of dissociable uptake and binding components of Ca^{2+}, then whether inhibitors act to alter different agonist-mobilized components can be ascertained.

Measurement of effects of inhibitors on tension responses to NE or K^+ provides initial insight into the existing mechanism of action. An inhibitory effect on the K^+-induced response would indicate a possible blockade of low-affinity Ca^{2+} uptake, whereas an inhibition of the NE-induced response might result from an interference with one or more of the actions of NE on release or uptake of Ca^{2+}. A differential inhibition of K^+ or NE would indicate specificity of action, presumably at a point in the excitation-contraction coupling sequence preceding elevation of the level of free intracellular Ca^{2+}.

Initially the actions of two vasodilators, nitroprusside (NP) and methoxyverapamil (D 600), were compared. In rabbit aortic smooth muscle (and mesenteric artery), D 600 inhibited Ca^{2+}-induced contractile responses in K^+-depolarized preparations at concentrations several magnitudes lower than those needed to inhibit NE-induced contractile responses (38). Conversely, NP decreased the contractile response to high concentrations of NE but had little effect on the K^+-induced contractile response in canine renal arteries (16) and rabbit aortic smooth muscle (H. Karaki and G. B. Weiss, unpublished data). Effects of these two inhibitors on Ca^{2+} retention are summarized in Table 1.

TABLE 1. *Effects of D 600 and NP on Ca^{2+} retention in the presence and absence of K^+ and NE in rabbit aortic smooth muscle*

^{45}Ca Incubation Conditions*	Ca^{2+} Retention, nmol/g muscle ± SE	Change in Ca^{2+} Retention, %
Control	301.9 ± 10.7	
D 600	293.6 ± 33.0	−2.8
NP	295.7 ± 16.3	−2.1
NE	299.4 ± 15.0	−0.8
K^+	605.0 ± 26.5	100.4
K^+ + NP	639.2 ± 30.5	111.7
K^+ + D 600	298.9 ± 8.6	−1.0
K^+ + NE	785.0 ± 33.0	160.0
K^+ + NE + NP	580.8 ± 28.9	92.4
K^+ + NE + D 600	397.1 ± 19.0	31.5
K^+ + NE + D 600 + NP	307.1 ± 28.0	1.7

* All muscles were washed out for 60 min with isosmotic La^{3+} solution at 0.5°C after a 60-min incubation with ^{45}Ca in 1.5-mM Ca^{2+} solution. Concentrations of agents used: NE, 10^{-6} M; K^+, 160 mM; D 600, 5×10^{-7} M; and NP, 10^{-6} M. [Adapted from Karaki and Weiss (27).]

Neither NP nor D 600 alter low-affinity Ca^{2+} retention (after 60-min washout in 80.8 mM La^{3+} solution at 0.5°C) in control rabbit aortic smooth muscle. However, Ca^{2+} retention is doubled by 160 mM K^+, and this increase is blocked by D 600 (but not by NP). The additional increase in Ca^{2+} retention obtained with 10^{-6} M NE in the presence of 160 mM K^+ was about 180 nmol/g of muscle above the value obtained with a 160-mM K^+ solution alone. This additional increment of ^{45}Ca retention was inhibited by NP. Although the combined effects of K^+ and NE were not totally inhibited by either NP or D 600, they were blocked when both D 600 and NP were present. Thus each of the two inhibitory agents appears to block a different specific component of ^{45}Ca uptake.

Another group of agents, the aminoglycoside antibiotics, are of therapeutic interest and appear to be strong Ca^{2+} antagonists (see 42). Measurement of the effects of neomycin on contractile responses to K^+ and several stimulatory agents including NE (13) indicated that the initial rate of induction of the contractile response (measured as half time of rise to maximum effect) was decreased significantly (and markedly) only for K^+-induced responses. The total equilibrated uptake of ^{45}Ca was also reduced by neomycin, but the rate of uptake was not inhibited (equilibrated ^{45}Ca tissue:medium ratios in the presence of neomycin were lower than control ^{45}Ca tissue:medium ratios). The results were interpreted as showing that neomycin prevents uptake and/or binding of Ca^{2+} at a portion of the superficially located Ca^{2+} sites in rabbit aortic smooth muscle.

Further support for this view was obtained more recently by a different technique—measurement of Ca^{2+} uptake and binding in vascular smooth muscle microsomal membranes. Initially canine aortic media-intimal layer microsomes were used and the ATP-dependent accumulation of ^{45}Ca was characterized (30). More recently (32) conditions affecting the binding and release of ^{45}Ca from membrane sites were examined and Scatchard coordinate plots were obtained in the presence of adenosine triphosphate (ATP) (but in the absence of oxalate). These plots indicated the presence of two types of Ca^{2+}-binding sites (high and low affinity) similar to those observed in canine (32) and rabbit (43, 44) aortic strips. Removal of ATP inhibited accumulation of Ca^{2+} but not superficial binding of Ca^{2+}. Thus comparison of effects of different inhibitors on ATP-dependent Ca^{2+} uptake (active accumulation plus binding) and ATP-independent Ca^{2+} uptake (binding) yields additional information about specific effects of these agents on Ca^{2+} accumulation and binding. However, since the microsomal preparation is an anatomically mixed preparation and the two parameters measured are not completely separate, the dissociation obtained is only approximate. The major differences observed (31) are summarized in Table 2. Clearly the strongest effects of Sr^{2+} are on high-affinity Ca^{2+} uptake and binding, whereas those of high concentrations of neomycin and gentamicin (heart muscle is much more sensitive to these agents and of greater clinical significance in this respect than vascular muscle) are on membrane (ATP-independent) binding. This latter action supports the earlier conclusion (13) that agents such as neomycin interfere with binding of Ca^{2+} at superficially located Ca^{2+} sites in vascular smooth muscle. The only effect of NP in this series of measurements was to decrease ATP-dependent high-affinity Ca^{2+} uptake. In even greater contrast, D 600 had no effect on any Ca^{2+} uptake or binding parameter. Thus this type of preparation may provide a satisfactory method for differentiating between agents that compete with Ca^{2+} at membrane sites (e.g., aminoglycoside antibiotics) and agents that act to prevent movement of Ca^{2+} through membrane channels (e.g., D 600).

MODEL FOR CALCIUM RELEASE AND ENTRY

A schematic model illustrating the routes of entry and types of binding sites for Ca^{2+} in vascular smooth muscle is shown in Figure 4. Superimposed on this are the action of K^+ on the voltage-sensitive channel and the multiple

TABLE 2. *Effects of Sr^{2+}, gentamicin, neomycin, NP, and D 600 on Ca^{2+} uptake and binding in canine aortic smooth muscle microsomes*

Experimental Conditions*	Change in Ca^{2+} Uptake, %			
	0.03 mM Ca^{2+}		3.0 mM Ca^{2+}	
	ATP	no ATP	ATP	no ATP
Sr^{2+}, 5.0 mM	−87%	−56%	−21%	−31%
Neomycin, 7 mM	−32%	−71%	−57%	−80%
Gentamicin, 6 mM	−21%	−70%	−52%	−80%
NP, 0.1 mM	−14%	NS	NS	NS
D 600, 5×10^{-5} M	NS	NS	NS	NS

* Microsomes were loaded with ^{45}Ca for 60 min in experimental solutions. Inhibitory agents were present during the ^{45}Ca incubation period and for 1 min before simultaneous addition of nonradioactive Ca^{2+} (0.03 mM or 3.0 mM) and ^{45}Ca. NS, not significant. [Adapted from Kutsky and Weiss (31).]

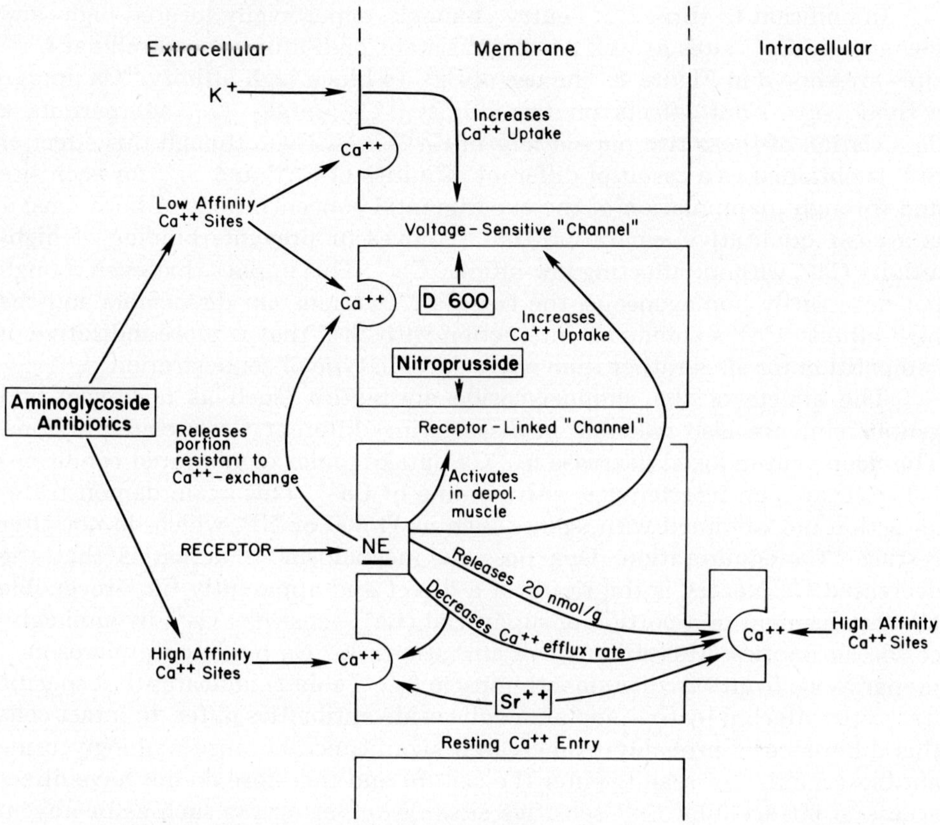

FIG. 4. Model for sites of action of different inhibitory agents on Ca^{2+}-uptake pathways and associated Ca^{2+}-binding sites in vascular smooth muscle.

actions of NE under different experimental conditions. The sites at which the inhibitors Sr^{2+}, D 600, NP, and aminoglycoside antibiotics apparently act are also indicated. Clearly the different specific effects of the inhibitors on entry and release of ^{45}Ca components help dissociate the postulated sites and channels. For example, the specific inhibitory effects of D 600 and NP indicate the presence of at least three separate entry mechanisms: a resting (La^{3+}-resistant control) Ca^{2+} entry unaffected by either D 600 or NP (25, 27), a voltage-sensitive Ca^{2+} entry obtained with a high level of K^+ (15, 25–27, 47) or depolarizing concentrations of NE (15, 25, 26) and blocked differentially by D 600 (25, 27), and a receptor-linked entry obtained in K^+-depolarized muscle with NE (26–28) and inhibited by NP. This last entry mechanism superficially resembles the receptor-operated channel for Ca^{2+} entry postulated by Bolton (4). Our studies (25, 26), however, indicate that different types of voltage-sensitive ^{45}Ca entry (whether elicited with K^+ or NE) are blocked by D 600 and may well be mediated by the same channels, whereas the additional NE-induced ^{45}Ca uptake in K^+-depolarized muscle is specifically sensitive to inhibition by NP.

In addition to three Ca^{2+} entry channels, superficially located high- and low-affinity Ca^{2+} sites as well as La^{3+}-resistant high-affinity intracellular Ca^{2+} sites are shown in Figure 4. The use of Sr^{2+} to block high-affinity ^{45}Ca uptake without significant effects on low-affinity ^{45}Ca uptake (28, 44) permits a dissociation of these two populations of Ca^{2+} sites. Even though this effect of Sr^{2+} is obtained as a result of different affinities of Ca^{2+} and Sr^{2+} for each site and through manipulation of the experimental concentrations of both ions, it is a clear qualitative separation that removes or prevents binding of high-affinity Ca^{2+} without affecting low-affinity Ca^{2+}. This implies that even though not necessarily homogeneous, the two Ca^{2+} fractions are dissociable and the high-affinity Ca^{2+} exhibits an interaction with Sr^{2+} that is more indicative of competition for sites rather than a less specific type of sequestration.

The effects of the aminoglycoside antibiotics (such as neomycin and gentamicin) are also valuable in dissociating different Ca^{2+}-related actions. The neomycin-induced decrease in ^{45}Ca uptake under equilibrated conditions (13) indicates an interference with binding of Ca^{2+}. This again demonstrates an action not obtained with agents such as D 600 or NP, which do not alter resting ^{45}Ca equilibration. One possible mechanism of action is that the decreased Ca^{2+} entry is the result of a direct and apparently Ca^{2+}-reversible (1) displacement of a portion of superficial (La^{3+}-sensitive) Ca^{2+} by aminoglycoside antibiotics. Measurement of alterations in ^{45}Ca binding in microsomal preparations from vascular smooth muscle (see Table 2) indicates that specific Ca^{2+} sites affected by Sr^{2+} and aminoglycoside antibiotics differ. In intact cells this difference is probably of increased significance because aminoglycoside antibiotics may not readily enter the cell (9) and therefore do not have direct access to intracellular Sr^{2+}-sensitive sites. Whether agents such as neomycin and gentamicin also block Ca^{2+} entry channels or, conversely, whether they block a surface-bound Ca^{2+} fraction that is subsequently mobilized through channels by stimulatory agents is not clear.

CONCLUSION

The use of different types of Ca^{2+}-related inhibitors has facilitated dissociation of specific components of Ca^{2+} with distinct but different roles in agonist-induced mobilization of Ca^{2+}. In the vascular smooth muscle cell, there are multiple Ca^{2+} entry channels as well as populations of binding sites differing in affinity and/or location. The concept that different types of inhibitors previously termed nonspecific may well have divergent but clearly defined cellular actions as Ca^{2+} antagonists helps us focus on mechanism-related questions. Measurements of ^{45}Ca uptake under conditions favoring either high- or low-affinity binding of Ca^{2+} as well as the use of Sr^{2+}, La^{3+}, and low temperature to remove or prevent removal of specific Ca^{2+} fractions have resulted in the development of a more quantitative model for patterns of Ca^{2+} binding and mobilization than has previously been possible in vascular smooth muscle.

Within this model the specific actions of Ca^{2+} voltage-sensitive channel blockers such as D 600 may well be of particular physiological and pharmacological significance. Other Ca^{2+}-channel blockers chemically related to nifedipine or cinnarizine also have cellular actions similar to those of D 600 (G. B. Weiss, unpublished data). Clearly agents acting in this manner would not alter equilibration of resting Ca^{2+} or resting free intracellular Ca^{2+} distribution. Increases in uptake of Ca^{2+}, however, would be blocked and the corresponding loss of increased intracellular Ca^{2+} would be unaffected. Thus at least in a superficial sense, this group of agents might well be anticipated to inhibit increased responsiveness or tension in vascular smooth muscle in a particularly specific manner. The therapeutic implications of this for vascular smooth muscle and, hopefully, for other smooth muscle systems, could be considerable.

The experimental work from my laboratory was supported by National Institutes of Health Grant HL-14775. Much of this work was performed in collaboration with Drs. Hideaki Karaki, Phyllis Kutsky, R. Kelly Hester, and Frank R. Goodman.

REFERENCES

1. ADAMS, H. R., F. R. GOODMAN, AND G. B. WEISS. Alteration of contractile function and calcium ion movements in vascular smooth muscle by gentamicin and other aminoglycoside antibiotics. *Antimicrob. Agents Chemother.* 5: 640–646, 1974.
2. BARELA, T. D., S. BURCHETT, AND D. E. KIZER. Terbium as a molecular probe of magnesium binding to ribosomes and ribosomal RNA. *Proc. 12th Rare Earth Conf.* Oak Ridge, TN: U.S. At. Energy Comm. Tech. Inf. Cent., 1976, vol. I, p. 223–230.
3. BOHR, D. F. Contraction of vascular smooth muscle. *Can. Med. Assoc. J.* 90: 174–179, 1964.
4. BOLTON, T. B. Mechanisms of action of transmitters and other substances on smooth muscle. *Physiol. Rev.* 59: 606–718, 1979.
5. BURTON, J., AND T. GODFRAIND. Sodium-calcium sites in smooth muscle and their accessibility to lanthanum. *J. Physiol. London* 241: 287–298, 1974.
6. DETH, R. C. Effect of lanthanum and reduced temperature on ^{45}Ca efflux from rabbit aorta. *Am. J. Physiol.* 234 (*Cell Physiol.* 5): C139–C145, 1978.
7. FREEMAN, D. J., AND E. E. DANIEL. Calcium movement in vascular smooth muscle and its detection using lanthanum as a tool. *Can. J. Physiol. Pharmacol.* 51: 900–913, 1973.
8. GODFRAIND, T. Calcium exchange in vascular smooth muscle, action of norepinephrine and lanthanum. *J. Physiol London* 260: 21–35, 1976.
9. GOODMAN, F. R. Distribution of ^{14}C-gentamicin in vascular smooth muscle. *Pharmacology* 16: 17–25, 1978.
10. GOODMAN, F. R., AND G. B. WEISS. Dissociation by lanthanum of smooth muscle responses to potassium and acetylcholine. *Am. J. Physiol.* 220: 759–766, 1971.
11. GOODMAN, F. R., AND G. B. WEISS. Effects of lanthanum on ^{45}Ca movements and on contractions induced by norepinephrine, histamine and potassium in vascular smooth muscle. *J. Pharmacol. Exp. Ther.* 177: 415–425, 1971.
12. GOODMAN, F. R., AND G. B. WEISS. Lanthanum-induced alterations in ^{45}Ca and ^{147}Pm distribution in vascular smooth muscle. In: *Proc. 11th Rare Earth Conf.* Oak Ridge, TN: U.S. At. Energy Comm. Tech. Inf. Cent., 1974, vol. II, p. 716–722.
13. GOODMAN, F. R., G. B. WEISS, AND H. R. ADAMS. Alterations by neomycin of ^{45}Ca movements and contractile responses in vascular smooth muscle. *J. Pharmacol. Exp. Ther.* 188: 472–480, 1974.
14. GOODMAN, F. R., G. B. WEISS, M. N. WEINBERG, AND S. D. POMARANTZ. Effects of added or substituted potassium ion on ^{45}Ca movements in rabbit aortic smooth muscle. *Circ. Res.* 31: 672–681, 1972.
15. HESTER, R. K., AND G. B. WEISS. Comparison of degree of dependence of canine renal arteries and veins on high and low affinity calcium for responses to norepinephrine and potassium. *J. Pharmacol. Exp. Ther.* 216: 239–246, 1981.
16. HESTER, R. K., G. B. WEISS, AND W. J. FRY. Differing actions of nitroprusside and D-600 on tension and ^{45}Ca fluxes in canine renal arteries. *J. Pharmacol. Exp. Ther.* 208: 155–160, 1979.
17. HESTER, R. K., G. B. WEISS, AND J. T. WILLERSON. Basis of pH-independent inhibitory effects of lactate on ^{45}Ca movements and responses to KCl and PGF_2 in canine coronary arteries. *Circ. Res.* 46: 771–779, 1980.
18. HINKE, J. A. M. Calcium requirements for noradrenaline and high potassium ion contraction in arterial smooth muscle. In: *Muscle*, edited by W. M. Paul, E. E. Daniel, C. M. Kay, and G. Monckton. New York: Pergamon, 1965, p. 269–284.
19. HIRAOKA, M., S. YAMAGISHI, AND T. SANO. Role of calcium ions in the contraction of vascular smooth muscle. *Am. J. Physiol.* 214: 1084–1089, 1968.
20. HUDGINS, P. M. Some drug effects on calcium movements in aortic strips. *J. Pharmacol. Exp. Ther.* 170: 303–310, 1969.
21. HUDGINS, P. M., AND G. B. WEISS. Differential effects of calcium removal upon vascular smooth muscle contraction induced by norepinephrine, histamine and potassium. *J. Pharmacol. Exp. Ther.* 159: 91–97, 1968.
22. HUDGINS, P. M., AND G. B. WEISS. Characteristics of ^{45}Ca binding in vascular smooth muscle. *Am. J. Physiol.* 217: 1310–1315, 1969.
23. HUDGINS, P. M., AND G. B. WEISS. Effects of Ba, Sr and stimulatory agents on Ca movements and contraction in vascular smooth muscle. *Federation Proc.* 28: 541, 1969.

24. ISHIYAMA, Y., H. YABU, AND E. MIYAZAKI. Changes in contractility and calcium binding of guinea pig taenia coli by treatment with enzymes which hydrolyze sialic acid. *Jpn. J. Physiol.* 25: 719–732, 1975.
25. KARAKI, H., R. K. HESTER, AND G. B. WEISS. Cellular basis of nitroprusside-induced relaxation of graded responses to norepinephrine and potassium in canine renal arteries. *Arch. Int. Pharmacodyn. Ther.* 245: 198–210, 1980.
26. KARAKI, H., AND G. B. WEISS. Alterations in high and low affinity binding of ^{45}Ca in rabbit aortic smooth muscle by norepinephrine and potassium after exposure to lanthanum and low temperature. *J. Pharmacol. Exp. Ther.* 211: 86–92, 1979.
27. KARAKI, H., AND G. B. WEISS. Effects of stimulatory agents on mobilization of high and low affinity site ^{45}Ca in rabbit aortic smooth muscle. *J. Pharmacol. Exp. Ther.* 213: 450–455, 1980.
28. KARAKI, H., AND G. B. WEISS. Dissociation of varied actions of norepinephrine on ^{45}Ca uptake and release at different sites in rabbit aortic smooth muscle. *J. Pharmacol. Exp. Ther.* 215: 363–368, 1980.
29. KARAKI, H., AND G. B. WEISS. Qualitative differences in ^{45}Ca efflux from membrane sites in vascular smooth muscle when washout conditions are varied. *Gen. Pharmacol.* 11: 483–489, 1980.
30. KUTSKY, P., AND F. R. GOODMAN. Calcium incorporation by canine aortic smooth muscle microsomes. *Arch. Int. Pharmacodyn. Ther.* 231: 4–20, 1978.
31. KUTSKY, P., AND G. B. WEISS. Specific alterations in binding and uptake of high and low affinity Ca^{++} in canine aortic microsomes. *Blood Vessels* in press.
32. KUTSKY, P., G. B. WEISS, AND H. KARAKI. Delineation of high and low affinity ^{45}Ca incorporation and of ^{45}Ca efflux in canine aortic smooth muscle microsomes. *Gen. Pharmacol.* 11: 475–481, 1980.
33. LETTVIN, J. Y., W. F. PICKARD, W. S. McCULLOCH, AND W. PITTS. A theory of passive ion flux through axon membranes. *Nature London* 202: 1338–1339, 1964.
34. MALVEY, J. E., F. P. J. DIECKE, AND B. A. SCHOTTELIUS. Effects of lanthanum ions on calcium-45 efflux from vascular smooth muscle. *Federation Proc.* 33: 436, 1974.
35. MEKATA, F. Rectification in the smooth muscle cell membrane of rabbit aorta. *J. Physiol. London* 258: 269–278, 1976.
36. REUTER, H., M. P. BLAUSTEIN, AND G. HAEUSLER. Na-Ca exchange and tension development in arterial smooth muscle. *Philos. Trans. R. Soc. London Ser. B* 264: 87–94, 1973.
37. SCATCHARD, G. The attractions of proteins for small molecules and ions. *Ann. NY Acad. Sci.* 51: 660–671, 1949.
38. SCHUMANN, H. J., B. D. GORLITZ, AND J. WAGNER. Influence of papaverine, D600, and nifedipine on the effects of norepinephrine and calcium on the isolated aorta and mesenteric artery of the rabbit. *Naunyn-Schmiedeberg's Arch. Pharmacol.* 289: 409–418, 1975.
39. VAN BREEMEN, C. Blockade of membrane calcium fluxes by lanthanum in relation to vascular smooth muscle contractility. *Arch. Int. Physiol. Biochim.* 77: 710–716, 1969.
40. VAN BREEMEN, C., B. R. FARINAS, P. GERBA, AND E. D. McNAUGHTON. Excitation-contraction coupling in rabbit aorta studied by the lanthanum method for measuring cellular calcium influx. *Circ. Res.* 30: 44–54, 1972.
41. VAN BREEMEN, C., AND E. McNAUGHTON. The separation of cell membrane calcium transport from extracellular calcium exchange in vascular smooth muscle. *Biochem. Biophys. Res. Commun.* 39: 567–574, 1970.
42. WEISS, G. B. Calcium and contractility in vascular smooth muscle. In: *Advances in General and Cellular Pharmacology*, edited by T. Narahashi and C. P. Bianchi. New York: Plenum, 1977, vol. II, p. 71–154.
43. WEISS, G. B. Approaches to delineation of differing calcium binding sites in smooth muscle. In: *Excitation-Contraction Coupling in Smooth Muscle*, edited by R. Casteels, T. Godfraind, and J. C. Rüegg. Amsterdam: Elsevier, 1977, p. 253–260.
44. WEISS, G. B. Quantitative measurement of binding sites and washout components for calcium ion in vascular smooth muscle. In: *Calcium in Drug Action*, edited by G. B. Weiss. New York: Plenum, 1978, p. 57–74.
45. WEISS, G. B., AND F. R. GOODMAN. Effects of lanthanum on contraction, calcium distribution and Ca^{45} movements in intestinal smooth muscle. *J. Pharmacol. Exp. Ther.* 169: 46–55, 1969.
46. WEISS, G. B., AND F. R. GOODMAN. Interactions between several rare earth ions and calcium ion in vascular smooth muscle. *J. Pharmacol. Exp. Ther.* 195: 557–564, 1975.
47. WHEELER, E. S., AND G. B. WEISS. Effects of prostaglandin E_1 on contractility and ^{45}Ca release in rabbit aortic smooth muscle. *Prostaglandins* 19: 761–778, 1980.
48. YAMASHITA, K., T. TAKAGI, AND K. HOTTA. Mobilization of cellular calcium and contraction-relaxation of vascular smooth muscle. *Jpn. J. Physiol.* 27: 551–564, 1977.

7

Calcium Influx and Receptor-Response Coupling

T. GODFRAIND

Laboratoire de Pharmacodynamie Générale et de Pharmacologie, Université Catholique de Louvain, Brussels, Belgium

THE ROLE OF CALCIUM as a second messenger in smooth muscle receptor-response coupling has been proposed on the basis that Ca^{2+} withdrawal from perfusion solutions reduces or even abolishes the contraction evoked by receptor activation (1, 13). Recent biochemical findings have characterized the preeminent role of Ca^{2+} in the activation of the contractile machinery (see 6, 18). The general biochemical mechanism responsible for smooth muscle contraction is myosin phosphorylation catalyzed by myosin light-chain kinase. Relaxation is achieved by a dephosphorylation catalyzed by myosin light-chain phosphatase and occurs at a lower rate than the contraction reaction. The phosphorylation of smooth muscle myosin is apparently essential for activation of myosin ATPase activity by actin; dephosphorylated myosin is not activated by actin. Myosin light-chain kinase activity depends on Ca^{2+} and calmodulin (Ca^{2+}-dependent regulator protein). Calcium ions first bind to calmodulin to form a Ca_n^{2+}-calmodulin complex, which in turn binds to the inactive myosin light-chain kinase to form a catalytically active phosphotransferase. Some drugs, including phenothiazine antipsychotic agents, bind to the Ca_n^{2+}-calmodulin complex and prevent it from activating myosin light-chain kinase.

Myosin light-chain kinase is also regulated by cyclic adenosine 3′,5′-monophosphate (cyclic AMP). An increase in cyclic AMP (by β-adrenergic stimulation or by phosphodiesterase inhibition) activates cyclic AMP-dependent protein kinase that in turn catalyzes the phosphorylation of myosin light-chain kinase. The affinity of phosphorylated myosin light-chain kinase for the Ca_n^{2+}-calmodulin complex is 20-fold lower than nonphosphorylated myosin light-chain kinase and therefore will not be as readily activated. Subsequently the phosphorylation of myosin is inhibited and the smooth muscle relaxes. Indeed cyclic nucleotide-dependent myosin phosphorylation could lead to a Ca^{2+}-independent smooth muscle contraction. Although such a mechanism might account for the residual contraction observed after Ca^{2+} depletion, its physiological role remains to be documented.

CELLULAR CALCIUM METABOLISM

It is generally accepted that in resting muscle the cytoplasmic free-Ca^{2+} concentration is not more than 10^{-7} M, so that the ratio of extracellular Ca^{2+} to intracellular Ca^{2+} exceeds 10^4. The Ca^{2+} electrochemical gradient is oriented inward and is approximately 60 mV. This low pCa is due to a low membrane permeability to Ca^{2+} (10, 20) and also to cellular buffering processes, which are intracellular and plasmalemmal. Cytoplasmic proteins such as pavalbumins have Ca^{2+}-binding properties. Mitochondria are able to take up large amounts of Ca^{2+} and may serve as Ca^{2+} sinks. They can also release Ca^{2+} rapidly in the presence of metabolic inhibitors or Na^+. These properties are required for mitochondria to regulate cellular Ca^{2+} metabolism. A role for mitochondria in regulation of tension is not yet established, however (see 16). Sarcoplasmic reticulum (SR) appears to be more efficient than mitochondria for normal regulation of Ca^{2+} levels in the sarcoplasm. This is illustrated by experiments performed in visceral smooth muscle showing that SR vesicles have a higher affinity for Ca^{2+} than mitochondria, whereas the capacity of mitochondria is higher than that of SR vesicles (Fig. 1).

Although intracellular binding of Ca^{2+} may rapidly buffer changes in the free Ca^{2+}, the constancy of the intracellular Ca^{2+} concentration over long periods of time requires that Ca^{2+} efflux exactly balances Ca^{2+} influx. It is therefore likely that Ca^{2+} pumps are located in the cell membrane. At least two types of mechanisms are feasible. The first is a Ca^{2+} pump that actively extrudes Ca^{2+} at the expense of metabolic energy. The second is a Ca^{2+}

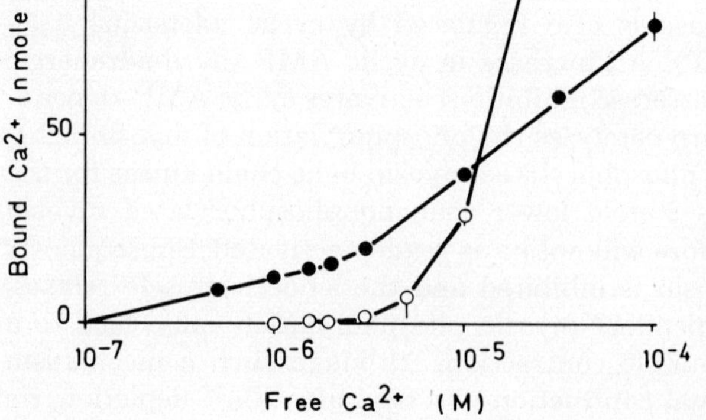

FIG. 1. Microsomal and mitochondrial Ca^{2+} binding as function of free-Ca^{2+} concentrations in medium. Microsomal ^{45}Ca content (●) was measured after 30 min incubation at 37°C in presence of 20 mM Tris-maleate buffer (pH 6.6), 5 mM $MgCl_2$, 3 mM ATP, 100 mM KCl, and various concentrations of Ca^{2+} buffered by calcium EGTA (ethylene glycol-bis(β-aminoethylether)-*N,N'*-tetraacetic acid). Each value is the mean of 6 determinations ± SE. Mitochondrial ^{45}Ca content (○) was estimated as the azide-sensitive Ca^{2+} binding of the mitochondrial fraction. It was measured after 30 min incubation at 37°C in presence of 20 mM KCl, 0 or 10 mM NaN_3, and various concentrations of Ca^{2+} buffered by calcium EGTA. Each result is mean of 9 determinations ± SE. [From Verbeke et al. (21).]

exchange with some other particle moving down its electrochemical gradient. The latter mechanism has been postulated to operate in smooth muscle as an Na^{+}-Ca^{2+} exchange. Whether such a regulatory mechanism exists in smooth muscle is controversial (5, 20). It is clear that the total intracellular calcium content and the rate of Ca^{2+} exchange across the smooth muscle membrane are affected by the intracellular sodium content. It has been demonstrated that the exchangeable calcium fraction increases when the intracellular Na^{+} concentration increases (4). It nevertheless seems that the maintenance of the inward Ca^{2+} gradient depends more on cellular ATP than on the Na^{+} gradient. When cellular ATP is depleted by metabolic inhibition with iodoacetic acid and dinitrophenol, cellular Ca^{2+} increases linearly, presumably due to inhibition of the Ca^{2+} pump (20). In resealed red cell ghosts, it is well established that an active extrusion of Ca^{2+} is accomplished by a membrane ATPase that depends on Ca^{2+}, requires Mg^{2+}, is insensitive to Na^{+} and K^{+}, and is activated by calmodulin. This is a prominent feature of plasmalemmal Ca^{2+} pumps, as demonstrated not only in red blood cells, but also in brain cells and in adipocytes (N. Morel et al., unpublished data). Calcium translocation in inside-out microsomal vesicles prepared from rat aorta has recently been studied. The ATP-driven Ca^{2+} transport is enhanced 2.5-fold in the presence of calmodulin, which increases the apparent affinity of the transport system for Ca^{2+} (Fig. 2). The vesicles that take up Ca^{2+} are distributed like plasmalemmal marker enzymes when the microsomal fraction is subfractionated by density-gradient centrifugation. This indicates that the calmodulin-stimulated Ca^{2+} pump associated with the microsomal fraction is located in the plasma mem-

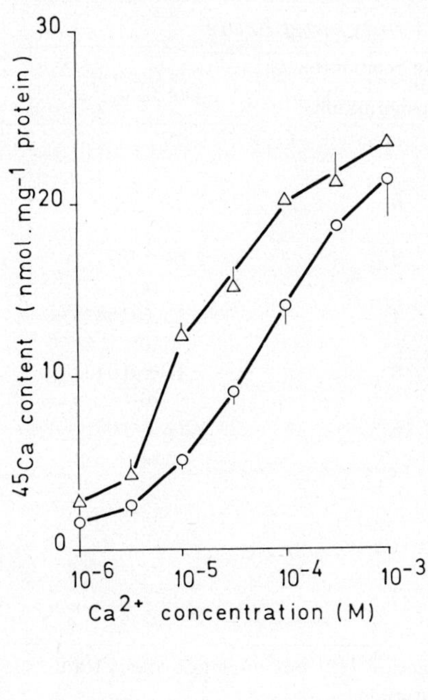

FIG. 2. Effect of calmodulin on ATP-dependent Ca^{2+} uptake by rat aorta microsomes. A microsomal pellet prepared from rat aorta was suspended in a chilled solution containing 10 mM Tris-HCl buffer (pH 7.4), 0.2 mM dithiothreitol, and 1 mM EGTA. After centrifugation at 10^5 g for 60 min, pellet was resuspended in 0.25 M sucrose buffered at pH 7.4 with 5 mM Tris-HCl. The EGTA-treated microsomes were incubated at 37°C in the presence (△) or absence (○) of calmodulin (6 μg/ml) with 20 mM Tris maleate (pH 6.6), 100 mM KCl, 5 mM $MgCl_2$, 3 mM ATP, 10 mM creatine phosphate, creatine kinase (3U/ml), and Ca^{2+} as indicated. Up to 10^{-4} M Ca^{2+}, medium contained 0.2 mM $^{45}CaCl_2$ and various amounts of EGTA to achieve the required free-Ca^{2+} concentration. After incubation for 15 min in absence of ATP, ^{45}Ca accumulation was initiated by addition of ATP. After 20 min, mixtures were filtered on Sartorius filters (0.45 μm). Filters were rapidly washed with 20 ml of chilled 0.25 M sucrose and then dissolved in a scintillation cocktail. The ATP-dependent Ca^{2+} accumulation was obtained after subtraction of the ^{45}Ca taken up in the absence of ATP. Long vertical bars indicate SE for 3–9 determinations on 2 microsomal preparations. (N. Morel, M. Wibo, and T. Godfraind, unpublished data).

brane and might be at least partly responsible for active Ca^{2+} extrusion from the cell.

CALCIUM FLUXES DURING MUSCLE ACTIVATION

The origin of activator Ca^{2+} was first established from indirect evidence, namely studies on contractions evoked by various agents in media with or without Ca^{2+}. Edman and Schild (7) showed that in Ca^{2+}-free medium, the contraction of the rat uterus evoked by K^+ depolarization declined faster than the contraction evoked by acetylcholine. Godfraind and Kaba (12) observed that vascular smooth muscle bathed in Ca^{2+}-free medium was still contracted by epinephrine, whereas K^+ depolarization was no longer able to activate the contraction. These results, indicating that α-adrenergic stimulation acted on two Ca^{2+} pools (one extracellular and the other intracellular), are consistent with other reports (1). This suggests that depolarization of the membrane is not required for α-adrenergic activation of vascular smooth muscle. More recent experiments with microelectrodes have shown that norepinephrine (NE) evokes a contraction without changes in membrane potential (5). Direct estimation of transmembrane Ca^{2+} fluxes in smooth muscle is obscured by exchange occurring at extracellular sites. Table 1 reports data obtained from rat aorta. The accompanying contraction of the muscle in response to NE shows a component sensitive to extracellular Ca^{2+}, suggesting that NE stimulates Ca^{2+} influx. The rate of ^{45}Ca uptake by the tissue, however, does not seem to be influenced by NE.

TABLE 1. *^{45}Ca content of rat aortas incubated in radioactive physiological solution at 37°C and in solution with 10^{-5} M norepinephrine*

Time in Radioactive Physiological Solution, min	^{45}Ca Content, mmol/kg wet wt	Time in Solution + 10^{-5} M Norepinephrine, min	^{45}Ca Content, mmol/kg wet wt
2	1.62 ± 0.055 (6)		
5	2.08 ± 0.046 (6)	0	
7	2.17 ± 0.022 (6)	2	2.19 ± 0.047 (4)
9	2.07 ± 0.030 (6)	4	2.22 ± 0.047 (6)
13	2.21 ± 0.052 (6)	8	2.31 ± 0.040 (6)
20	2.22 ± 0.025 (20)	15	2.24 ± 0.051 (14)
21	2.15 ± 0.033 (6)		
120	2.31 ± 0.022 (28)		
240	2.43 ± 0.023 (8)		

Norepinephrine (10^{-5} M) evokes maximum contraction (10) but does not affect total ^{45}Ca content. Numbers in parentheses are number of aortas studied.

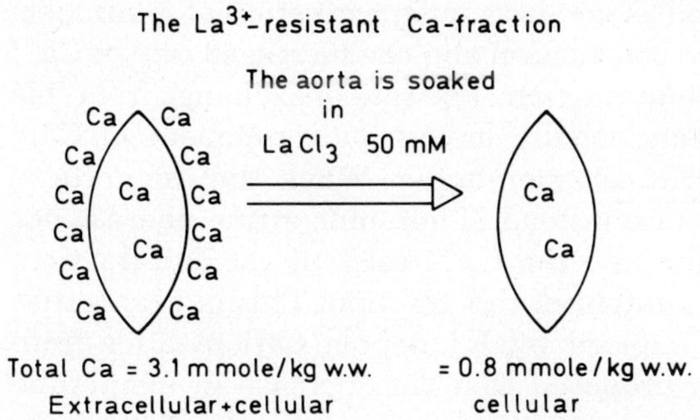

FIG. 3. Schematic illustration of method used to measure ^{45}Ca fluxes in vascular smooth muscle. [From Godfraind (10).]

Several attempts have been made to measure changes in the level of cytoplasmic Ca^{2+} during excitation of smooth muscle. Lanthanum replaces Ca^{2+} at superficial binding sites, and in the presence of La^{3+}, Ca^{2+} does not penetrate the cell (22). Because La^{3+} might block transmembrane fluxes of Ca^{2+}, it has been proposed that the calcium content of a muscle washed in an La^{3+} solution could provide an estimate of cellular Ca^{2+} (21). Evidence that La^{3+} blocks binding of Ca^{2+} to superficial sites as well as Ca^{2+} influx is based on the observation that the Ca^{2+} diffusion space in the presence of La^{3+} is similar to the inulin diffusion space. As shown for several smooth muscles, however, La^{3+} does not completely block Ca^{2+} efflux (4, 10).

There have been recent attempts to estimate Ca^{2+} fluxes across the smooth muscle cell membrane by measuring Ca^{2+} turnover in the La^{3+}-resistant calcium fraction (10). This fraction was defined as that portion not displaced when the tissue was washed in 50 mM $LaCl_3$. In this solution most of the Ca^{2+} displacement occurred during the first 5 min, and Ca^{2+} efflux from intracellular stores were apparently delayed (Fig. 3). This allowed the La^{3+}-resistant calcium fraction to be measured. In resting tissue, the rate of uptake of ^{45}Ca into this fraction was 6 μmol $^{45}Ca \cdot kg^{-1} \cdot min^{-1}$. In the presence of 10^{-5} M NE, it increased to 26 μmol $^{45}Ca \cdot kg^{-1} \cdot min^{-1}$ (Table 2). There was no net gain in tissue calcium, and the Ca^{2+} efflux increased in a similar manner (Table 2). The increase in Ca^{2+} uptake was dose dependent. In the presence of phentolamine, the dose-effect curves for the action of NE on ^{45}Ca uptake were displaced in a manner characteristic of competitive antagonism. The pA_2 for phentolamine was 7.8 for both Ca^{2+} uptake and contraction. In view of the similarity of the pA_2

TABLE 2. *Action of flunarizine on norepinephrine-dependent Ca^{2+} influx and efflux*

	Flunarizine, M				
	0	10^{-7}	3×10^{-7}	10^{-6}	10^{-5}
Ca^{2+} influx,	20 ± 0.9	17 ± 2	12 ± 2	9 ± 1	2 ± 1
$\mu mol \cdot kg^{-1} \cdot min^{-1}$	(80)	(6)	(6)	(12)	(12)
Ca^{2+} efflux,	29 ± 1	25 ± 4			23 ± 5
$\mu mol \cdot kg^{-1} \cdot min^{-1}$	(60)	(9)			(9)

Aortas were preincubated for 90 min in presence of flunarizine. Each value is the difference ± SE between changes in ^{45}Ca content of La^{3+}-resistant calcium fraction of control rat aortas and of aortas stimulated with 10^{-5} M norepinephrine. Numbers in parentheses are number of pairs.

values when phentolamine was used as an antagonist, activation of α-adrenergic receptors may underlie both the contraction and the increased rate of Ca^{2+} exchange in the La^{3+}-resistant calcium fraction. The rate of exchange from this fraction thus appears to be the functionally important component of Ca^{2+} exchange across the smooth muscle cell membrane. When the rat aorta is stimulated by K^+-depolarizing solution instead of norepinephrine, there is not only an increased rate of ^{45}Ca influx into the La^{3+}-resistant calcium fraction, but also a net increase of the ^{45}Ca content of this fraction. This indicates that membrane permeability changes induced by K^+ depolarization differ from those evoked by NE. It has been proposed that the increase in membrane permeability for Ca^{2+} might be due to the opening of membrane Ca^{2+} channels. Those associated with receptors are called receptor-operating channels (2), whereas others were postulated to be sensitive only to cell membrane potential. Analysis of the action of Ca^{2+} entry blockers may permit characterization of some properties of Ca^{2+} channels.

ACTION OF CALCIUM ENTRY BLOCKERS

The simplest experimental procedure to assess the ability of a pharmacological compound to inhibit Ca^{2+} entry is to preincubate a preparation in Ca^{2+}-free physiological solution and then depolarize it with a KCl-rich solution and gradually increase the Ca^{2+} concentration in the bathing solution. This will result in an increase in tension dependent on Ca^{2+} concentration. When this procedure is repeated in the presence of a putative Ca^{2+} entry blocker (Fig. 4), the contractile responses are depressed in a dose-dependent manner.

Figure 4 illustrates the Ca^{2+} dose-response curves obtained with rabbit mesenteric arteries. The preparations are very sensitive to increases in Ca^{2+} concentrations, and 20 μM of this ion can trigger a contraction. The response to Ca^{2+} is attenuated by cinnarizine concentrations as low as 10^{-9} M. If cinnarizine is added during a Ca^{2+}-evoked contraction, a dose-dependent relaxation is noted (Fig. 5). In the same preparations, cinnarizine reduces the contraction evoked by epinephrine in the presence of Ca^{2+}, but not the residual response obtained with the catecholamine in Ca^{2+}-free depolarizing solutions; this residual response is thought to be due to the release of intracellularly bound Ca^{2+}. Similar results have been obtained in rat aorta not only with cinnarizine, but also with flunarizine and methoxyverapamil (D 600). The results indicate that all these drugs reduce the contraction by decreasing the availability of Ca^{2+} by reducing its entry.

Figure 6 illustrates experiments comparing the effect of 10^{-6} M cinnarizine on arteries from different parts of the body. For inhibition of contraction evoked by depolarization, sensitivity to the action of the drug increased in the following order: thoracic aorta < abdominal aorta < renal artery < mesenteric arteries. For the mesenteric arteries, sensitivity was inversely proportional to the outside diameter, and the response of the smaller mesenteric arteries was completely inhibited. Analogous results were obtained with epinephrine: the

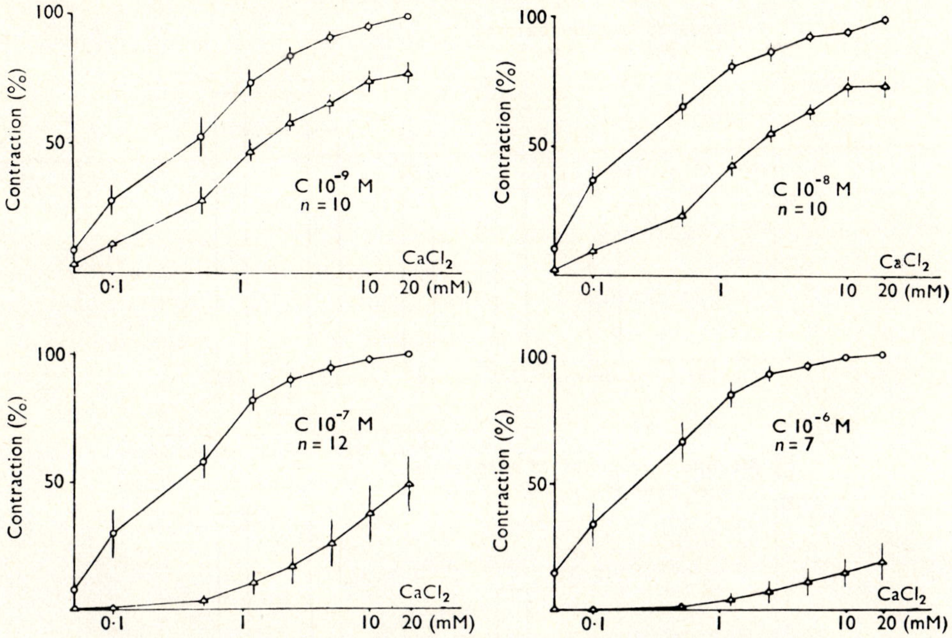

FIG. 4. Action of cinnarizine (C) on contraction evoked by Ca^{2+} in depolarized arteries. Dose-effect curves of $CaCl_2$ on depolarized rabbit mesenteric arteries in absence of cinnarizine (○) and 90 min after its addition to medium (Δ). [From Godfraind et al. (14).]

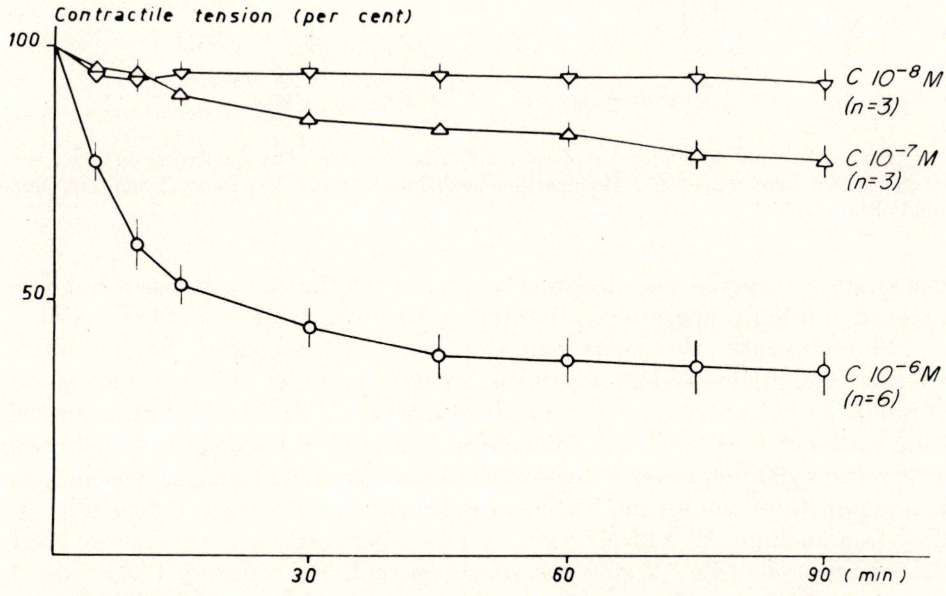

FIG. 5. Relaxation by cinnarizine (C) of rabbit mesenteric arteries contracted by K^+ depolarization. [From Godfraind et al. (14).]

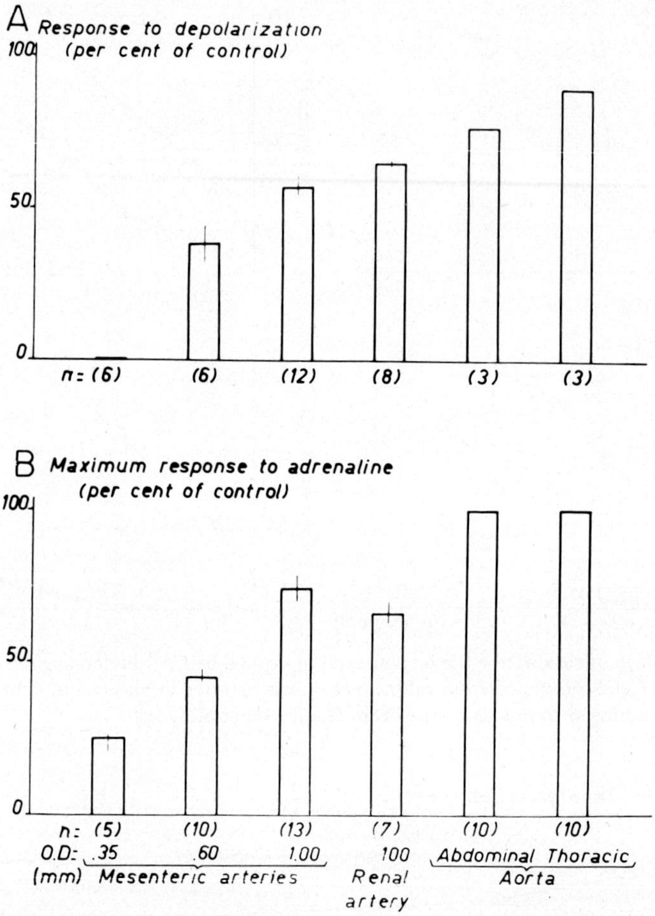

FIG. 6. Inhibition by cinnarizine of contractile response evoked by depolarization by K^+ and by epinephrine (adrenaline); 10^{-6} M cinnarizine was added 90 min before assay. [From Godfraind and Polster (15).]

maximum contractile response was unaltered for the thoracic aorta but was reduced in other preparations according to the decreasing size of the vessel.

More recently Broekaert and Godfraind (3) compared the inhibitory effects of cinnarizine and papaverine on the NE- and Ca^{2+}-induced contraction of rabbit thoracic aorta and mesenteric artery. They observed that the cinnarizine concentration inhibiting the contraction evoked by Ca^{2+} in depolarized arteries was 200-fold lower in mesenteric artery than in aorta. Papaverine was less potent than cinnarizine and showed no tissue selectivity. For arteries in Ca^{2+}-free medium, 10^{-5} M NE evoked a fast, nonsustained contraction; after the readmission of Ca^{2+}, a slow sustained contraction developed. The action of papaverine and cinnarizine on the two components of the contractile response was studied in mesenteric artery at concentrations of inhibitor that reduced the maximum response evoked by NE 50%. Cinnarizine selectively reduced the extracellular Ca^{2+}-dependent response, whereas papaverine inhibited the initial fast response more than the extracellular Ca^{2+}-dependent response. This action of papaverine was confirmed in aorta where cinnarizine was ineffective. Such results suggest that cinnarizine and papaverine antagonize NE action by different mechanisms. As noted above, cinnarizine seems to act by reducing membrane permeability to extracellular Ca^{2+}. Papaverine could inhibit the contraction by an inhibition of phosphodiesterase (see above) and to a lesser degree by an action similar to that of cinnarizine. For the latter, tissue selectivity was observed with cinnarizine but not with papaverine.

I have examined the action of cinnarizine and flunarizine on the NE- and depolarization-dependent Ca^{2+} fluxes using rat aorta and rat superior mesenteric artery (T. Godfraind, unpublished data). These experiments show that both flunarizine and cinnarizine reduce in a dose-dependent manner the NE-dependent Ca^{2+} entry, whereas the NE-dependent Ca^{2+} efflux is relatively insensitive to these drugs (Table 2). Since flunarizine depresses in a dose-dependent manner the sustained contraction evoked by NE, the IC_{50} (mean inhibitory concentration) values for contraction and for Ca^{2+} entry can be compared. As shown in Table 3, the two values are in good agreement. The contraction evoked by KCl depolarization of rat aorta is sustained, but in preparations pretreated with flunarizine, the initial peak tension is not sustained and a secondary low plateau is reached after 20 min (Fig. 7). I have measured ^{45}Ca influx after 2 min and after 35 min in K^+-rich solution with and without flunarizine. Data reported in Table 3 show that the IC_{50} value for the flunarizine was 10 times lower after 35 min than after 2 min, but that IC_{50} values for both contraction and Ca^{2+} entry were similar.

These data show that the interactions of Ca^{2+} entry blockers with receptor-operated channels and with membrane potential–sensitive channels differ. Blockade of receptor-operated channels reaches maximum level immediately after the receptors are activated; blockade of membrane potential–sensitive channels, on the other hand, is increased during the depolarization. Since a preincubation time of 90 min in physiological solution containing flunarizine is

TABLE 3. *Comparison of concentrations of flunarizine producing a half saturation of saturable binding sites, 50% diminution of ^{45}Ca entry in stimulated preparations (I_{50}), and 50% maximum reduction of contraction (IC_{50})*

	Flunarizine, M	
Half saturation	10^{-7}	
Norepinephrine-dependent processes		
IC_{50}	5×10^{-7}	
I_{50}	7×10^{-7}	
Depolarization-dependent processes	*After 2 min*	*After 35 min*
IC_{50}	2×10^{-7}	2×10^{-8}
I_{50}	2×10^{-7}	2×10^{-8}

sufficient to attain a steady effect, the observation for K^+-depolarizing solution may result from a time-dependent process not observed with NE (which can evoke contraction without a change in membrane potential). Such a process has been characterized in heart muscle for the interaction of local anesthetics with the fast Na^+ channels (17) and the interaction of verapamil with slow Ca^{2+} channels (8). It is interesting that membrane potential–sensitive Ca^{2+} channels in vascular smooth muscle behave like the potential-sensitive channels in cardiac tissue. The greater sensitivity to flunarizine blockade shown by the depolarization-dependent contraction in mesenteric artery is characterized not only by a lower IC_{50} value against the peak contraction, but also by a higher responsiveness of the time-dependent process as shown by a ratio of IC_{50} (2 min) to IC_{50} (35 min) of near 10 in aorta and 20 in superior mesenteric artery. This is consistent with the greater sensitivity to flunarizine blockade of the NE-dependent contraction and Ca^{2+} influx in mesenteric artery.

Another major difference between the action of Ca^{2+} entry blockers in rat aorta and in rat mesenteric artery is the amplitude of the residual response evoked by NE after a maximum inhibitory effect of the blocker. In aorta, 50% of the response is unaltered; in mesenteric artery, this value is only 10%. These values are similar to those obtained in Ca^{2+}-free solution. As shown by the data in Table 4, Ca^{2+} efflux evoked by NE in Ca^{2+}-free solution is insensitive to flunarizine. It is likely that this intracellularly released Ca^{2+} is responsible for the sustained contraction evoked by NE when the NE-dependent Ca^{2+} influx is blocked. This suggests that flunarizine does not alter the refilling of

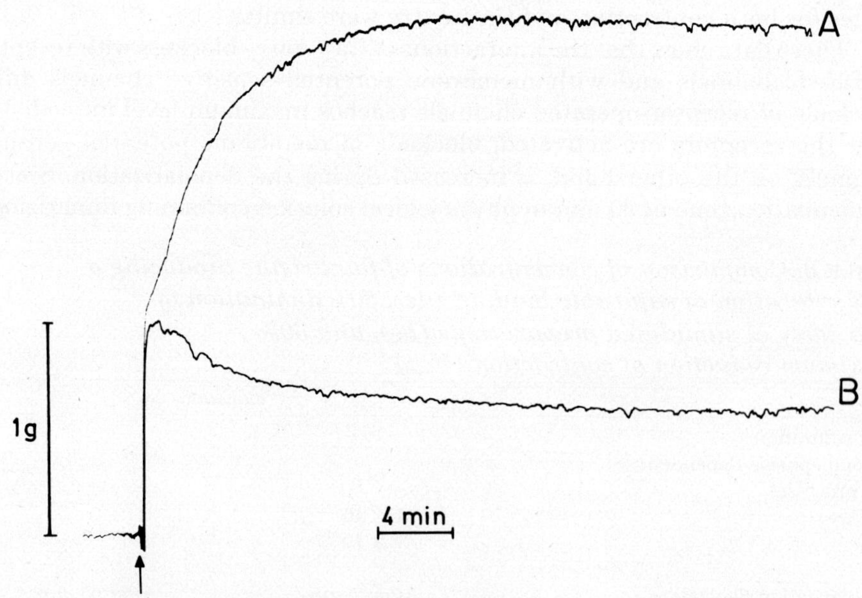

FIG. 7. Contraction in rat aorta evoked by KCl before (*A*) and after (*B*) 90 min incubation with 10^{-7} M flunarizine. (T. Godfraind, unpublished data.)

TABLE 4. *Norepinephrine-dependent Ca^{2+} efflux in rat aorta*

	Ca^{2+} Efflux, $\mu mol \cdot kg^{-1} \cdot min^{-1}$
Physiological solution	29 ± 1 (60)
Ca^{2+}-free solution	19 ± 2 (17)
Ca^{2+}-free solution + 10^{-6} M flunarizine	18 ± 1 (6)
Ca^{2+}-free solution + 5×10^{-6} M flunarizine	15 ± 4 (17)

Aortas were preincubated for 2 h in ^{45}Ca physiological solution (10) at 37°C with or without flunarizine. They were transferred into Ca^{2+}-free solution (plus 2×10^{-4} M EGTA) for 5 min before addition of 10^{-5} M norepinephrine. Numbers in parentheses are number of aortas studied.

the intracellular Ca^{2+} stores sensitive to NE. The Ca^{2+} stores could be replenished by a Ca^{2+}-Ca^{2+} exchange occurring through a leaky membrane channel.

CELLULAR LOCALIZATION OF THE SITE OF ACTION OF CALCIUM ENTRY BLOCKERS

We have attempted (T. Godfraind and N. Morel, unpublished data) to characterize the sites of flunarizine action in the membrane by studying the binding sites of [^{3}H]flunarizine in isolated rat arota. For concentrations ranging between 10^{-8} M and 5×10^{-7} M, uptake by the tissue reaches a steady-state level after 90 min of incubation. The uptake of [^{3}H]flunarizine by the aorta consists of two components—a nonsaturable uptake, which has a clearance of 150 liters/kg wet wt and a saturable uptake, which is characterized by a maximum binding capacity of 14 μmol flunarizine/kg wet wt. The concentration for half saturation is 0.1 μM, and the Hill coefficient of the binding is greater than 1. This saturable uptake is reversible. Since the saturable binding occurs within the range of flunarizine concentrations at which the effect on contraction and Ca^{2+} entry also occur, we examined how [^{3}H]flunarizine is taken up by different subfractions of rat aorta. A microsomal fraction was prepared by differential centrifugation. It was incubated at 37°C in a 20-mM tris-(hydroxymethyl)aminomethane (Tris)–maleate buffer (pH 7.4) with or without [^{3}H]flunarizine. The reaction was stopped by filtration on glass-fiber filters (0.45-μm pore diam), and the radioactivity retained by the filter was counted by liquid scintillation. The maximum flunarizine uptake was achieved after 4 min incubation and was partly reversible. The reversible binding showed saturation. The binding capacity of this saturable process was equal to about 400 pmol/mg protein. The concentration of flunarizine required for 50% saturation was 70 nM, and the Hill coefficient of the binding was greater than 1. In nuclear and mitochondrial fractions, no saturable binding could be detected. Since the saturable binding follows the distribution of 5′ nucleotidase, it probably occurs on the plasma membrane. It is also likely that the saturable binding occurs on sites responsible for the action of flunarizine, since the concentration for half saturation was close to the concentration producing half-maximum effects (Table 3).

CONCLUSIONS

In a Ca^{2+}-free medium, NE evokes only a phasic contraction of arterial smooth muscle; the presence of Ca^{2+} in the extracellular fluid is required to observe a sustained contraction. This indicates that α-adrenoceptor stimulation mobilizes two calcium pools: an intracellular one responsible for the phasic contraction and an extracellular one responsible for the tonic contraction. Mobilization of the pools is inhibited to the same extent by competitive α-blockers, whereas some noncompetitive antagonists appear to inhibit selectively one of these two processes of activation. The sustained contraction is specifically blocked by cinnarizine, flunarizine, and other Ca^{2+} entry blockers (Table 5), and the phasic contraction is inhibited by chlorpromazine and papaverine.

Studies have shown that there is a selective sensitivity to Ca^{2+} entry blockers among smooth muscles and that some of these compounds do not affect heart contractility. This observation is of great interest since it indicates that it is possible to select a Ca^{2+} entry blocker specific for a given tissue and therefore potentially active for a given disease. Indeed a drug with a specific action on arteriolar and venous smooth muscle could have great potential in treatment of hypertension and heart failure.

Calcium antagonists have been recognized on the basis of their ability to block the contraction evoked by Ca^{2+} in depolarized smooth muscle and to interfere with slow inward current in heart muscle. Direct estimates of Ca^{2+} fluxes indicate that drugs such as cinnarizine and flunarizine block NE-dependent Ca^{2+} influx when NE-dependent Ca^{2+} efflux is unaltered. This observation is the basis for identifying a drug as a Ca^{2+} entry blocker.

TABLE 5. *Uses of putative Ca^{2+} entry blockers*

Neuroleptic drug	Chlorpromazine
Antibiotic with curarizing actions	Neomycin
Antianginal drugs	Diltiazem Lidoflazine Nifedipine
Antiarrhythmic drug	Verapamil
Peripheral anti-ischemic drugs	Cinnarizine Flunarizine
Smooth muscle relaxants	Papaverine Diazoxide
Experimental tools	D 600 La^{3+}

See References 9 and 11.

Comparisons have been made between the activation by NE (which does not change membrane potential) and that by high-K^+ solutions (which depolarize the membrane). It has been observed that a Ca^{2+} entry blocker causes a concentration- and a time-dependent reduction of the contraction and of the Ca^{2+} influx evoked by depolarization, whereas with NE the reduction is only concentration dependent.

This indicates that the interaction of Ca^{2+} entry blockers with Ca^{2+} channels differs when the channels are opened by depolarization or by α-adrenoceptor activation.

REFERENCES

1. BOHR, D. F. Vascular smooth muscle updated. *Circ. Res.* 32: 665–672, 1973.
2. BOLTON, T. B. Mechanisms of action of transmitters and other substances on smooth muscle. *Physiol. Rev.* 59: 606–718, 1979.
3. BROEKAERT, A., AND T. GODFRAIND. A comparison of the inhibition of cinnarizine and papaverine of the noradrenaline evoked contraction of isolated rabbit vessels in presence and absence of extracellular calcium. *Eur. J. Pharmacol.* 53: 281–288, 1979.
4. BURTON, J., AND T. GODFRAIND. Sodium-calcium sites in smooth muscle and their accessibility to lanthanum. *J. Physiol. London* 241: 287–298, 1974.
5. CASTEELS, R. Electro- and pharmacomechanical coupling in vascular smooth muscle. *Chest* 78: 150–156, 1980.
6. EBASHI, S. Muscle contraction and pharmacology. *Trends Pharmacol.* 1: 29–31, 1979.
7. EDMAN, K. A. P., AND H. O. SCHILD. The need for calcium in the contractile responses induced by acetylcholine and potassium in the uterno. *J. Physiol. London* 161: 424–441, 1962.
8. EHARA, T., AND R. DAUFMANN. The voltage- and time-dependent effects of (−)-verapamil on the slow inward current in isolated cat ventricular myocardium. *J. Pharmacol. Exp. Ther.* 207: 49–55, 1978.
9. FLECKENSTEIN, A. Specific pharmacology of calcium in myocardium, cardiac pacemakers and vascular smooth muscle. *Annu. Rev. Pharmacol. Toxicol.* 17: 149–166, 1977.
10. GODFRAIND, T. Calcium exchange in vascular smooth muscle, action of noradrenaline and lanthanum. *J. Physiol. London* 260: 21–35, 1976.
11. GODFRAIND, T. Cellular metabolism of calcium and control of smooth muscle function. *Prog. Pharmacol.* 2: 57–68, 1978.
12. GODFRAIND, T., AND A. KABA. Blockade or reversal of contraction induced by calcium and adrenaline in depolarized arterial smooth muscle. *Br. J. Pharmacol.* 36: 549–560, 1969.
13. GODFRAIND, T., AND A. KABA. The role of calcium in the action of drugs on vascular smooth muscle. *Arch. Int. Pharmacodyn. Ther.* 196, Suppl.: 35–49, 1972.
14. GODFRAIND, T., A. KABA, AND P. POLSTER. Differences in sensitivity of arterial smooth muscles to inhibition of their contractile response to depolarization by potassium. *Arch. Int. Pharmacodyn. Ther.* 172: 235 -239, 1968.
15. GODFRAIND, T., AND P. POLSTER. Étude comparative de médicaments inhibant la réponse contractile de vaisseaux isolés d'origine humaine ou animale. *Therapie* 23: 1209–1220, 1968.
16. GODFRAIND-DE BECKER, A., AND T. GODFRAIND. Calcium transport system: a comparative study in different cells. *Int. Rev. Cytol.* 67: 141–170, 1980.
17. HONDEGHEM, L. M., AND B. G. KATZUNG. Time- and voltage-dependent interactions of antiarrhythmic drugs with cardiac sodium channels. *Biochem. Biophys. Acta* 472: 373–398, 1977.
18. STULL, J. T., D. K. BLUMENTHAL, AND R. COOKE. Regulation of contraction by myosin phosphorylation. A comparison between smooth and skeletal muscles. *Biochem. Pharmacol.* 29: 2537–2543, 1980.
19. VAN BREEMEN, C., P. AARONSON, R. LOUTZENHISER, AND K. MEISHERI. Ca^{2+} movements in smooth muscle. *Chest* 78: 157–165, 1980.
20. VAN BREEMEN, C., B. R. FARINAS, R. CASTEELS, R. GERBA, F. WUYTACK, AND R. DETH. Factors controlling cytoplasmic Ca^{2+} concentration. *Philos. Trans. R. Soc. London* 265: 57–71, 1973.
21. VERBEKE, N., N. MOREL, AND T. GODFRAIND. Microsomal and mitochondrial calcium pumps in smooth muscle. In: *Excitation-Contraction Coupling in Smooth Muscle*, edited by J. C. Rüegg. Amsterdam: Elsevier, 1977, p. 219–224.
22. WEISS, G. B., AND F. R. GOODMAN. Effect of lanthanum on contraction, calcium distribution and ^{45}Ca movements in intestinal smooth muscle. *J. Pharmacol. Exp. Ther.* 169: 46–55, 1969.

8

Differential Effects of Calcium Entry Blockers on Vascular Smooth Muscle

PAUL M. VANHOUTTE

Division of Pharmacology, Faculty of Medicine, Universitaire Instelling Antwerpen, University of Antwerp, Wilrijk, Belgium

ALTERATION IN THE DEGREE OF contraction of vascular smooth muscle cells affects cardiovascular function in different ways, depending on the anatomical location of the blood vessel. Thus under physiological conditions, contraction of the cells present in the wall of the aorta and in the large conduit arteries may primarily help adjust the impedance of the cardiovascular system, whereas constriction of the arterioles governs the local control of blood flow as well as total peripheral resistance and passive mobilization of blood from the postcapillary blood vessels. Constriction of the systemic veins (especially those of the splanchnic area) is the dynamic controller of the total vascular capacitance and thus of the filling of the heart. In the skin, cutaneous veins play a major role in thermoregulatory adjustments (see 22–25). In pathological conditions, exaggerated contraction of the smooth muscle cells of the large arteries (especially in the cerebral and coronary vascular beds) may cause an abnormal reduction in vascular diameter and thus endanger tissue perfusion (see 15, 39, 42). Contraction of the arteriolar smooth muscle plays an important role in the genesis and maintenance of the increased peripheral resistance observed in several forms of hypertension.

Activation of the contractile proteins in vascular smooth muscle cells depends on an increase in cytoplasmic Ca^{2+} concentration (Fig. 1; see 2, 3, 16, 23, 25, 26). The source of the activator ion, however, may differ among vascular smooth muscle cells depending on their anatomical origin and the substance that triggers contraction. Hence it is not surprising that Ca^{2+} entry blockers differ in their ability to interfere with the function of vascular smooth muscles from different anatomical origins or under different experimental conditions. The purposes of this chapter are to focus briefly on the heterogeneity of vascular effects of these agents and to suggest why this may be important in understanding the beneficial effects of Ca^{2+} entry blockers in the treatment of cardiovascular failure.

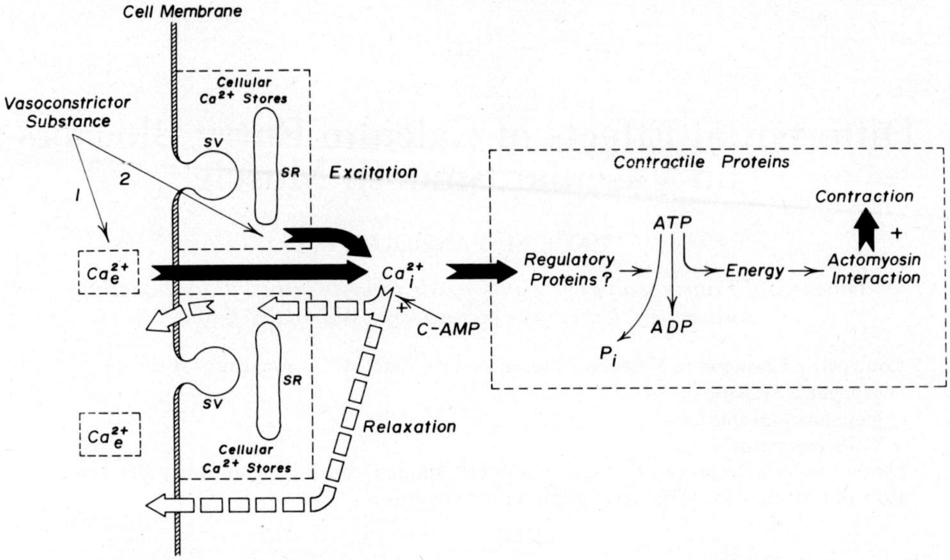

FIG. 1. Major steps in excitation and contraction processes in vascular smooth muscle cells. ADP, adenosine diphosphate; ATP, adenosine triphosphate; C-AMP, adenosine 3′,5′-monophosphate; e, extracellular; i, free intracellular; P_i, inorganic phosphate; SR, sarcoplasmic reticulum; SV, surface vesicles. [From Shepherd and Vanhoutte (25).]

CONTRACTILE RESPONSE OF VASCULAR SMOOTH MUSCLE CELLS TO CALCIUM ENTRY BLOCKERS

Myogenic Activity

Certain vascular smooth muscle cells are able to develop myogenic tone. This can be sustained (as at the arteriolar level) or phasic (as seen most typically in portal-mesenteric veins). The development of myogenic tone is generally attributed to spontaneous depolarization of the cell membrane of the vascular smooth muscle cell. This triggers the influx of extracellular Ca^{2+} (see 23, 28, 31). Hence it is not surprising that Ca^{2+} entry blockers markedly inhibit myogenic activity in isolated portal veins (see 4, 13, 19, 23) and decrease arteriolar resistance (see 4, 32, 33, 40, 42). Marked differences in potency exist among the available drugs, however (Fig. 2; 43, 44).

Depolarizing Solutions

When exposed to solutions of increasing K^+ concentrations, isolated blood vessels contract. In adrenergically innervated blood vessels, part of the contractile response to high K^+ levels is due to release of endogenous norepinephrine (see 23, 31, 34, 36). The direct effect of high K^+ on the vascular smooth muscle cell is sustained cell membrane depolarization, which results in an augmented influx of extracellular activator Ca^{2+} (see 2, 3, 16, 22, 23, 26, 31). In

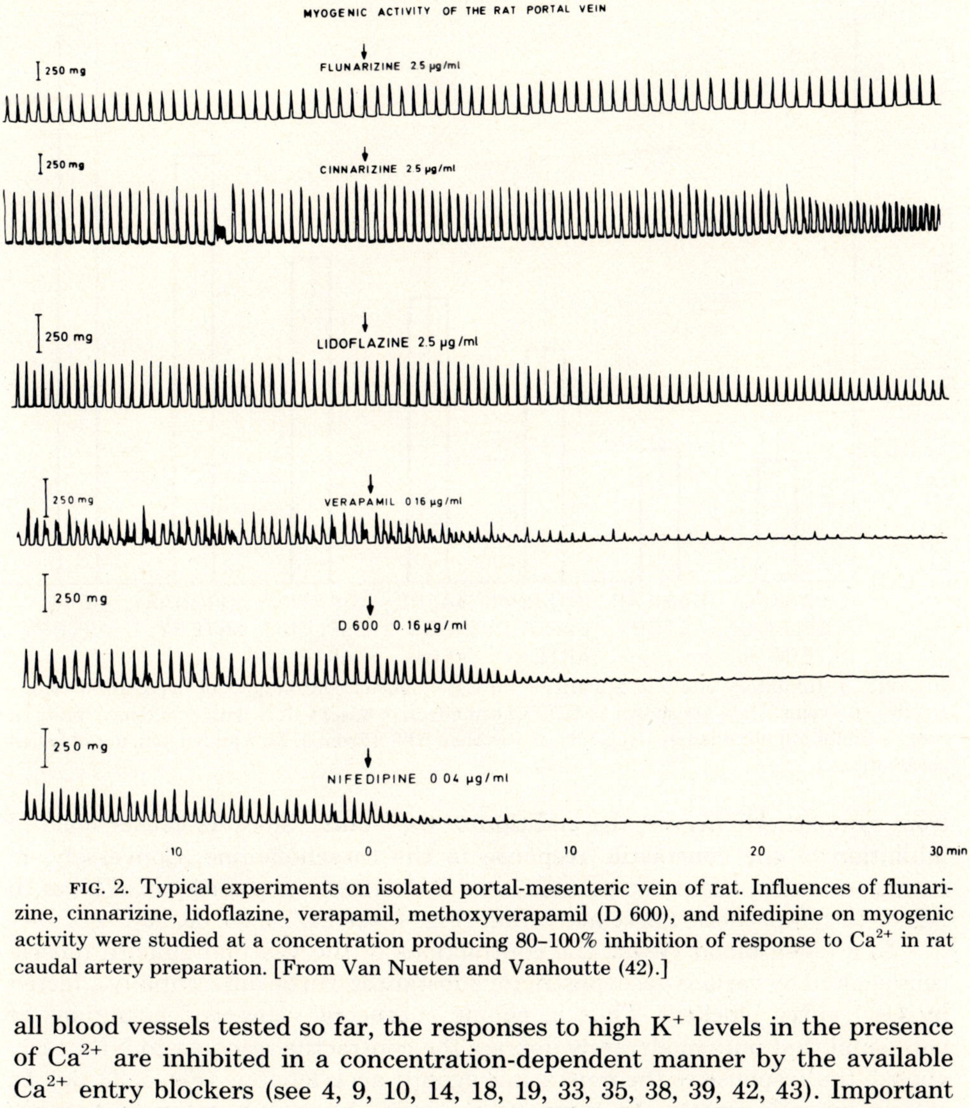

FIG. 2. Typical experiments on isolated portal-mesenteric vein of rat. Influences of flunarizine, cinnarizine, lidoflazine, verapamil, methoxyverapamil (D 600), and nifedipine on myogenic activity were studied at a concentration producing 80–100% inhibition of response to Ca^{2+} in rat caudal artery preparation. [From Van Nueten and Vanhoutte (42).]

all blood vessels tested so far, the responses to high K^+ levels in the presence of Ca^{2+} are inhibited in a concentration-dependent manner by the available Ca^{2+} entry blockers (see 4, 9, 10, 14, 18, 19, 33, 35, 38, 39, 42, 43). Important differences, however, exist in the potency of a given Ca^{2+} entry blocker to antagonize Ca^{2+}-induced contractions in various blood vessels of the same species (Fig. 3).

Vasoconstrictor Drugs

For a given vasoconstrictor substance (e.g., catecholamines), the relative contribution of the two sources of activator ions (influx of extracellular Ca^{2+} and cellular mobilization) varies among different blood vessels (see 2, 3, 16, 18, 26, 27, 29). Thus when arterial or splanchnic venous smooth muscle is contracted with norepinephrine (NE), whether exogenously added or released

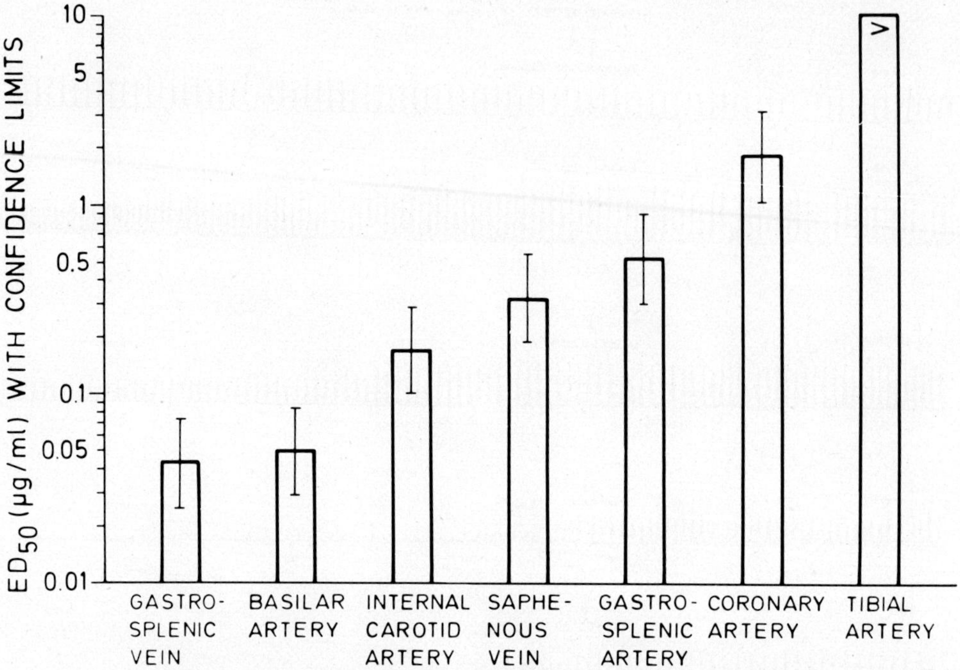

FIG. 3. Inhibitory effect of flunarizine on Ca^{2+}-induced contractions of depolarized canine arteries and veins. Data are shown as ED_{50} (mean effective dose) values with confidence limits ($n = 6$). > Inhibition obtained at 10 μg/ml was less than 50%. (From J. M. Van Nueten, unpublished observations.)

from sympathetic nerves, the addition of Ca^{2+} entry blockers causes marked inhibition of the contractile response to the catecholamine. Conversely, in cutaneous veins, these compounds have little effect on the responsiveness to NE or to sympathetic nerve activation (Fig. 4; see 21, 23, 30, 33, 35).

In a given blood vessel, the contractions of the vascular smooth muscle cells elicited by various vasoconstrictor substances can be differentially affected by Ca^{2+} entry blockers. Thus in canine cutaneous veins, concentrations of verapamil that only moderately depress the contractile response to NE nearly abolish the response to the muscarinic agonist acetylcholine (Fig. 5; 23, 30). In arterial smooth muscle, the contractions caused by most vasoconstrictors are markedly sensitive to the inhibitory effect of the Ca^{2+} entry blockers (see 1, 9–12, 14, 18, 19, 21, 33, 35, 37, 38, 42, 43). In particular, experiments on isolated coronary arteries demonstrate that Ca^{2+} entry blockers inhibit contractions caused by several possible mediators of coronary vasospasm (Fig. 6) and by the diagnostically used coronary vasoconstrictor ergonovine maleate (Fig. 7; 9, 33, 39, 41).

HETEROGENEOUS RESPONSE OF VASCULAR SMOOTH MUSCLE CELLS TO CALCIUM ENTRY BLOCKERS

There is no fully satisfactory explanation for the heterogeneous sensitivity of vascular smooth muscle cells to the inhibitory effect of the Ca^{2+} entry

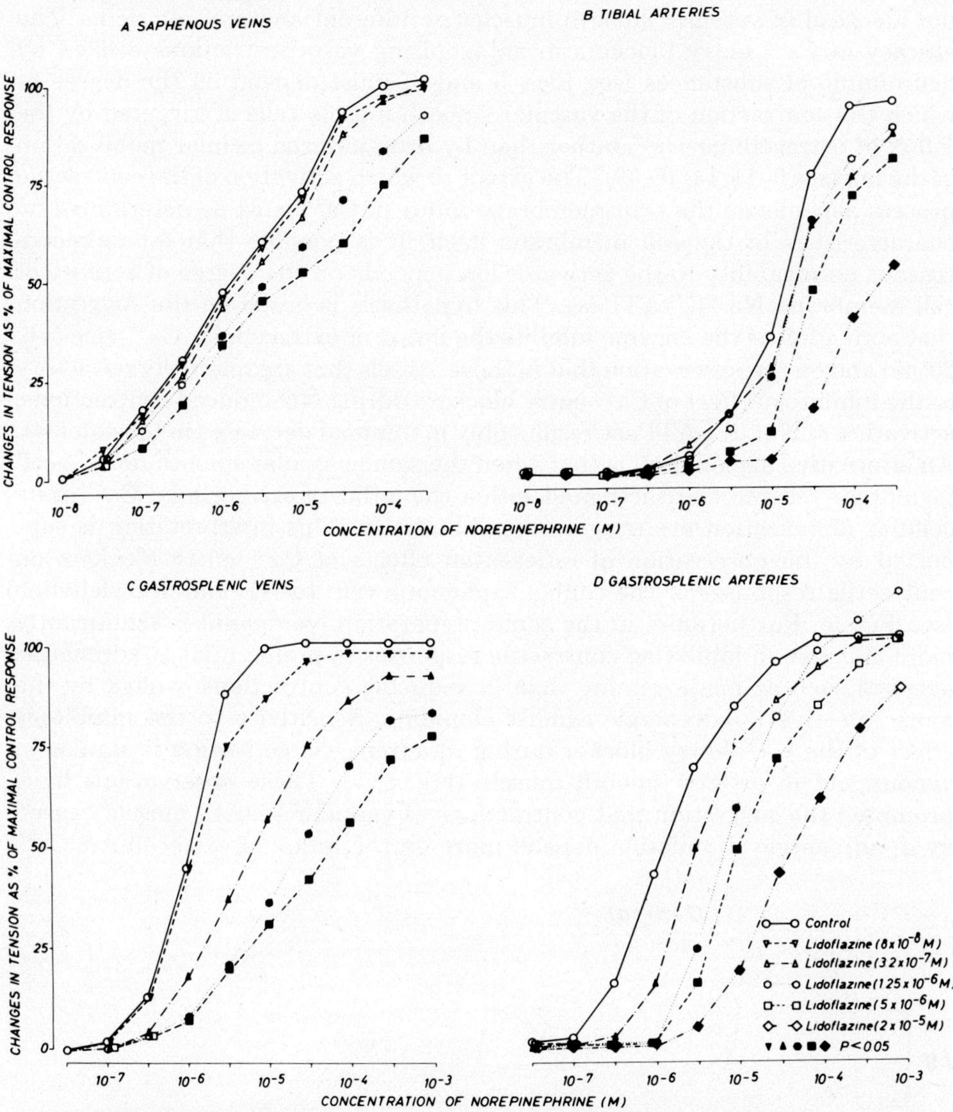

FIG. 4. Influence of increasing concentrations of lidoflazine on dose-response curves for norepinephrine in canine saphenous veins, tibial arteries, gastrosplenic veins, and gastrosplenic arteries. Values are means. Solid symbols indicate that values obtained in presence of lidoflazine are significantly different from control values. Note that Ca^{2+} entry blocker has little effect in cutaneous veins. [From Vanhoutte et al. (35).]

blockers (see 42). Differences in the efficacy of various compounds in inhibiting a specific effect in a given blood vessel (see Fig. 2) must reflect the kinetic properties of their interaction with, and their access to the Ca^{2+} channels of the vascular smooth muscle cell membrane. Tissue selectivity for the effects of a given Ca^{2+} entry blocker on contractions that are most likely triggered by similar events in all blood vessels tested (see Fig. 3) must mean that access to the Ca^{2+} channels or the configuration of the channels in the cell membrane is not identical in vascular smooth muscles of different anatomical origins. The efficacy of Ca^{2+} entry blockers in antagonizing vasoconstrictions evoked by neurohumoral substances (see Figs. 5 and 6) must depend on the degree to which the contraction of the vascular smooth muscle cells is triggered by the influx of extracellular Ca^{2+} rather than by drug-induced cellular mobilization of the ion (see 9–11, 14, 18, 19). The extent to which activation of the contractile process depends on the transmembrane influx of Ca^{2+} must be determined by characteristics of the cell membrane itself. It is possible that for unknown reasons permeability to the activator ion depends on the degree of activity of cell membrane Na^+-K^+-ATPase. This hypothesis is based on the suggestion that activation of the enzyme inhibits the influx of extracellular Ca^{2+} (see 5, 6, 20, 45) and on the observation that in those vessels that are relatively refractory to the inhibitory effect of Ca^{2+} entry blockers during NE-induced contractions, activation of Na^+-K^+-ATPase results only in minimal decreases in tension (34). An alternative explanation is that when the same vascular smooth muscle cell membrane receptors are activated, either the influx of extracellular Ca^{2+} or its cellular mobilization are triggered preferentially. This interpretation is supported by the observation of differential effects of Ca^{2+} entry blockers on contractile responses of the canine saphenous vein to NE and acetylcholine (see Fig. 5). Furthermore in the same preparation, verapamil is significantly more effective in inhibiting contractile responses to preferential α_1-adrenergic agonists such as methoxamine than in reducing contractions evoked by the more selective α_2-adrenergic agonist clonidine. Sensitivity to the inhibitory effect of the Ca^{2+} entry blocker during α_1-adrenergic activation is similar in venous and in arterial smooth muscle (Fig. 8; 7). These observations have prompted the suggestion that contractions of vascular smooth muscle caused by α_1-adrenergic stimulation depend more on the influx of extracellular Ca^{2+}

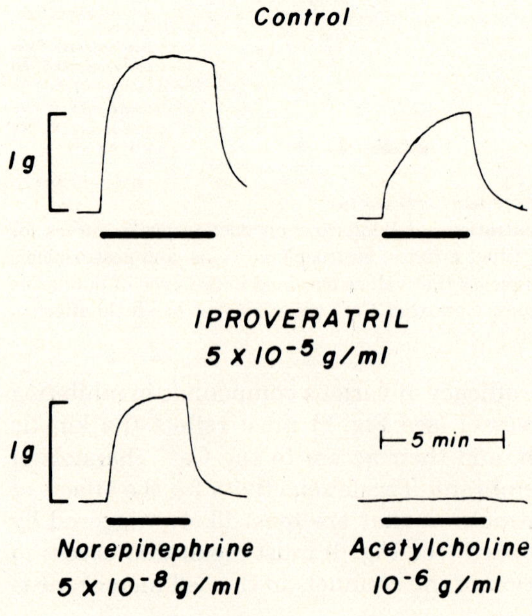

FIG. 5. Response of same canine saphenous vein strip to norepinephrine (*left*) and acetylcholine (*right*) before (*top*) and after (*bottom*) inhibition of Ca^{2+} entry with verapamil (iproveratril). [From Shepherd and Vanhoutte (23).]

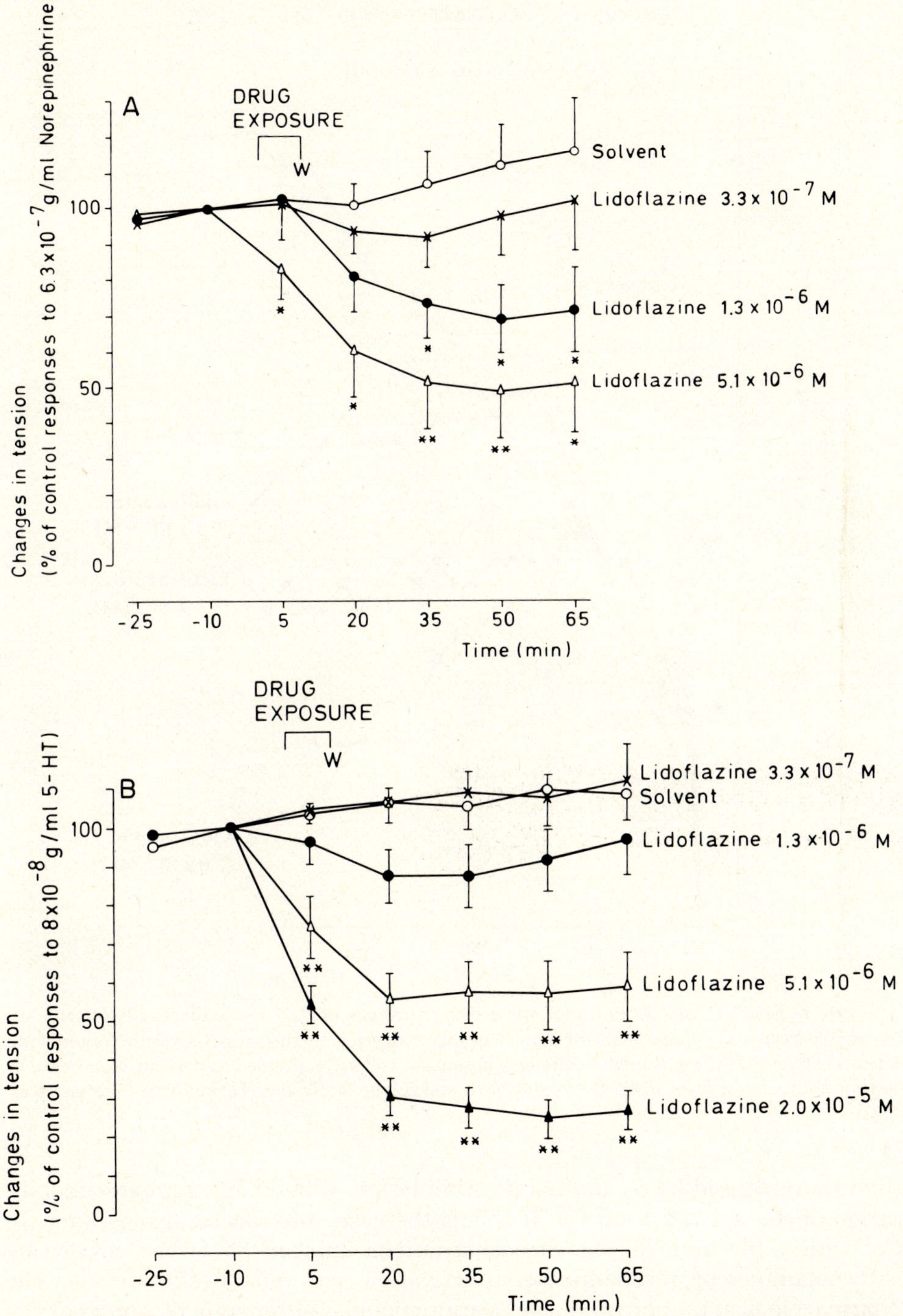

FIG. 6. Effect of solvent and Ca^{2+} entry blocker lidoflazine on contractions of isolated canine coronary arteries evoked by norepinephrine (*A*) and 5-hydroxytryptamine (5-HT, *B*). W, washout of lidoflazine; *, ** differences from control are statistically significant. [From Van Nueten et al. (39).]

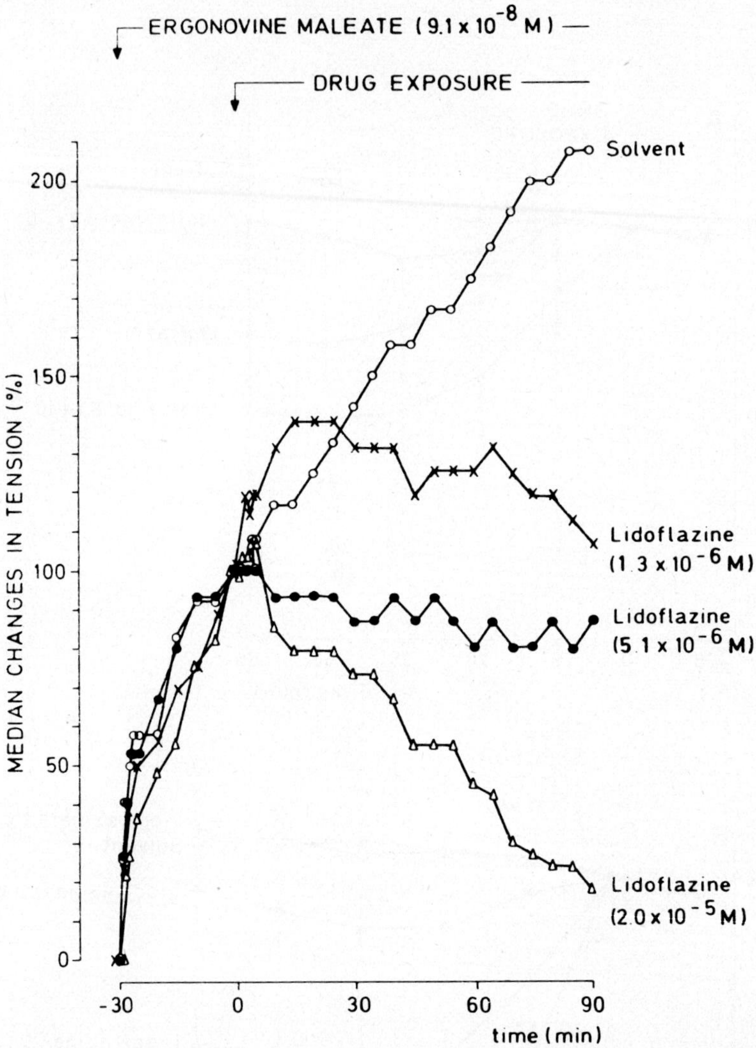

FIG. 7. Effect of solvent and increasing concentrations of Ca^{2+} entry blocker lidoflazine on contractile response of segments of same coronary arteries to the vasoconstrictor ergonovine maleate. Data are shown as median changes in tension and are expressed as percent of increase in tension above basal level upon treatment with solvent or lidoflazine. [From Van Nueten et al. (39).]

than those caused by α_2-adrenergic stimulation, which causes greater mobilization of the activator ion (7). If this is the case, differences in sensitivity to Ca^{2+} entry blockers during vasoconstrictions induced by either circulating catecholamines or adrenergic nerve activation may reflect differences in the pharmacological properties of the postjunctional α-adrenergic receptors of the vascular smooth muscle cells (see 36).

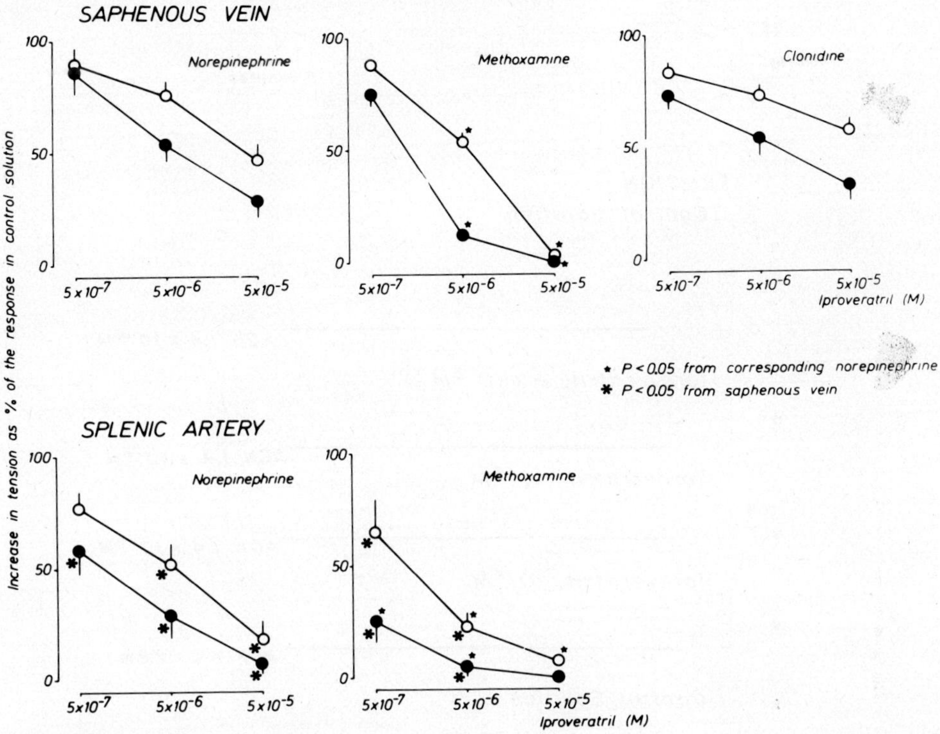

FIG. 8. Effect of Ca^{2+} entry blocker verapamil (iproveratril) on responses to α-adrenergic agonists in canine saphenous veins (*top*) and splenic arteries (*bottom*). Data are for 2 concentrations of norepinephrine (*left*), methoxamine (*middle*), and clonidine (*right*). Values are means ± SE and are expressed as percents of increase in tension obtained in control solution with individual doses of agonists tested. ★ Difference from corresponding response to norepinephrine is significant; * difference from saphenous veins is significant. [From De Mey and Vanhoutte (7) by permission of the American Heart Association, Inc.]

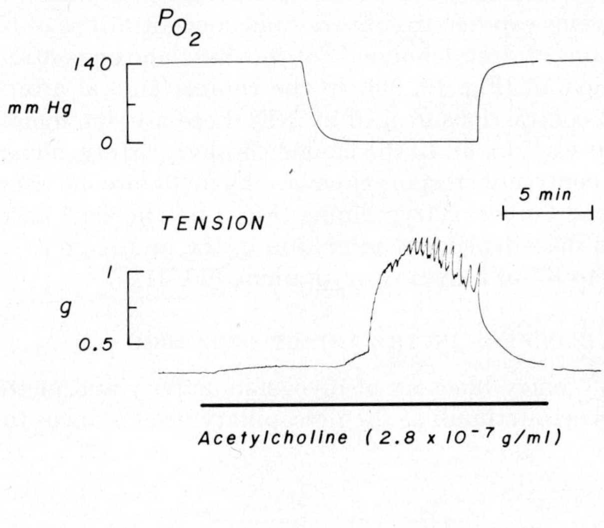

FIG. 9. Isometric tension recording from helical saphenous vein strip (*bottom*) and continuous recording of bath P_{O_2} (*top*). Acetylcholine causes only a weak contraction in normal oxygenation; contraction becomes strong during anoxia. At maximum increase in tension caused by anoxia, preparation becomes spontaneously active. Return to control oxygenation stops rhythmic activity and causes immediate relaxation. [From Shepherd and Vanhoutte (23).]

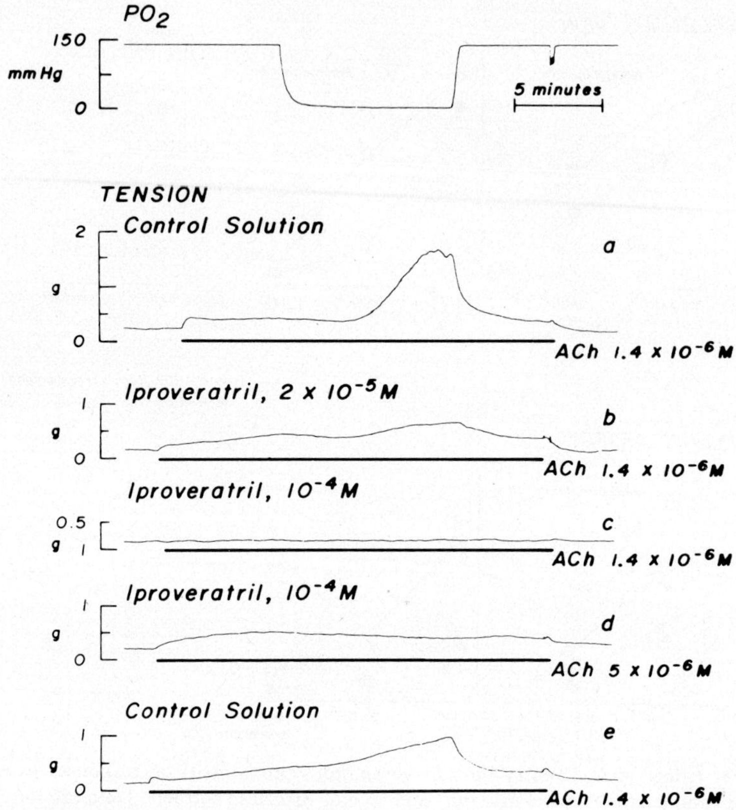

FIG. 10. Inhibition by verapamil (iproveratril) of potentiating effect of anoxia during acetylcholine (ACh)-induced contraction of canine saphenous vein. [From Vanhoutte (30).]

The ability of a given vasoconstrictor substance to augment the influx of Ca^{2+} also depends on the local conditions to which the vascular smooth muscle cells are exposed. Thus acute anoxia causes marked increases in tension in isolated canine saphenous veins exposed to subthreshold concentrations of K^+ or to threshold concentrations of acetylcholine (Fig. 9); these anoxic contractions are inhibited by verapamil (Fig. 10; 30). In the canine femoral artery, anoxia augments sustained contractions evoked by NE; these are due mainly to the influx of extracellular Ca^{2+} (5, 6). In the canine coronary artery, anoxia considerably augments the contractile responses caused by moderate increases in K^+ concentration, NE, and 5-hydroxytryptamine (Fig. 11). The Ca^{2+} entry blocker lidoflazine causes a dose-dependent inhibition of the anoxic contractions seen during exposure to K^+ or 5-hydroxytryptamine (30, 41).

ROLE OF CALCIUM ENTRY BLOCKERS IN THE INTACT ORGANISM

The depression by Ca^{2+} entry blockers of myogenic activity and of the responsiveness to vasoconstrictor stimuli at the precapillary level reduces the

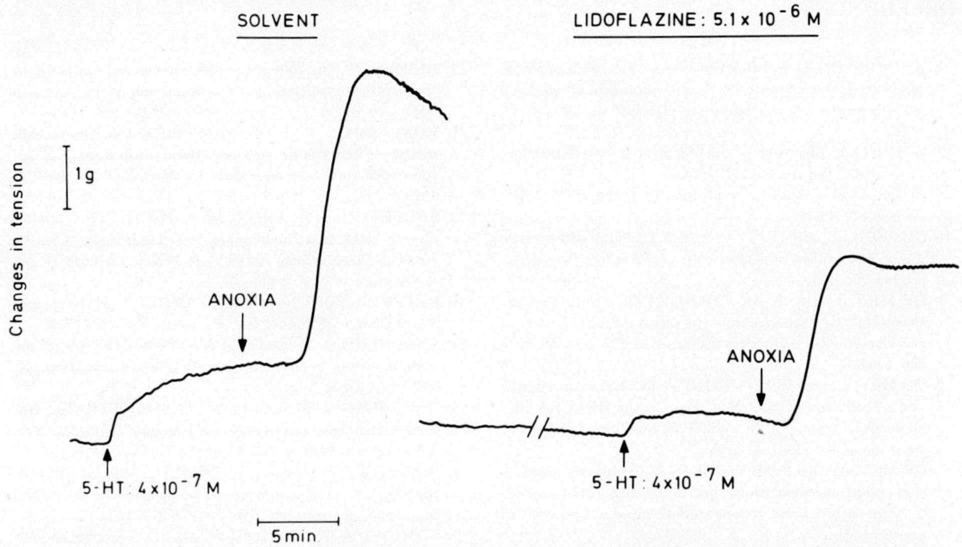

FIG. 11. In solvent (*left*), anoxia markedly augments contractile response of canine coronary artery to 5-hydroxytryptamine (5-HT). The Ca^{2+} entry blocker lidoflazine (*right*) reduces both normoxic and anoxic components of response to 5-HT. [From Van Nueten and Vanhoutte (41).]

systemic peripheral resistance (a major dynamic determinant of the circulatory impedance) and thus the afterload of the left ventricle. These actions help explain the improvement in cardiac function seen in patients with angina pectoris or heart failure as well as the antihypertensive properties of the compounds that are most effective at the arteriolar level (e.g., nifedipine and verapamil). The inhibitory effect of Ca^{2+} entry blockers on contractile responses of vascular smooth muscle cells of large arteries results in the reduction or abolition of vasospastic episodes; these are particularly important for patients with angina pectoris. In this condition, spastic constriction may be the primary cause of myocardial ischemia (Prinzmetal's angina) or may exacerbate the restriction in blood flow due to partial atherosclerosis of the coronary lumen. This may also occur in other vascular beds. By inhibiting the constriction of the capacitance vessels to sympathetic nerve activation, which is the main controller of venous function, the Ca^{2+} entry blockers may reduce the preload of the heart. The experiments performed on isolated veins suggest that they do this primarily by affecting the splanchnic veins, which dynamically regulate return of blood to the right heart, whereas the responses of cutaneous veins are only slightly affected. This differential effect of Ca^{2+} entry blockers ensures that the thermoregulatory role of the cutaneous veins and the orthostatic pooling of blood in the lower extremities are not essentially affected by these compounds (see 4, 8, 9, 15, 17, 23–25, 32, 33, 40, 42, 43).

Present address of P. M. Vanhoutte: Department of Physiology and Biophysics, Mayo Clinic, Rochester, MN 55901.

REFERENCES

1. ANDERSSON, K. E. Effects of calcium and calcium antagonists on the excitation-contraction coupling in striated and smooth muscle. *Acta Pharmacol. Toxicol. Suppl.* 1: 5–14, 1978.
2. BOHR, D. F. Electrolytes and smooth muscle contraction. *Pharmacol. Rev.* 16: 85–111, 1964.
3. BOHR, D. F. Vascular smooth muscle updated. *Circ. Res.* 32: 665–672, 1973.
4. CHURCH, J., AND T. T. ZSOTER. Calcium antagonistic drugs. Mechanism of action. *Can. J. Physiol. Pharmacol.* 58: 254–264, 1980.
5. DE MEY, J., AND P. M. VANHOUTTE. Interaction between Na^+, K^+ exchanges and the direct inhibitory effect of acetylcholine on canine femoral arteries. *Cir. Res.* 46: 826–836, 1980.
6. DE MEY, J., AND P. M. VANHOUTTE. Effect of removal and substitution of potassium ions on the adrenergic and cholinergic reactivity in canine femoral artery. *Eur. J. Pharmacol.* 67: 159–164, 1980.
7. DE MEY, J., AND P. M. VANHOUTTE. Uneven distribution of postjunctional $alpha_1$ and $alpha_2$-like adrenoceptors in canine arterial and venous smooth muscle. *Circ. Res.* in press.
8. EKELUND, L. G. Ca-blockers and peripheral circulation—physiological viewpoints. *Acta Pharmacol. Toxicol. Suppl.* 1: 5–14, 1978.
9. FLECKENSTEIN, A. Specific pharmacology of calcium in myocardium, cardiac pacemakers, and vascular smooth muscle. *Annu. Rev. Pharmacol. Toxicol.* 17: 149–166, 1977.
10. GODFRAIND, T., AND A. KABA. Blockade or reversal of the contraction induced by calcium and adrenaline in depolarized arterial smooth muscle. *Br. J. Pharmacol.* 36: 549–560, 1969.
11. GODFRAIND, T., AND A. KABA. The role of calcium in the action of drugs on vascular smooth muscle. *Arch. Int. Pharmacodyn. Ther.* 196: 35–49, 1972.
12. GODFRAIND, T., AND N. MOREL. Inhibitors of calcium influx. In: *Mechanisms of Vasodilatation*, edited by P. M. Vanhoutte and I. Leusen. Basel: Karger, 1978, p. 144–151.
13. GOLENHOFEN, K., AND E. LAMMEL. Selective suppression of some components of spontaneous activity in various types of smooth muscle by iproveratril (verapamil). *Pfluegers Arch.* 331: 233–243, 1972.
14. HAYASHI, S., AND N. TODA. Inhibition by Cd^+, verapamil and papaverine of Ca^{2+}-induced contractions in isolated cerebral and peripheral arteries of the dog. *Br. J. Pharmacol.* 60: 35–43, 1977.
15. HILLIS, L. D., AND E. BRAUNWALD. Coronary-artery spasm. *N. Engl. J. Med.* 299: 695–702, 1978.
16. JOHANSSON, B. Processes involved in vascular smooth muscle contraction and relaxation. *Circ. Res. Suppl.* 1: 14–20, 1978.
17. MASERI, A., A. L'ABBATE, S. CHIERCHIA, O. PARODI, S. SEVERI, A. BIAGINI, A. DISTANTE, M. MARZILLI, AND A. M. BALLESTRA. Significance of spasm in the pathogenesis of ischemic heart disease. *Am. J. Cardiol.* 44: 788–792, 1979.
18. MIKKELSEN, E., K. E. ANDERSSON, AND O. LEDERBALLE PEDERSEN. The effect of nifedipine on isolated human peripheral vessels. *Acta Pharmacol. Toxicol.* 43: 291–298, 1978.
19. RAHWAN, R. G., M. F. PIASCIK, AND D. T. WITIAK. The role of calcium antagonism in the therapeutic action of drugs. *Can. J. Physiol. Pharmacol.* 57: 443–460, 1979.
20. REINER, O. The role of the electrogenic sodium pump in the potassium relaxation of the rabbit ear artery. *Naunyn-Schmiedebergs Arch. Pharmacol.* 303: 213–220, 1978.
21. ROBINSON, B. F., AND J. G. COLLIER. Vascular smooth muscle. Correlations between basic properties and responses of human blood vessels. *Br. Med. Bull.* 35: 305–312, 1979.
22. SHEPHERD, J. T., AND P. M. VANHOUTTE. Skeletal-muscle blood flow: neurogenic determinants. In: *The Peripheral Circulations*, edited by R. Zelis. New York: Grune & Stratton, 1975, p. 3–55.
23. SHEPHERD, J. T., AND P. M. VANHOUTTE. *Veins and Their Control.* Philadelphia: Saunders, 1975, p. 1–269.
24. SHEPHERD, J. T., AND P. M. VANHOUTTE. Role of the venous system in circulatory control. *Mayo Clin. Proc.* 53: 247–255, 1978.
25. SHEPHERD, J. T., AND P. M. VANHOUTTE. *The Human Cardiovascular System. Facts and Concepts.* New York: Raven, 1979, p. 1–351.
26. SOMLYO, A. P., AND A. V. SOMLYO. Vascular smooth muscle: I. Normal structure, pathology, biochemistry and biophysics. *Pharmacol. Rev.* 20: 197–272, 1968.
27. SOMLYO, A. P., AND A. V. SOMLYO. Vascular smooth muscle: II. Pharmacology of normal and hypertensive vessels. *Pharmacol. Rev.* 22: 249–353, 1970.
28. UCHIDA, E., AND D. F. BOHR. Myogenic tone in isolated perfused vessels: occurrence among vascular beds and along vascular trees. *Circ. Res.* 25: 549–555, 1969.
29. VAN BREEMEN, C. Calcium requirement for activation of intact aortic smooth muscle. *J. Physiol.* London 272: 317–329, 1977.
30. VANHOUTTE, P. M. Effects of anoxia and glucose depletion on isolated veins of the dog. *Am. J. Physiol.* 230: 1261–1268, 1976.
31. VANHOUTTE, P. M. Heterogeneity of vascular smooth muscle. In: *Microcirculation*, edited by G. Kaley and B. M. Altura. Baltimore: University Park, 1978, vol. II, p. 181–309.
32. VANHOUTTE, P. M. Calcium entry blockers and cardiovascular failure. *Federation Proc.* in press.
33. VANHOUTTE, P. M., AND J. M. VAN NUETEN. The pharmacology of lidoflazine. *Proc. R. Soc. Med.* 29: 61–77, 1980. (Int. Congr. Symp. Ser.)
34. VANHOUTTE, P. M., T. J. VERBEUREN, AND R. R. LORENZ. Effects of potassium ions on nerve endings and effector cells in adrenergically innervated vascular smooth muscle. In: *Smooth Muscle Pharmacology and Physiology*, edited by M. Worcel and G. Vassort. Paris: INSERM, 1976, p. 425–441.
35. VANHOUTTE, P. M., T. J. VERBEUREN, AND J. M. VAN NUETEN. Effect of lidoflazine on systemic arteries and veins. *Angiology* 31: 581–593, 1980.
36. VANHOUTTE, P. M., T. J. VERBEUREN, AND R. C. WEBB. Local modulation of the adrenergic neuroeffector interaction in the blood vessel wall. *Physiol. Rev.* 61: 151–247, 1981.
37. VAN NUETEN, J. M. Vasodilatation or inhibition of vasoconstriction? In: *Mechanisms of Vasodilatation*, edited by P. M. Vanhoutte and I. Leusen. Basel: Karger, 1978, p. 137–143.
38. VAN NUETEN, J. M., J. VAN BEEK, AND P. A. J. JANSSEN. Effect of flunarizine on calcium-induced responses of peripheral vascular smooth muscle. *Arch. Int. Pharmacodyn. Ther.* 232: 42–52, 1978.

39. VAN NUETEN, J. M., J. VAN BEEK, AND P. M. VANHOUTTE. Inhibitory effect of lidoflazine on contractions of isolated canine coronary arteries caused by norepinephrine, 5-hydroxytryptamine, high potassium, anoxia and ergonovine maleate. *J. Pharmacol. Exp. Ther.* 213: 179–187, 1980.
40. VAN NUETEN, J. M., AND P. M. VANHOUTTE. Improvement of tissue perfusion with inhibitors of calcium ion influx. *Biochem. Pharmacol.* 29: 479–481, 1980.
41. VAN NUETEN, J. M., AND P. M. VANHOUTTE. Effect of the Ca^{2+} antagonist lidoflazine on normoxic and anoxic contractions of canine coronary arterial smooth muscle. *Eur. J. Pharmacol.* 64: 173–176, 1980.
42. VAN NUETEN, J. M., AND P. M. VANHOUTTE. Calcium entry blockers and vasospasm. In: *Vasodilatation*, edited by P. M. Vanhoutte and I. Leusen. New York: Raven, in press.
43. VAN NUETEN, J. M., AND P. M. VANHOUTTE. Calcium entry blockers and vascular smooth muscle heterogeneity. *Federation Proc.* in press.
44. VAN NUETEN, J. M., AND D. WELLENS. Tissue specificity of calcium antagonistic properties of lidoflazine. *Arch. Int. Pharmacodyn. Ther.* 242: 329–331, 1979.
45. WEBB, R. C., P. M. VANHOUTTE, AND D. F. BOHR. Calcium efflux in rat tail artery during potassium-induced relaxation. *Proc. Soc. Exp. Biol. Med.* 164: 252–256, 1980.

9

Receptor-Activated Calcium Mechanisms and Their Antagonism in Cerebrovascular Muscle

JOHN A. BEVAN, THOMAS A. McCALDEN, AND ROBERT M. RAPOPORT

Department of Pharmacology, School of Medicine, and Brain Research Institute, University of California, Los Angeles, California

SOME OF THE CHARACTERISTICS of the Ca^{2+} dependence of agonist-induced contractions of a systemic mid-sized artery are presented in this chapter. Against this background, some unique features of these mechanisms in cerebral blood vessels are discussed as a prelude to a description of the actions of Ca^{2+} antagonists. The effects of these drugs on vascular tone and their variability can only be understood when related to these physiological characteristics and their differences in different vessels.

It should be emphasized that our present perspective and understanding of the action of Ca^{2+} antagonists on blood vessels is probably limited by the lack of a systematic study of the Ca^{2+} dependence of tone development in a variety of blood vessels. Only when the full spectrum of vascular heterogeneity is appreciated will a balanced viewpoint be possible.

BIPHASIC RESPONSES OF BLOOD VESSELS TO AGONISTS

A biphasic pattern of response to most antagonists, including norepinephrine (NE), is commonly seen in studies of mid-sized blood vessels in vitro, whether arteries or veins, or rings or strips, immersed in a tissue bath or perfused. Although the response of an elastic vessel can be resolved into two components (6), the complex nature of the contraction can more easily be resolved in smaller vessels. In the rabbit ear artery, for example, there is a rapid initial contraction that reaches its maximum after about 12 s. This is followed by a slower contraction that reaches a plateau after 1–2 min. The two phases of response appear to depend on two different mechanisms of activation (3, 5, 19). The first phase probably stems from the action of the transmitter on superficial cells and from these excitation spreads by myogenic propagation through the thickness of the media. It is *1*) triggered within a few seconds of exposure to NE, *2*) associated with mass electrical activity, and *3*) abolished by prior depolarization. In 1974 Bevan and Ljung (4) demonstrated that the longitudinal myogenic propagation of excitation that takes place in the ear artery after exposure to NE occurs only during this first phase of contraction. The second phase is slower and has a time course similar to the rate of entry of the agonist into the media. It is probably caused by the local direct action of the agonist on individual smooth muscle cells.

Since the first phase of contraction is resistant to removal of Ca^{2+} from the environment of the vessel, activator Ca^{2+} is thought to be held at sites not easily depleted by this procedure (i.e., it is sequestered). In contrast, the second plateau phase is highly susceptible to removal of extracellular Ca^{2+} and thus is presumed to depend on Ca^{2+} entry into the cell from the extracellular space.

This concept was recently enlarged when patterns of contraction were studied with adjacent segments of the rabbit ear artery in vitro, mounted either with their adventitial surface facing outward (the normal in vitro configuration) or with the vessel everted so that the intimal surface was exposed (11). Drugs added to the tissue bath entered the blood vessel wall predominantly through the outer and inner surfaces, respectively. Everted vessels were more sensitive to NE than noneverted vessels. Although some of the differences in the smooth muscle contraction were abolished by blockade of neuronal and extraneuronal catecholamine disposition, the relative size of the first transient response compared with the maintained component of contraction remained greater for intimal than adventitial entry. These observations have been interpreted as showing that the arterial wall is functionally asymmetric, at least for the ease with which the transient contractile component can be elicited and the increased sensitivity of its smooth muscle cells.

A difference between the inner and outer vascular smooth muscle cell layers was first established by Keatinge and colleagues (see, e.g., 8), who claimed that the inner smooth muscle cell layers in the sheep carotid artery are more sensitive to certain agonists than the outer ones. A difference in the dynamic response depending on the surface of entry of the drugs has also been demonstrated in the rabbit thoracic aorta (14, 15), where the latency of contraction was shorter and the initial velocity faster for a variety of drugs entering at the intimal surface. A similar specialization also occurs in veins (A. Gonzales and J. A. Bevan, unpublished data). This asymmetry of vascular response is consistent with electrophysiological observations of Mekata and Keatinge (13), who showed that the outer and inner cell layers of the media of the sheep carotid artery had different electrical characteristics.

It should be emphasized that the functional asymmetry of the vessel wall (related to the surface of entry of the agonist) was observed in the rabbit ear artery with either NE or serotonin (5-HT), but not with KCl. It is generally agreed that the response of the vessel to KCl depends exclusively on external sources of Ca^{2+}, whereas other agonists rely on both external and internal sources. The experimental results suggest that the rapid phasic contraction is *1*) preferentially obtained from the intimal surface, *2*) associated with mobilization of sequestered Ca^{2+} stores, and *3*) largely responsible for the asymmetry.

One of the functional implications of these findings is that blood-borne vasoactive substances may initially generate responses from the blood vessel wall (from the inner layers of smooth muscle) that are fast because they are myogenically propagated. This would not be the case if the drug effect were solely the consequence of its movement through the wall resulting in a maintained plateau effect due to an exclusively local action.

BIPHASIC CONTRACTION PATTERN OF CEREBRAL BLOOD VESSELS

In the rabbit basilar artery and some arterial segments derived from the vertebral and internal carotid arteries, only transient contractile responses to particular concentrations of NE have been noted (2). The contraction is fast and reaches a peak after only several seconds. It then reverses rapidly and relaxes with a half time of approximately 15 s. This transient contraction is small compared to the maximum contractile capacity of the tissue. The concentration of NE producing the maximum transient response was approximately 10^{-7} M. In general the proportion of segments showing the transient response increased as the anatomical site of origin of the vascular segment along the course of the vertebral and internal carotid arteries became more cranial. No transient responses were observed in the brachiocephalic and common carotid artery segments.

The short-lived response mediated by an α-like receptor was observed after chronic reserpinization but was not seen after exposure to α-adrenergic receptor blocking agents. Since it was seen with concentrations of NE similar to those that caused the first-phase contraction in systemic vessels, it is tempting to conclude that these two responses are equivalent features in both vessels.

In contrast to the first phase, the concentrations of NE required to produce a steady-state contraction in the cerebral vessel are more than 2 times higher than in the ear artery. There is no reason to question that the α-adrenoceptors associated with the two modes of contraction in a particular vessel are not the same. Thus differences in the α-adrenoceptors between the two vessels do not account for this effect. It must be concluded that the coupling between the α-adrenoceptor and the two Ca^{2+} sources is independent and that the relative efficacy or effectiveness of these coupling systems differs in different vessels.

Two recent in vivo studies have demonstrated that cerebral blood flow escapes the immediate effects of sympathetic stimulation (9, 10). It is interesting to speculate that the transient responses to NE observed are the in vitro counterpart of vascular escape.

INDEPENDENCE OF THE TWO COMPONENTS OF CONTRACTION OF RABBIT EAR ARTERY

The possibility that the relative effectiveness of coupling of the α-adrenoceptor to sequestered and extracellular Ca^{2+} sources differs in different vessels is supported by recent studies suggesting that these coupling systems can be manipulated independently. Direct electrical stimulation (DES) of the rabbit ear artery in vitro reduces the size of subsequent contractions to equieffective concentrations of NE, histamine, and 5-HT by similar amounts; the contractile response to KCl is unimpaired by DES (16). When ear arteries that had received DES were exposed to Ca^{2+}-free Krebs solution for 5 min and then challenged by an agonist such as histamine, the resultant transient responses (presumed to be equivalent to the first-phase response) did not differ from rested controls (17). In contrast, the subsequent sustained contraction resulting from the addition of Ca^{2+} in the continued presence of histamine (assumed to be equivalent to the second or sustained phase of contraction) was less in those tissues that had received DES. This reduction was exactly equivalent to the loss of responsiveness or subsensitivity determined in normal Krebs solution after DES compared to rested controls. In other words, DES appeared to alter the second sustained phase of agonist-induced contraction but not the first, and the response to KCl was unaltered.

These results are further supported by experiments performed on everted ear artery segments. Everted vessels that were directly stimulated responded to subsequent agonist exposure with initial transient responses equivalent in magnitude to those of rested controls; however, the secondary plateau contractions were depressed. A similar type of experiment (18) showed that prior exposure of the rabbit ear artery to a submaximum concentration of agonist increases responsiveness and maximum contractility of the vessels. This effect was considered nonspecific (independent of the particular agonist used), although this conclusion was complicated by an additional effect of receptor desensitization in specific instances. The increased responsiveness was shown to depend on the contraction actually occurring and was independent of the agonist-receptor interaction per se. Again using everted ear artery segments in which the two phases of response could be clearly delineated and accurately measured, the effect of prior agonist exposure on subsequent responses was shown to be confined to the second plateau phase of response.

These results are consistent with the idea that DES or prior contraction can induce sub- or supersensitivity by changing the efficiency or effectiveness of the coupling between receptor activation and the mechanism altering permeability of the cell membrane to Ca^{2+}. This can be achieved without influencing the interaction between adrenoceptor occupation and sequestered Ca^{2+} sources. Since the response to KCl is not effected by DES, however, these findings suggest that it is the coupling of the receptor to the Ca^{2+} channel in the smooth muscle cell that is altered. There are, however, other less likely possibilities.

If the tonic phase of contraction resulting from receptor occupation can be independently increased or decreased relative to the phasic component, this would suggest an independent linkage between the receptor and the two Ca^{2+} pools. The physiological variation in the relative coupling effectiveness between the receptor and these two pools (for example, as in the ear and basilar arteries) emphasizes this independence and supports this conclusion.

COMPARISON OF CALCIUM SOURCES AND EFFECT OF CALCIUM ANTAGONISTS ON AGONIST-INDUCED RESPONSES OF CEREBRAL AND SYSTEMIC MUSCULAR ARTERIES

A recent comparison of the effect of incubation in Ca^{2+}-free physiological saline solution on the responses of the rabbit basilar artery versus the ear artery (12) showed that the KCl-induced contraction in both vessels entirely depends on extracellular Ca^{2+}. The response of the ear artery decreased with a half time of about 2 min, whereas the half time for the basilar artery was 0.7 min. These rates are much slower than would be predicted if they were determined by the rate of loss of Ca^{2+} from the extracellular space. Thus other rate-limiting phenomena must be considered. Although the initial rates of decline of contractions induced by 5-HT and NE in both vessels could not be distinguished from those for KCl, a subsequent slower rate of decline with a half time of approximately 20 min was observed in the ear artery. Corresponding rates for the basilar artery are difficult to define, but are probably about several minutes. By extrapolation it is quite clear that the contribution of Ca^{2+} associated with the slower rate of decline (presumably from sequestered sources) is much smaller for the cerebral than the systemic vessel. The data suggest that most of the activator Ca^{2+} in the basilar artery cannot be distinguished on the basis of washout rate from the exclusively extracellular pool utilized in the contractile response to high KCl. In the ear artery, Ca^{2+} from both sources plays a significant role.

The Ca^{2+} antagonist (Ca^{2+}-uptake inhibitor) methoxyverapamil (D 600) completely prevented the response of the rabbit basilar artery to KCl and blocked 80% of the response of the ear artery. The IC_{50} values (concentrations inhibiting contraction of arteries 50%) for D 600 against the KCl-induced contractions in the two vessels were similar (3.93×10^{-9} M and 5.06×10^{-9} M for basilar and ear arteries, respectively). When D 600 was tested against contractions induced by NE and 5-HT, considerably higher concentrations of NE were required to elicit equivalent responses in the basilar artery (i.e., the

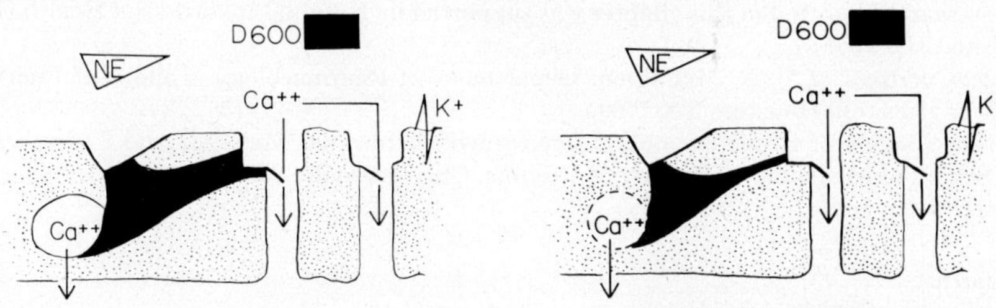

FIG. 1. Diagram of some differences between the coupling of the α-adrenoceptor to Ca^{2+} systems in rabbit basilar and ear arteries. Norepinephrine (NE) binds to a corresponding site in the membrane. Receptor occupation results in release of sequestered Ca^{2+} (shown in membrane), which is a smaller and less tightly bound fraction in the basilar artery. It also leads to the opening of a transmembrane Ca^{2+} channel. Efficacy of the coupling is considered lower in the basilar than in the ear artery. D 600 more easily blocks the entry of extracellular Ca^{2+} in the cerebral vessel.

effective concentrations of D 600 on this vessel were considerably lower than on the ear artery). In addition the maximum effect of D 600 on the basilar artery was greater than on the ear artery. These findings are consistent with earlier observations (1, 7).

These results suggest that there is a fundamental difference in the Ca^{2+}-related contractile mechanisms in the two vessels. Several of these differences (the greater utilization of extracellular Ca^{2+} and the increased sensitivity to and efficacy of D 600 in the basilar artery) result in a considerably greater effect of this drug and other Ca^{2+} antagonists (J. A. Bevan, unpublished data) on the rabbit cerebral artery versus the systemic ear artery. Some of the differences in the α-adrenoceptor system in the basilar and ear arteries are illustrated in Figure 1.

SUMMARY

In contrast to a systemic vessel such as the rabbit ear artery, where the concentrations of NE required to elicit phasic and tonic components of contraction are similar, concentrations of NE required to elicit phasic contractions in a cerebral vessel like the rabbit basilar artery are considerably lower than those that induce the tonic phase of contraction. This independence of phasic and tonic components of contraction suggests different and separate coupling mechanisms between the α-adrenoceptor and the two Ca^{2+} pools responsible for the two phases of contraction. This conclusion is strengthened by studies with the ear artery that show that the size of the second phase of contraction (but not the first) in response to α-adrenoceptor occupation can be altered experimentally.

The rabbit basilar artery contractile response to NE is more readily blocked than that of the ear artery by the Ca^{2+} antagonist D 600 because it depends more on external Ca^{2+}. Thus it is more sensitive to D 600 (based on IC_{50} values), and D 600 has a greater maximum inhibitory effect.

The research reported in this chapter was supported by National Institutes of Health Grants HL-15805 and HL-20581.

Present address of T. A. McCalden: Department of Pharmacology, College of Pharmacy, University of Houston, Houston, TX 77004.

Present address of R. M. Rapoport: Department of Internal Medicine and Pharmacology, Box 419, School of Medicine, University of Virginia, Charlottesville, VA 22901.

REFERENCES

1. ALLEN, G. S., AND S. B. BANGHART. Cerebral arterial spasm (part 9): in vitro effects of nifedipine on serotonin, phenylephrine, and potassium induced contractions of canine basilar and femoral arteries. *Neurosurgery* 4: 37–42, 1979.
2. BEVAN, J. A. Transient responses of rabbit cerebral blood vessels to norepinephrine: correlation with intrinsic myogenic tone. *Circ. Res.* 45: 566–572, 1979.
3. BEVAN, J. A., W. GARSTKA, C. SU, AND M. SU. The bimodal basis of the contractile response of the rabbit ear artery to norepinephrine and other agonists. *Eur. J. Pharmacol.* 22: 47–53, 1973.
4. BEVAN, J. A., AND B. LJUNG. Longitudinal propagation of myogenic activity in rabbit arteries and in the rat portal vein. *Acta Physiol. Scand.* 90: 703–715, 1974.
5. BEVAN, J. A., AND J. G. WATERSON. Biphasic constrictor response of the rabbit ear artery. *Circ. Res.* 28: 655–661, 1971.

6. BOHR, D. F. Vascular smooth muscle: dual effect of calcium. *Science* 139: 597, 1963.
7. FLECKENSTEIN, A. Specific pharmacology of calcium in myocardium, cardiac pacemakers, and vascular smooth muscle. *Annu. Rev. Pharmacol. Toxicol.* 17: 149–166, 1977.
8. KEATINGE, W. R., AND C. TORRIE. Action of sympathetic nerves on inner and outer muscle of sheep carotid artery, and effect of pressure on nerve distribution. *J. Physiol. London* 257: 699–712, 1976.
9. LACOMBE, P., R. SERCOME, A.-M. REYNIER-RUBUSSEL, N. LUSAMVUKU, E. PINARD, AND J. SEYLAZ. Time course of the effects of cervical sympathetic stimulation on local CBF: evidence for an escape phenomenon. *Acta Neurol. Scand. Suppl.* 64: 40–41, 1977.
10. MARCUS, M. L., P. M. GROSS, AND D. D. HEISTAD. Temporal and regional characteristics of cerebral vascular responses to sympathetic nerve stimulation (abstr.). *Circ. Res.* 58: 68, 1978.
11. McCALDEN, T. A., AND J. A. BEVAN. Asymmetry of the contractile response of the rabbit ear artery to exogenous amines. *Am. J. Physiol.* 238 (*Heart Circ. Physiol.* 7): H618–H624, 1980.
12. McCALDEN, T. A., AND J. A. BEVAN. Sources of activator calcium in the rabbit basilar artery. *Am. J. Physiol.* 241 (*Heart Circ. Physiol.* 10): H129–H133, 1981.
13. MEKATA, F., AND W. R. KEATINGE. Electrical behavior of inner and outer smooth muscle of sheep carotid artery. *Nature London* 258: 534–535, 1975.
14. PASCUAL, R., AND J. A. BEVAN. Variation in contractile response characteristics of rabbit aorta strips with surface of drug entry. *J. Pharmacol. Exp. Ther.* 210: 51–55, 1979.
15. PASCUAL, R., AND J. A. BEVAN. Asymmetry of consequences of drug disposition mechanisms in the wall of the rabbit aorta. *Circ. Res.* 46: 22–28, 1980.
16. RAPOPORT, R. M., AND J. A. BEVAN. Vascular smooth muscle contraction induced by direct electrical stimulation and its subsequent effect on reactivity to agonists. *J. Pharmacol. Exp. Ther.* in press.
17. RAPOPORT, R. M., AND J. A. BEVAN. Mechanism of electrical stimulation—induced decreased responsiveness. *J. Pharmacol. Exp. Ther.* in press.
18. RAPOPORT, R. M., AND J. A. BEVAN. Prior contraction results in increased responsiveness and maximum contractility of vascular smooth muscle in vitro. *Blood Vessels* in press.
19. STEINSLAND, O. S., R. F. FURCHGOTT, AND S. M. KIRPEKAR. Biphasic vasoconstriction of the rabbit ear artery. *Circ. Res.* 32: 49–58, 1973.

10

General Anesthetics and Magnesium Ions as Calcium Antagonists on Vascular Smooth Muscle

BURTON M. ALTURA AND BELLA T. ALTURA

Department of Physiology, State University of New York, Downstate Medical Center, Brooklyn, New York

IT HAS LONG BEEN KNOWN that systemic administration of general anesthetics as well as Mg^{2+} can rapidly lower blood pressure and induce peripheral vasodilation (1, 41, 46, 48, 55, 61, 64, 66). At the turn of this century, it was reported that systemic administration of Mg^{2+} produced general anesthesia accompanied by hypotension (55).

This chapter reviews recent evidence that suggests that both general anesthetics and excess Mg^{2+} lower blood pressure and induce peripheral vasodilation by their actions on movement and/or translocation of Ca^{2+} across vascular smooth muscle cell membranes and intracellularly. If the latter is true, one must then pose the question: are certain general anesthetics and Mg^{2+} "true" Ca^{2+} antagonists?

INFLUENCE OF GENERAL ANESTHETICS AND MAGNESIUM IONS ON SPONTANEOUS MECHANICAL ACTIVITY AND BASAL TENSION OF ARTERIES AND VEINS

General Anesthetics

In vitro, most of the commonly used general anesthetics (e.g., halothane, diethyl ether, chloroform, barbiturates, ketamine, urethane, propranidid, steroids, alcohol, and even acetaldehyde) cause a dose-dependent inhibition of

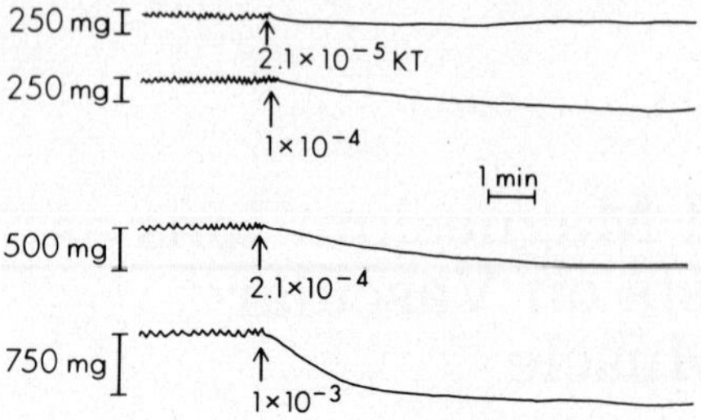

FIG. 1. Influence of various concentrations (in mol/liter) of ketamine hydrochloride on development of spontaneous mechanical activity in a single isolated rat aortic strip. Arrows indicate points at which tissue was exposed to ketamine. Values on *left* represent tension. [From Altura et al. (20).]

spontaneous mechanical activity in venous and arterial smooth muscles and a decreased resting tension in arteries (Fig. 1) and arterioles capable of developing basal tone (20, 21, 23–25, 27–29, 34, 39, 42). In all cases the greater the concentration of anesthetic agent, the more rapid the inhibition and the more the resting tension is lowered. Venous smooth muscle seems more sensitive to these inhibitory actions than arterial smooth muscle (27–29).

Most of these inhibitory actions are observed with the same concentrations of anesthetics achieved during surgery (Table 1). The relative order of potency for inhibition of vascular myogenic tone is similar to that needed to induce anesthesia, e.g., thiopental > secobarbital ≅ pentobarbital > amobarbital > phenobarbital > barbital (28); and ketamine > thiopental > halothane > diethyl ether > ethanol > urethane (Table 1). These concentration-dependent changes in mechanical activity are not tachyphylactic and are not all-or-none responses (20, 21, 23–25, 27–29, 34, 39, 42). None of these anesthetic-induced

TABLE 1. *General anesthetics known to inhibit spontaneous contractile activity in mammalian arteries and veins*

Anesthetic	*Inhibitory In Vitro Concentrations, M*
Intravenous	
Thiopental*	2.5×10^{-5}–2×10^{-3}
Thiamylal*	5×10^{-5}–5×10^{-3}
Pentobarbital	1×10^{-4}–2×10^{-3}
Secobarbital	1×10^{-4}–2×10^{-3}
Amobarbital	2.5×10^{-4}–1×10^{-3}
Hexobarbital	4×10^{-4}–5×10^{-3}
Phenobarbital	5×10^{-4}–5×10^{-3}
Ketamine	5×10^{-6}–1×10^{-3}
Althesin	1×10^{-4}–1×10^{-3}
Ethanol	8.5×10^{-3}–2×10^{-1}
Propanidid	1×10^{-3}–5×10^{-1}
Acetaldehyde†	6.5×10^{-4}–1.7×10^{-3}
Magnesium ions	2×10^{-3}–1×10^{-2}
Volatile	
Halothane	5×10^{-4}
Diethyl ether	2×10^{-3}–7×10^{-3}
Chloroform	2.5×10^{-3}–8.5×10^{-3}
Urethane	1×10^{-2}–1×10^{-1}

* Usually contracts arteries but inhibits spontaneous activity in veins. †Usually enhances amplitude in veins but inhibits spontaneous activity in arteries. (Data from 4, 6, 8, 12, 14, 16–21, 23, 25, 27–31, 39, 42; also from unpublished observations.)

inhibitory (relaxation) responses (or those noted below) can be mimicked, attenuated, or prevented with known pharmacological antagonists, nor are they due to release of prostaglandins or to actions on β-adrenergic receptors (20, 21, 23–25, 27–29, 34, 39, 42). Such data suggest that the vasodepressant effects of the anesthetic agents may be due to some direct actions on the vascular smooth muscle cells (see below). Interestingly, the effects of these anesthetics on basal tension and spontaneous activity rather closely resemble the effects of removal of extracellular Ca^{2+} on these vessels (Fig. 2).

If venous smooth muscle cells are more sensitive to these inhibitory actions than arterial smooth muscle cells, this may help to explain why, in particular, barbiturates, alcohol, halothane, urethane, and ether compromise the control of the cardiovascular system (35, 41, 48, 72), produce hypotension and venous pooling, and exacerbate the effects of circulatory shock (35, 38, 52).

Magnesium Ions

Like general anesthetics, Mg^{2+} appears to be able to directly alter base-line tension or tone (4–8, 11–14, 16–19, 30–32, 40, 44, 47, 54, 63, 68, 71). Decrements in $[Mg^{2+}]_o$ result, in a dose-dependent manner, in rapid elevations in tension development in a variety of mammalian arteries (e.g., coronary, cerebral, umbilical, rat aorta; Figs. 3 and 4) and arterioles that are either spontaneously active or exhibit tone. Similar results have also been reported in ileal smooth muscle (47). Raising the $[Mg^{2+}]_o$ above normal physiological levels (i.e., > 1.2 mM) inhibits spontaneous mechanical activity and lowers base-line tension (Figs. 3 and 4); none of these findings are due to changes in osmolarity or the release (or inhibition of release) of any known vasoactive substance from the vascular walls (5, 6, 8, 13, 14, 16, 19, 43, 44, 71). The elevations in mechanical activity when $[Mg^{2+}]_o$ is lowered are markedly reduced and rapidly disappear as $[Ca^{2+}]_o$ is lowered or chelated; addition of calcium ethylenediaminetetraacetic acid (calcium EDTA) to the bath fluid potentiates these contractile responses, whereas ethylene glycol-bis(β-aminoethylether)-*N,N'*-tetraacetic acid (EGTA) promotes rapid relaxations, suggesting that influx of Ca^{2+} is necessary for these contractile responses (4, 11, 17). Conversely, lowering base-line tone with excess extracellular Mg^{2+} can be reversed by elevating the $[Ca^{2+}]_o$ (4, 6, 8, 11, 16, 19).

In vitro studies using spontaneously active venous smooth muscle (e.g., portal veins) indicate that withdrawal of extracellular Mg^{2+} results in a rapid enhancement of the spontaneously evoked mechanical responses and increases

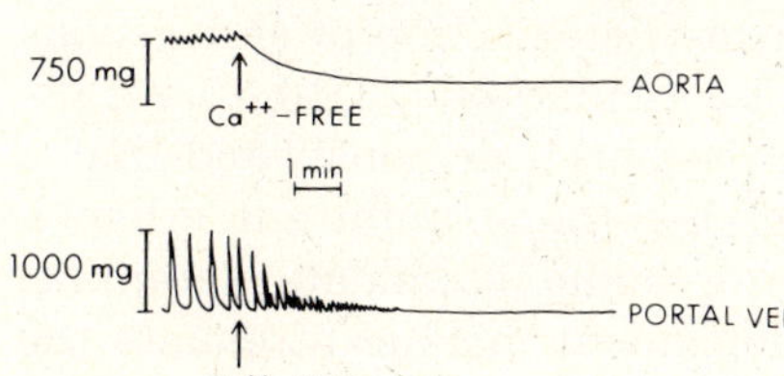

FIG. 2. Influence of removal of Ca^{2+} on spontaneous mechanical activity and base-line tension in rat aorta and portal vein. [From Altura et al. (20).]

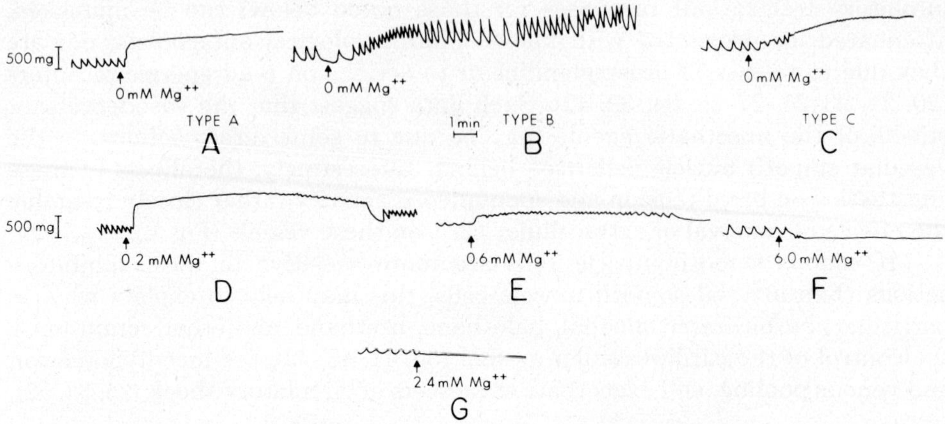

FIG. 3. Influence of $[Mg^{2+}]_o$ on spontaneous mechanical activity and resting tension in isolated rat aortic strips. *Panels A–C* represent 3 different types of contractile responses observed with 0 mM Mg^{2+}. [Adapted from Altura and Altura (8).]

in rhythmic contractility (Fig. 5; 4, 8, 12, 14, 16, 17, 19, 33, 63). Elevations in $[Mg^{2+}]_o$ above the physiological level lower, in a dose-dependent manner, the frequency and the contractile tension (Fig. 5; Table 2). Addition of approximately 10 mM Mg^{2+} usually obliterates development of all spontaneous mechanical events (Fig. 5; Table 2); these findings are not the results of hyperosmolarity or release of any known vasoactive agent from the venous walls. Like the arteries and arterioles, the magnitude of each mechanical spike elevation, when $[Mg^{2+}]_o$ is lowered, is directly proportional to $[Ca^{2+}]_o$ (4, 8, 12, 14, 16, 17, 19, 33). The attenuation of spike activity by elevations in $[Mg^{2+}]_o$ can be surmounted by an elevation in $[Ca^{2+}]_o$ or by the addition of extracellular Sr^{2+} (4, 8, 12, 14, 16, 17, 19, 33). Divalent cations such as Ni^{2+}, Co^{2+}, Cd^{2+}, and Mn^{2+}, which inhibit uptake of Ca^{2+} in excitable tissues, can completely block the rise in tensions when $[Mg^{2+}]_o$ is lowered, both in venous and arterial smooth muscles (4, 6, 11, 16, 17).

INFLUENCE OF GENERAL ANESTHETICS AND MAGNESIUM IONS ON CONTRACTIONS ELICITED BY NEUROHUMORAL AGENTS AND POTASSIUM IONS

General Anesthetics

In addition to their direct actions on myogenic tone, general anesthetics may cause vasodilation by inhibition of the constrictor action of endogenous neurohumoral substances that help to maintain vascular tone (3, 10, 20, 21, 23–25, 27–29, 36, 39, 42, 53, 65). For example, it has been demonstrated that a variety of general anesthetics (e.g., barbiturates, halothane, isoflurane, ethanol, urethane, steroids, trichloroethylene, ether, propanidid, etomidate, ketamine, and acetaldehyde) markedly and in a dose-dependent manner attenuate the

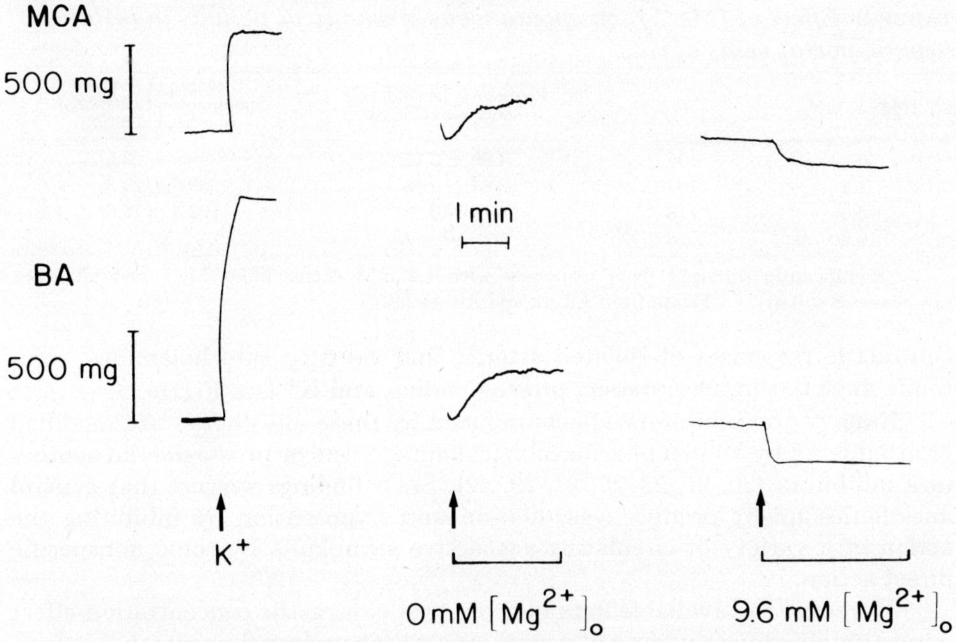

FIG. 4. Influence of $[Mg^{2+}]_o$ on basal tension of a middle cerebral artery (MCA) and basilar artery (BA). *Left panel* indicates contractile tension developed in response to addition of 30 mM KCl in normal Krebs-Ringer bicarbonate solution (1.2 mM Mg^{2+}). *Middle panel* shows tension developed (in absence of any vasoactive agent) in a Mg^{2+}-free environment. *Right panel* indicates that basal tension is lowered by elevated $[Mg^{2+}]$ (i.e., 9.6 mM). [Adapted from Altura and Altura (31).]

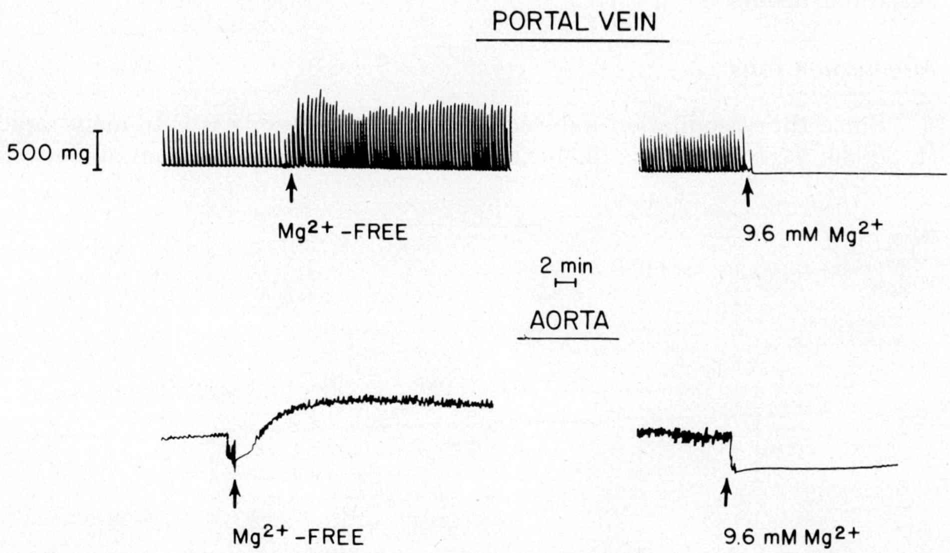

FIG. 5. Influence of $[Mg^{2+}]_o$ on spontaneous mechanical activity of rat portal vein and aorta.

TABLE 2. *Effect of [Mg^{2+}]$_o$ on spontaneous mechanical activity in isolated hepatic-portal veins of rats*

[Mg^{2+}]$_o$, mM	*n*	Frequency, mechanical responses/min ± SE	Strength of Average Spontaneous Contraction, mg tension ± SE
0	104	4.99 ± 0.12*	379.9 ± 23.5*
1.2	104	3.64 ± 0.16	239.5 ± 21.3
5.0	18	1.16 ± 0.12*	102.4 ± 16.6*
10.0	18	0*	0*

* Significantly different from responses with 1.2 mM extracellular Mg^{2+} and all other responses ($P < 0.01$). [Data from Altura and Altura (16).]

contractile responses of isolated arteries and veins to catecholamines, serotonin, angiotensin, vasopressin, prostaglandins, and K^+ (see Fig. 6).

None of the inhibitory effects induced by these anesthetics are modified or attenuated by known pharmacological antagonists or prostaglandin synthetase inhibitors (20, 21, 23–25, 27, 29, 42). Such findings suggest that general anesthetics might produce vasodilation and hypotension by inhibiting the action of a variety of circulating vasoactive stimulants by some nonspecific direct action.

Moreover the available data indicate that contractile concentration-effect curves obtained for a variety of neurohumoral substances (i.e., catecholamines, serotonin, angiotensin II, vasopressin, and prostaglandins) and K^+, in the presence of anesthetic agents, are shifted to the right to higher concentrations concomitant with dose-dependent reductions in maximum responses (10, 20, 21, 23–25, 27–29, 39, 42). In addition much of these data suggest that venous contractions may be more sensitive to the inhibitory actions of the general anesthetic agents.

Magnesium Ions

Since the vasodilation induced by hypermagnesemia is extremely rapid (1, 30, 46, 54, 55, 61), it is likely that it is due to some alteration at the cell

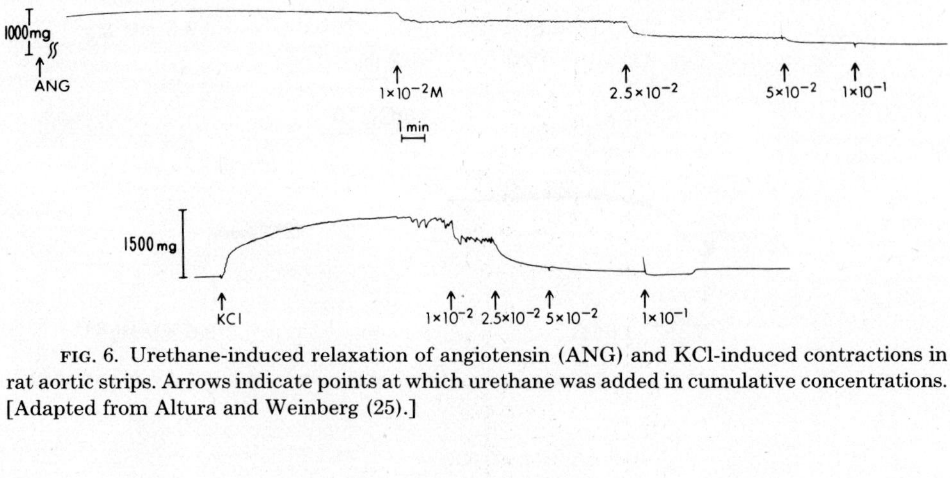

FIG. 6. Urethane-induced relaxation of angiotensin (ANG) and KCl-induced contractions in rat aortic strips. Arrows indicate points at which urethane was added in cumulative concentrations. [Adapted from Altura and Weinberg (25).]

surfaces of the arteries and arterioles. In vitro studies performed over the past decade clearly indicate that the contractile actions on isolated arterial smooth muscles of a variety of agonists, including neurohumoral agents (mentioned above), K^+, and ouabain, are depressed (e.g., Fig. 7) by even a slightly elevated $[Mg^{2+}]_o$ (i.e., > 1.2 mM). Such results have been noted for a wide variety of arteries (e.g., pulmonary, coronary, mesenteric, renal, femoral, cerebral, celiac, splenic, and umbilical) from rabbits, rats, dogs, piglets, guinea pigs, and humans (e.g., Table 3; 2–8, 12–19, 30, 37, 40, 44, 55, 58, 68, 71) as well as for single perfused rat mesenteric arterioles (17–22 μm ID; 15, 16, 30). These results cannot be attributed to hyperosmolarity, alteration of Na^+-K^+-ATPase activity, or release of any known vasoactive mediator from the smooth muscle or endothelial cells.

Conversely hypomagnesemia could elevate blood pressure and/or increase vascular resistance in certain vascular regions by potentiating the constrictor action of endogenous circulating neurohumoral substances (e.g., catecholamines, angiotensin II, acetylcholine, and serotonin) and ions (e.g., K^+) known to play important roles in regulation of blood flow (2, 3). Removal of Mg^{2+} from the medium bathing isolated arterial and venous vessels (e.g., coronary, pulmonary, and renal arteries, umbilical arteries and veins, and rat and rabbit

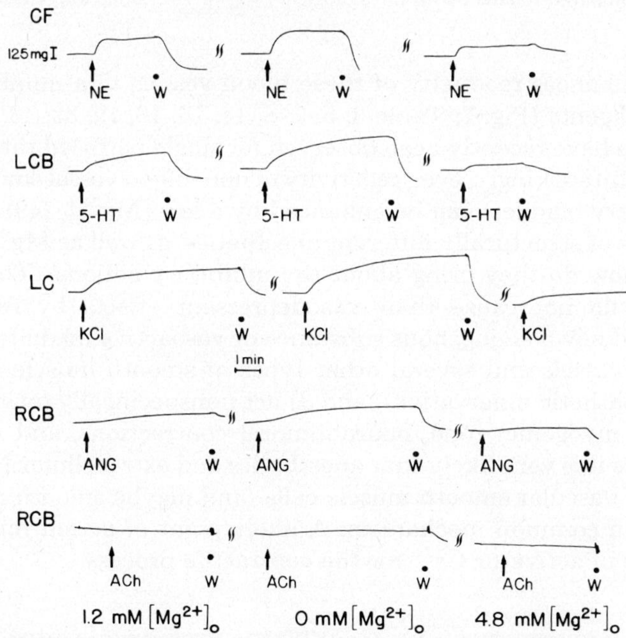

FIG. 7. Responses of canine circumflex (CF), left coronary branch (LCB), left coronary (LC), and right coronary branch (RCB) arterial strips to stimulation with norepinephrine bitartrate (NE; 50 ng/ml), serotonin creatinine sulfate (5-HT; 10 ng/ml), potassium chloride (KCl; 20 mM), angiotensin II amide (ANG; 5 ng/ml), and acetylcholine chloride (ACh; 0.4 mg/ml) in Krebs-Ringer bicarbonate solution containing 1.2, 0, and 4.8 mM Mg^{2+}. W, point at which strips were washed and relaxed in normal Krebs-Ringer bicarbonate solution. [Adapted from Turlapaty and Altura (68).]

TABLE 3. *Comparative effects of $[Mg^{2+}]_o$ on vasoactive drug-induced contractions of canine coronary, renal, and pulmonary arteries*

Agonist and Artery	n	Tension, mg $[Mg^{2+}]_o$, mM 1.2	0	4.8
Serotonin				
Left coronary	10	235.0 ± 47.5	447.5 ± 67.5*	142.5 ± 37.5*
Renal	6	879.2 ± 218	2125 ± 609*	691.7 ± 288
Intrapulmonary	6	891.7 ± 86.9	1100 ± 133*	416.7 ± 33.3*
Potassium				
Left coronary	8	270.0 ± 30.0	462.5 ± 55.0*	135.0 ± 18.7*
Renal	7	925 ± 80.7	1578.6 ± 269*	200.0 ± 31.3*
Intrapulmonary	6	1042 ± 171	1366.7 ± 223*	75.0 ± 38.1*
Angiotensin				
Left coronary	9	195.7 ± 58.5	292.5 ± 73.0*	152.8 ± 67.5
Renal	5	587.5 ± 157	1218.8 ± 371*	300 ± 91.3*
Intrapulmonary	6	1012 ± 168	1067 ± 205	733 ± 142*
Norepinephrine				
Left coronary†	8	100.0 ± 23.7	190.0 ± 40.0*	90.5 ± 29.5*
Renal	8	587.5 ± 100	972 ± 85.4	296.8 ± 31.1*
Intrapulmonary	6	650 ± 80.8	750 ± 120	375 ± 72.4

* $P < 0.02$, compared to effect with physiological $[Mg^{2+}]$. †Done in presence of propranolol.

portal veins) enhances reactivity of these blood vessels to a number of these neurohumoral agents (Fig. 7; Table 3; 5, 7, 8, 14, 16, 18, 19, 33, 43, 44, 68, 71). Similar findings have recently been observed for single perfused rat mesenteric arterioles (16). Interestingly, even reactivity of non–blood vessel smooth muscle cells (e.g., urinary bladder) can be enhanced by a low $[Mg^{2+}]_o$ (49).

If a variety of structurally different anesthetics, as well as Mg^{2+}, can exert such actions, how do they bring about the inhibitory actions? These diverse substances *1*) do not cause their vasodepressant effects by releasing (or inhibiting) any known endogenous substance or vasoactive mediator; *2*) act on isolated blood vessels and several other types of smooth muscle (even some devoid of sympathetic innervation); and *3*) act nonspecifically on spontaneous vasomotor (or myogenic) tone, neurohumoral contractions, and ion-induced responses. Thus it is very likely that anesthetics and extracellular Mg^{2+} inhibit contractility of vascular smooth muscle cells (and maybe smooth muscle cells in general) by a common mechanism. A likely point of action might involve the availability of activator Ca^{2+} for the contractile process.

INFLUENCE OF ANESTHETICS AND MAGNESIUM IONS ON CALCIUM-INDUCED CONTRACTIONS, CALCIUM EXCHANGE, AND CONTENT OF VASCULAR MUSCLE

General Anesthetics

Anesthetics might interfere with the mobility of Ca^{2+} both at the vascular membranes and intracellularly, since they penetrate both membranes and cells

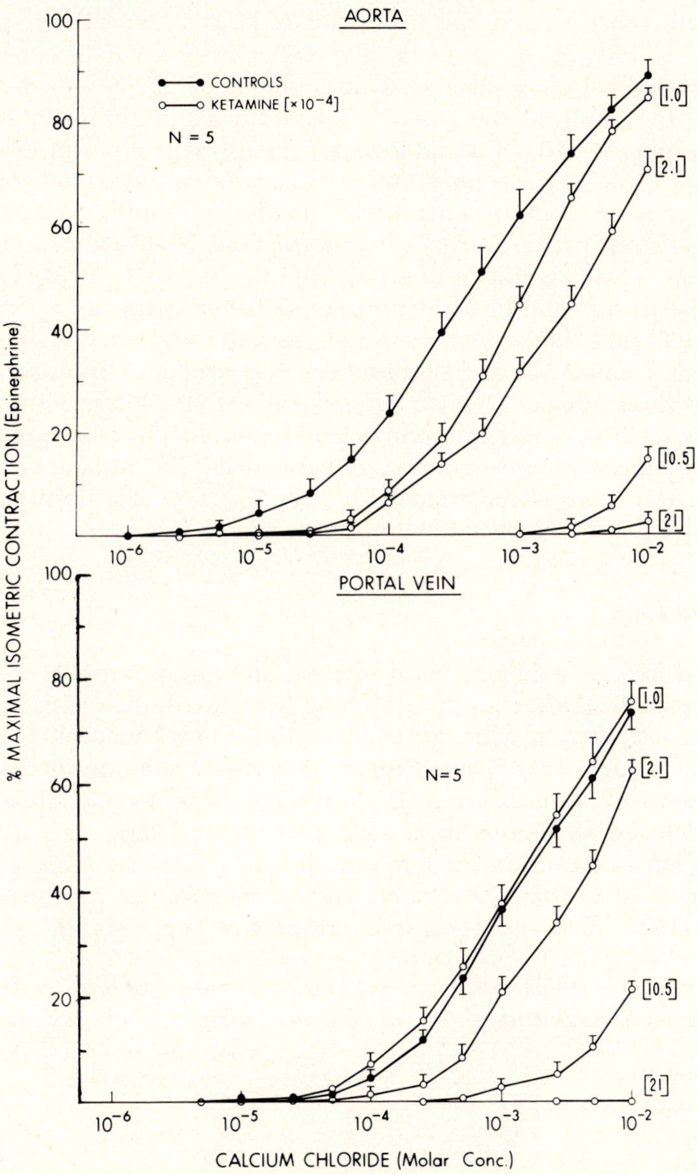

FIG. 8. Influence of ketamine HCl [$\times 10^{-4}$ M] on $CaCl_2$-induced contractions of K^+-depolarized rat aortas and portal veins; 100% isometric contractile tension for aortas = 1,572 ± 136 mg, 100% isometric contractile tension for portal veins = 2,242 ± 162 mg. [Adapted from Altura et al. (20).]

rapidly (41, 62, 72). If so, one would anticipate that Ca^{2+}-induced contractions of K^+-depolarized vascular muscles would be most sensitive to general anesthetic inhibition, as is exactly the case (e.g., Fig. 8; 10, 20, 21, 23–25, 27–29, 42). The observed shifts of the log concentration-response curves for the agonists (10, 20, 21, 23–25, 27–29, 39, 42), as well as for Ca^{2+}, together with the reduction in maximum tensions and the inhibition of $[Ca^{2+}]_o$-dependent spontaneous mechanical activity in the presence of anesthetics would all be consistent with the hypothesis that these diverse membrane-active agents interfere with the mobility of Ca^{2+} both at the vascular membranes and intracellularly. Recent direct experiments with ^{45}Ca and isolated rat aortas and portal veins indicate that at least three of these anesthetics (barbiturates, ketamine, and ethanol), when used in anesthetic concentrations, significantly inhibit membrane-bound and La^{3+}-resistant (intracellular) calcium fractions by 40–60%; total exchangeable calcium is also significantly attenuated (Table 4; 34, 70; B. T. Altura, P. D. M .V. Turlapaty, and B. M. Altura, unpublished data).

Price (60) and Nayler and Szeto (56), as well as others (57), have recently suggested that halothane and barbiturates may produce depression of cardiac muscle by interfering with Ca^{2+} at several cellular sites. Various volatile anesthetics, such as ether, chloroform, and halothane, have also recently been reported to depress microsomal and mitochondrial Ca^{2+} uptake in rat myocardium (50). Rat heart sarcolemmal Ca^{2+} binding has also been shown to be depressed by these three volatile anesthetic agents (51).

Magnesium Ions

Several lines of evidence, both indirect and direct, support the idea that $[Mg^{2+}]_o$ does indeed affect uptake, binding, and distribution of cellular Ca^{2+} in vascular smooth muscle. First, exposure of aortic smooth muscle to a Ca^{2+}-free versus a Ca^{2+}, Mg^{2+}–free Krebs-Ringer bicarbonate solution for 2 h, followed by a cumulative readmittance of Ca^{2+} subsequent to acetylcholine- or angiotensin II-induced residual contractions, results, after only Ca^{2+} depletion [> 90% reduction in tissue Ca; (9)], in 60% and 40% maximal responses, respectively. But after combined Ca^{2+} and Mg^{2+} depletion, one sees very dramatic 250% and 150% maximal contractile responses (e.g., Fig. 9; 16, 32). The

TABLE 4. *Effect of anesthetic concentrations of pentobarbital sodium and ethanol on total exchangeable and cellular calcium in rat aortas*

Treatment, M	Total Exchangeable Calcium	Calcium Fraction After 2-min La^{3+} Wash (Loosely Bound Ca^{2+})	Calcium Fraction After 5-min La^{3+} Wash (Intracellular Ca^{2+})
Control	4.11 ± 0.10 (48)	2.09 ± 0.09 (34)	0.88 ± 0.06 (31)
Pentobarbital			
5×10^{-4}	3.52 ± 0.14* (6)	1.54 ± 0.13* (8)	0.69 ± 0.04* (9)
2×10^{-3}	3.39 ± 0.34† (5)	1.34 ± 0.07* (7)	0.55 ± 0.06* (4)
Ethanol			
1.7×10^{-2}	3.46 ± 0.11* (6)	1.53 ± 0.14* (6)	0.78 ± 0.09 (6)

Values are mmol/kg wet wt (mean ± SE). Numbers in parentheses are number of different tissues used. *Significantly different from paired controls ($P < 0.01$). †Significantly different from paired controls ($P < 0.05$).

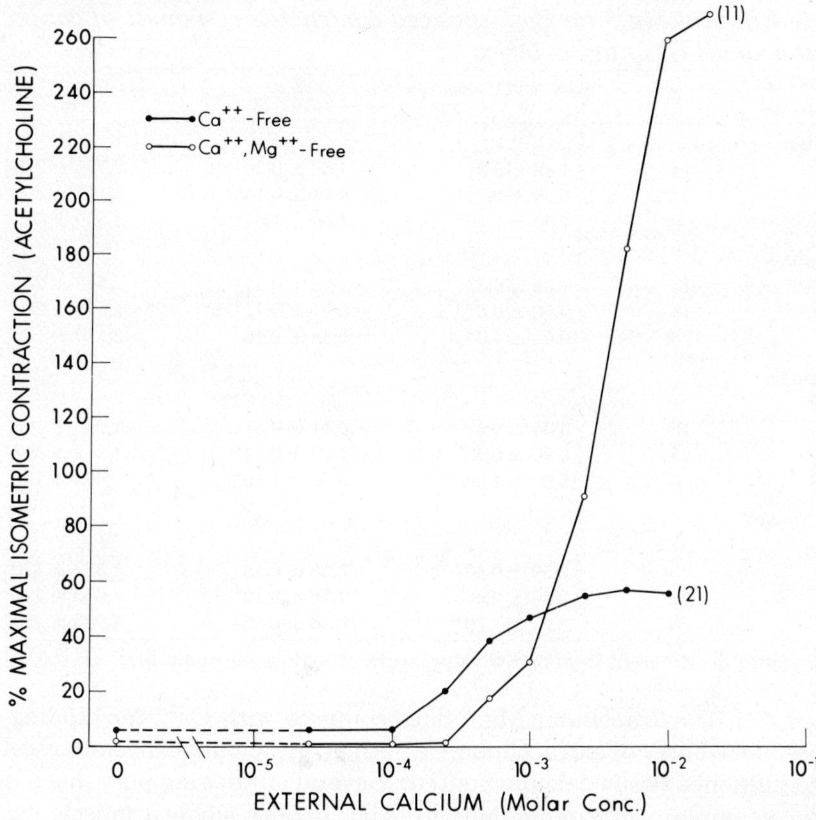

FIG. 9. Calcium restoration dose-response curves for acetylcholine-induced contractions of rabbit aortic strips. Numbers in parentheses are total number of aortic strips obtained and examined from different rabbits. Ordinate is % maximum isometric contraction in normal Krebs-Ringer bicarbonate solution. The SE for each point is ⩽ 5%. All paired points obtained with $CaCl_2$ ($> 10^{-4}$ M) are significantly different from one another ($P < 0.02$). [Adapted from Altura and Altura (16).]

potentiation of maximum responses in normal Krebs-Ringer bicarbonate solution containing 2.5 mM Ca^{2+} in the absence of Mg^{2+}, however, is only 30% and 10%, respectively (7). Enhanced entry or translocation of extracellular Ca^{2+} in the absence of both Ca^{2+} and Mg^{2+} (e.g., Fig. 9) could account for the increased maximal responses noted with acetylcholine- and angiotensin-induced contraction of vascular muscle. The inhibitory effects of elevated $[Mg^{2+}]_o$ (e.g., Fig. 7; Table 3) can also be explained as a result of a block of Ca^{2+} influx at the membrane. If this is true, one would expect that Ca^{2+}-induced contractions of K^+-depolarized vascular muscles would be *1*) sensitive to inhibition by an elevated $[Mg^{2+}]_o$ [i.e., increases in EC_{50} (mean effective concentration) values and threshold concomitant with reductions in maxima should be noted], as is exactly the case (e.g., Table 5; 6, 18, 19, 71) and *2*) potentiated in the absence of Mg^{2+} (i.e., decreases in EC_{50} and threshold values, possibly concomitant with increases in maxima), as is the case (e.g., Table 5;

TABLE 5. *Effect of Mg^{2+} on Ca^{2+}-induced contractile responses of canine large and small coronary arteries*

Artery and $[Mg^{2+}]_o$, mM	n	Threshold Concn, $\times 10^{-5}$ M	EC_{50}, $\times 10^{-4}$ M	Maximum Tension, mg
Circumflex				
1.2	14	1.42 ± 0.34	1.58 ± 0.39	2,775 ± 230
0	14	0.30 ± 0.02*	0.71 ± 0.14*	3,175 ± 200*
4.8	6	7.50 ± 1.40*	4.50 ± 1.07*	2,550 ± 625*
Left coronary				
1.2	15	1.53 ± 0.32	2.16 ± 0.41	2,567 ± 238
0	15	0.50 ± 0.07*	0.94 ± 0.11*	2,970 ± 230*
4.8	6	10.0 ± 1.0*	5.58 ± 0.16*	2,170 ± 78*
Left coronary branches				
1.2	15	3.30 ± 0.61	3.51 ± 0.61	1,046 ± 95
0	15	0.90 ± 0.30*	1.39 ± 0.14*	1,242 ± 97*
4.8	6	15.0 ± 4.20*	6.50 ± 1.19*	806 ± 196
Right coronary branches				
1.2	15	2.50 ± 0.40	3.58 ± 0.46	1,387 ± 120
0	15	0.56 ± 0.06*	1.16 ± 0.12*	1,660 ± 120*
4.8	6	8.0 ± 1.20*	6.70 ± 0.52*	1,080 ± 260*

* Significantly different from paired values with 1.2 mM extracellular Mg^{2+} ($P < 0.01$).

6, 18, 19, 71). If extracellular Mg^{2+} does compete with Ca^{2+} for binding sites and does block entry of extracellular Ca^{2+}, then one should expect extracellular Mg^{2+} to influence tissue calcium content. Several studies support this hypothesis. For example, acute reductions in $[Mg^{2+}]_o$ raise smooth muscle calcium content, whereas acute elevations in $[Mg^{2+}]_o$ lower calcium content in several types of smooth muscle (6, 7, 15, 19, 44, 45, 47, 59). These findings are supported by recent experiments with ^{45}Ca (18, 19, 44, 47, 67, 71). Collectively these experiments indicate that when $[Mg^{2+}]_o$ is lowered, ^{45}Ca influx (and intracellular content) is increased. The observation that high $[Mg^{2+}]_o$ decreases total exchangeable and membrane-bound calcium suggests that Mg^{2+} can displace and/or compete with Ca^{2+} at smooth muscle cell membranes for some functional binding sites. Other recent experiments indicate that extracellular Mg^{2+} can also prevent efflux of ^{45}Ca from arterial and venous smooth muscle (19). Overall these findings suggest that membrane Mg^{2+} sites may control and regulate content, entry, and exit of Ca^{2+} in smooth muscle. If this does occur, then one would have to consider that reciprocal changes in arterial (and arteriolar) and plasma Mg^{2+} and Ca^{2+} might take place pathophysiologically under a variety of conditions (5, 19, 22, 26, 31, 68, 69).

CONCLUSIONS

The data reviewed here suggest that the peripheral vasodilation and hypotension resulting from administration of general anesthetics or excess Mg^{2+} may largely be a consequence of the direct actions of these molecules on vascular smooth muscle cells. At least two mechanisms appear to contribute to this vasodilator effect: *1*) inhibition of the normal rhythm or vasomotion (spontaneous mechanical activity) of vascular smooth muscle and *2*) depression of the contractile responses to endogenous neurohumoral substances that act in maintaining vascular tone. The data acquired so far suggest that these dilator actions are related causally to interference with movement and/or translocation of Ca^{2+} across the vascular membranes and intracellularly. Extracellular Mg^{2+} and membrane Mg^{2+} may exert a regulatory role in vascular tone, vascular reactivity, or peripheral vascular resistance, and Mg^{2+} apparently has an important functional role in control of Ca^{2+} uptake, content, and distribution in smooth muscle cells. Overall the data are compatible with the notion that general anesthetics and Mg^{2+} may be true Ca^{2+} antagonists, at least in vascular smooth muscle.

This work was supported in part by National Institutes of Health Grants HL-18002, HL-18015, and DA-02339.

REFERENCES

1. AIKAWA, J. K. *The Relationship of Magnesium to Disease in Domestic Animals and in Humans.* Springfield: Thomas, 1971.
2. ALTURA, B. M. Chemical and humoral regulation of blood flow through the precapillary sphincter. *Microvasc. Res.* 3: 361–384, 1971.
3. ALTURA, B. M. Pharmacology of venular smooth muscle: new insights. *Microvasc. Res.* 16: 91–117, 1978.
4. ALTURA, B. M. Magnesium withdrawal and rhythmic contractility of arterial vs. venous smooth muscle. Differential effects of multivalent cations and EDTA. *Artery* 4: 512–517, 1978.
5. ALTURA, B. M. Sudden-death ischemic heart disease and dietary magnesium intake: is the target site coronary vascular smooth muscle? *Med. Hypotheses* 5: 843–849, 1979.
6. ALTURA, B. M. Magnesium and regulation of contractility of vascular smooth muscle. *Adv. Microcirc.* in press.
7. ALTURA, B. M., AND B. T. ALTURA. Influence of magnesium on drug-induced contractions and ion content in rabbit aorta. *Am. J. Physiol.* 220: 938–944, 1971.
8. ALTURA, B. M., AND B. T. ALTURA. Magnesium and contraction of arterial smooth muscle. *Microvasc. Res.* 7: 145–155, 1974.
9. ALTURA, B. M., AND B. T. ALTURA. Contractile actions of antihistamines on isolated arterial smooth muscle. *J. Pharmacol. Exp. Ther.* 191: 262–268, 1974.
10. ALTURA, B. M., AND B. T. ALTURA. Peripheral vascular actions of glucocorticoids and their relationship to protection in circulatory shock. *J. Pharmacol. Exp. Ther.* 190: 300–315, 1974.
11. ALTURA, B. M., AND B. T. ALTURA. Magnesium withdrawal and contraction of arterial smooth muscle: effects of EDTA, EGTA and divalent cations. *Proc. Soc. Exp. Biol. Med.* 151: 752–755, 1976.
12. ALTURA, B. M., AND B. T. ALTURA. Vascular smooth muscle and prostaglandins. *Federation Proc.* 35: 2360–2366, 1976.
13. ALTURA, B. M., AND B. T. ALTURA. Ouabain, membrane Na^+, K^+-ATPase and the extra-cellular action of magnesium ions in arterial smooth muscle. *Artery* 3: 72–83, 1977.
14. ALTURA, B. M., AND B. T. ALTURA. Extracellular magnesium and contraction of vascular smooth muscle. In: *Excitation-Contraction Coupling of Smooth Muscle,* edited by R. Casteels, T. Godfraind, and J. C. Ruegg. Amsterdam: North-Holland, 1977, p. 137–144.
15. ALTURA, B. M., AND B. T. ALTURA. Magnesium-neurohypophyseal hormone interactions in contraction of arterial smooth muscle. In: *Peptides: Chemistry, Structure and Biology,* edited by R. Walter and J. Meienhofer. Ann Arbor: Ann Arbor Science, 1978, p. 719–726.
16. ALTURA, B. M., AND ALTURA, B. T. Magnesium and vascular tone and reactivity. *Blood Vessels* 15: 5–16, 1978.
17. ALTURA, B. M., AND B. T. ALTURA. Differential interactions of divalent and trivalent cations with magnesium ions in rhythmic contractility of arterial versus venous smooth muscle. In: *Trace Substances in Environmental Health—XIII,* edited by D. D. Hemphill. Columbia: Univ. of Missouri, 1980, p. 408–416.
18. ALTURA, B. M., AND B. T. ALTURA. The role of external magnesium in contraction of pulmonary, renal, cerebral, and coronary blood vessels. *Magnesium Bull.* in press.
19. ALTURA, B. M., AND B. T. ALTURA. Magnesium ions and contraction of vascular smooth muscles: relationship to some vascular diseases. *Federation Proc.* in press.
20. ALTURA, B. M., B. T. ALTURA, AND A. CARELLA. Effects of ketamine on vascular smooth muscle function. *Br. J. Pharmacol.* 70: 257–267, 1980.
21. ALTURA, B. M., A. CARELLA, AND B. T. ALTURA. Acetaldehyde on vascular smooth muscle: possible role in vasodilator action of ethanol. *Eur. J. Pharmacol.* 52: 73–83, 1978.
22. ALTURA, B. M., A. CARELLA, AND B. T. ALTURA. Magnesium ions control prostaglandin reactivity of venous smooth muscle from spontaneously hypertensive rats. *Prostaglandins Med.* 4: 255–262, 1980.
23. ALTURA, B. M., H. EDGARIAN, AND B. T. ALTURA. Differential effects of ethanol and mannitol on contraction of arterial smooth muscle. *J. Pharmacol. Exp. Ther.* 197: 352–362, 1976.
24. ALTURA, B. M., A. OGUNKOYA, A. GEBREWOLD, AND B. T. ALTURA. Effects of ethanol on terminal arterioles and muscular venules: direct observations on the microcirculation. *J. Cardiovasc. Pharmacol.* 1: 97–113, 1979.
25. ALTURA, B. M., AND J. WEINBERG. Urethane and con-

traction of vascular smooth muscle. *Br. J. Pharmacol.* 67: 255–263, 1979.

26. ALTURA, B. T. Type-A behavior and coronary vasospasm: a possible role of hypomagnesemia. *Med. Hypotheses* 6: 753–758, 1980.
27. ALTURA, B. T., AND B. M. ALTURA. Pentobarbital and contraction of vascular smooth muscle. *Am. J. Physiol.* 229: 1635–1640, 1975.
28. ALTURA, B. T., AND B. M. ALTURA. Barbiturates and aortic and venous smooth-muscle function. *Anesthesiology* 43: 432–445, 1975.
29. ALTURA, B. T., AND B. M. ALTURA. Intravenous anesthetic agents and vascular smooth muscle function. In: *Mechanisms of Vasodilatation*, edited by P. M. Vanhoutte and I. Leusen. Basel: Karger, 1978, p. 165–172.
30. ALTURA, B. T., AND B. M. ALTURA. Factors affecting vascular responsiveness. In: *Microcirculation*, edited by G. Kaley and B. M. Altura. Baltimore: University Park, 1978, vol. II, p. 547–615.
31. ALTURA, B. T., AND B. M. ALTURA. Withdrawal of magnesium causes vasospasm while elevated magnesium produces relaxation of tone in cerebral arteries. *Neurosci. Lett.* 20: 323–327, 1980.
32. ALTURA, B. T., AND B. M. ALTURA. Influence of magnesium on contractile activity in isolated rat arterial and venous muscle. In: *Magnesium in Health and Disease*, edited by M. Cantin and M. S. Seelig. New York: Spectrum, 1980, p. 695–702.
33. ALTURA, B. T., AND B. M. ALTURA. Magnesium-calcium interactions and contraction of isolated arterial smooth muscle. In: *Magnesium in Health and Disease*, edited by M. Cantin and M. S. Seelig. New York: Spectrum, 1980, p. 703–711.
34. ALTURA, B. T., P. D. M. V. TURLAPATY, AND B. M. ALTURA. Pentobarbital sodium inhibits calcium uptake in vascular smooth muscle. *Biochim. Biophys. Acta* 595: 309–312, 1980.
35. BAEZ, S. Anesthetics and the microcirculation. In: *Effects of Anesthetics on the Circulation*, edited by H. L. Price and P. J. Cohen. Springfield, IL: Thomas, 1964, p. 182–198.
36. BREZENOFF, H. E. Cardiovascular responses to noradrenaline in the rat before and after administration of various anaesthetics. *Br. J. Pharmacol.* 49: 565–572, 1973.
37. CARRIER, O., JR., R. K. HESTER, H. A. JUREVICS, AND T. E. TENNER, JR. Influence of magnesium on calcium- and potassium-related responses in vascular smooth muscle. *Blood Vessels* 13: 321–337, 1976.
38. CHIEN, S. Hemodynamics in hemorrhage: influences of sympathetic nerves and pentobarbital anesthesia. *Proc. Soc. Exp. Biol. Med.* 136: 271–275, 1971.
39. CLARK, S. C., AND K. L. MACCANNELL. Vascular responses to anaesthetic agents. *Can. Anaesth. Soc. J.* 22: 20–33, 1975.
40. CROPP, G. J. A. Reduction of hypoxic pulmonary vasoconstriction by magnesium chloride. *J. Appl. Physiol.* 24: 755–760, 1968.
41. DUNDEE, J. W., AND G. M. WYANT. *Intravenous Anesthesia.* Edinburgh: Churchill-Livingstone, 1974.
42. EDGARIAN, H., AND B. M. ALTURA. Ethanol and contraction of venous smooth muscle. *Anesthesiology* 44: 311–317, 1976.
43. FUJIWARA, M., H. KITAGAWA, AND K. KURAHASHI. Effects of magnesium on contractile responses induced by electrical transmural stimulation and noradrenaline in rabbit thoracic aorta. *Br. J. Pharmacol.* 63: 51–56, 1978.
44. GOLDSTEIN, S., AND T. T. ZSOTÉR. The effect of magnesium on the response of smooth muscle to 5-hydroxytryptamine. *Br. J. Pharmacol.* 62: 507–514, 1978.
45. GOODFORD, P. J. The calcium content of the smooth muscle of guinea pig taenia coli. *J. Physiol. London* 142: 145–157, 1967.
46. HAZARD, R., AND L. WURMSER. Action de sels de magnésium sur les vasoconstricteurs rénaux. *C.R. Soc. Biol.* 110: 525–528, 1932.
47. HUDDART, H., AND K. H. M. SAAD. The effect of sodium and magnesium and their interaction with quinine and lanthanum on spontaneous activity and related calcium movements of rat ileal smooth muscle. *J. Comp. Physiol.* 126: 233–240, 1978.
48. HUG, C. C., JR. Pharmacology-anesthetic drugs. In: *Cardiac Anesthesia*, edited by J. A. Kaplan. New York: Grune & Stratton, 1979, p. 3–37.
49. JOHNS, A. The effect of magnesium-free solutions on the responses of the guinea pig urinary bladder to adenosine 5'-triphosphate and nerve stimulation. *Can. J. Physiol. Pharmacol.* 57: 1320–1323, 1979.
50. LEE, S. L., L. E. ALTO, AND N. S. DHALLA. Subcellular effects of some anesthetic agents on rat myocardium. *Can. J. Physiol. Pharmacol.* 57: 65–70, 1979.
51. LEE, S. L., L. E. ALTO, AND N. S. DHALLA. Effects of some volatile anesthetic agents on rat heart sarcolemma. *Life Sci.* 24: 1441–1446, 1979.
52. LONGNECKER, D. E., AND P. D. HARRIS. Anesthesia. In: *Microcirculation*, edited by G. Kaley and B. M. Altura. Baltimore: University Park, 1980, vol. III, p. 335–371.
53. LUNDY, P. M., C. W. GOWDEY, AND E. H. COLHOUN. Further experimental evidence for a smooth muscle depressant effect of ketamine. *Arch. Int. Pharmacodyn. Ther.* 216: 57–62, 1975.
54. MARIER, J. R., L. C. NERI, AND T. W. ANDERSON. Water hardness, human health, and the importance of magnesium. *Natl. Res. Counc. Can.* Publ. No. 17581: 1–119, 1979.
55. MORDES, J. P., AND W. E. C. WACKER. Excess magnesium. *Pharmacol. Rev.* 29: 273–300, 1978.
56. NAYLER, W. G., AND J. SZETO. Effect of sodium pentobarbital on calcium in mammalian heart muscle. *Am. J. Physiol.* 222: 339–344, 1972.
57. OHNISHI, S. T., C. A. DICAMILLO, M. SINGER, AND H. L. PRICE. Correlation between halothane-induced myocardial depression and decreases in La^{3+}-displaceable Ca^{2+} in cardiac muscle cells. *J. Cardiovasc. Pharmacol.* 2: 67–76, 1980.
58. OZAKI, H., AND N. URAKAWA. Na-Ca exchange and tension development in guinea-pig aorta. *Naunyn-Schmiedebergs Arch. Pharmacol.* 309: 171–178, 1979.
59. PALATÝ, V. Distribution of magnesium in the arterial wall. *J. Physiol. London* 218: 353–368, 1971.
60. PRICE, H. L. Calcium reverses myocardial depression caused by halothane. *Anesthesiology* 41: 576–579, 1974.
61. SEELIG, M. S., AND H. A. HEGGTVEIT. Magnesium interrelationships in ischemic heart disease: a review. *Am. J. Clin. Nutr.* 27: 59–79, 1974.
62. SEEMAN, P. The membrane actions of anesthetics and tranquilizers. *Pharmacol. Rev.* 24: 583–655, 1972.
63. SIGURDSSON, S. B., AND B. UVELIUS. The effects of variations in extracellular magnesium concentration on electrical and mechanical activity in rat portal vein. *Acta Physiol. Scand.* 99: 368–376, 1977.
64. SMITH, N. T., AND P. C. SMITH. Circulatory effects of modern inhalation anesthetic agents. In: *Handbook of Experimental Pharmacology. Vol. XXX. Modern Inhalation Anesthetics*, edited by M. B. Chenoweth. Berlin: Springer-Verlag, 1972, p. 149–241.
65. SPRAGUE, D. H., J. C. YANG, AND S. H. NGAI. Effects of isoflurane and halothane on contractility and the cyclic 3',5'-adenosine monophosphate system in the rat aorta. *Anesthesiology* 40: 162–167, 1974.
66. SU, J. Y., D. W. AMORY, M. P. SANDS, AND H. MOHRI. Effects of ether anesthesia and surface-induced hypother-

mia on regional blood flow. *Am. Heart J.* 97: 53–60, 1979.

67. TURLAPATY, P. D. M. V., AND B. M. ALTURA. Extracellular magnesium ions control calcium exchange and content of vascular smooth muscle. *Eur. J. Pharmacol.* 52: 421–423, 1978.
68. TURLAPATY, P. D. M. V., AND B. M. ALTURA. Magnesium deficiency produces spasms of coronary arteries: relationship to etiology of sudden death ischemic heart disease. *Science* 208: 198–200, 1980.
69. TURLAPATY, P. D. M. V., AND B. M. ALTURA. Magnesium ions and contractions of alloxan-diabetic vascular muscle. *Artery* 6: 375–384, 1980.
70. TURLAPATY, P. D. M. V., B. T. ALTURA, AND B. M. ALTURA. Ethanol reduces ^{45}Ca concentration in arterial and venous smooth muscle. *Experientia* 35: 639–640, 1979.
71. TURLAPATY, P. D. M. V., B. T. ALTURA, AND B. M. ALTURA. Influence of magnesium on responses of coronary arteries to vasoactive agents. *Federation Proc.* 40: 563, 1981.
72. WYLIE, W. D., AND H. C. CHURCHILL-DAVIDSON. *A Practice of Anesthesia* (3rd ed.). London: Lloyd-Luke, 1972.

11

Actions of Calcium Antagonists on Secretory Cells

RONALD P. RUBIN

Department of Pharmacology, Medical College of Virginia, Richmond, Virginia

THE MAIN OUTLINE OF EVENTS governing the role of Ca^{2+} in the secretory process has been well established (12, 29). Electrical activity in neurons induced by depolarizing current or by a high K^+ concentration is associated with an increase in ionic permeability and a downhill movement of Ca^{2+} into the neuron through ion-specific channels. This inward Ca^{2+} current couples the membrane events to neurotransmitter release. This general mechanism is shared by certain other secretory cells that have been designated as excitable because they respond to high K^+ and possess voltage-sensitive ion channels. The neurohypophysis and adrenal medulla are included among this group and respond to stimulation with an increase in membrane permeability and a transmembrane influx of Ca^{2+}.

Secretory cells also possess cellular Ca^{2+} buffers, including mitochondria, endoplasmic reticulum, and secretory granules, which sequester Ca^{2+} and thus maintain the very low intracellular concentration of free Ca^{2+} during the resting state (4, 6, 29). The intracellular organelles may also serve as the sources of the Ca^{2+} released during heightened secretory activity. As in muscle cells, the increase in the free-Ca^{2+} concentration in secretory cells to trigger the secretory response may result from either an increase in Ca^{2+} entry from the extracellular fluid or a mobilization from cellular binding sites (16, 29). In nonexcitable secretory systems, such as the exocrine pancreas, salivary glands, and adrenal cortex, cellular Ca^{2+} appears to play an important role in coupling the membrane events to the secretory response (4, 7, 29, 36). Although in such systems an increase in Ca^{2+} uptake may be demonstrable during enhanced secretory activity, it has not been determined how this increase in Ca^{2+} uptake is linked to the stimulation of secretion (36).

Knowledge about the identification and characterization of the various channels involved in Ca^{2+} entry has been advanced by the development of pharmacological agents to selectively block ion channels. This has given us a greater understanding of the nature of the events associated with stimulus-secretion coupling. This chapter compares and contrasts the effects of Ca^{2+} antagonists on the activity of excitable and nonexcitable secretory cells.

EXCITABLE SYSTEMS

Baker and colleagues (3) first demonstrated in squid giant axons two channels by which Ca^{2+} enters the axoplasm during the action potential. One is a fast Na^+ channel, which is blocked by tetrodotoxin (TTX); the second is a slow Ca^{2+} channel, which is blocked by Mg^{2+}, La^{2+}, and the organic Ca^{2+} antagonist methoxyverapamil (D 600). The properties of the slow channel suggest that it is the route by which Ca^{2+} enters the neuron to trigger neurotransmitter release. Indeed Katz and Miledi demonstrated that after blockage of the Na^+ and K^+ channels with TTX and tetraethylammonium, respectively, a local depolarizing current can still cause an enhanced neurotransmitter release from presynaptic terminals that is blocked by cations such as Mg^{2+} and La^{3+} (16, 23). More recent studies on brain synaptosomes have shown that neurotransmitter release evoked by high K^+ can be blocked by organic Ca^{2+} antagonists (24; see also chapter 18). Thus despite the reservation that D 600 may also block Na^+ channels (3, 24), we have clear evidence that neurotransmitter release is associated with augmented Ca^{2+} entry through a voltage-sensitive Ca^{2+} channel.

Adrenal Chromaffin Cells

The chromaffin cells of the adrenal medulla are derived from the neural crest and secrete catecholamines in response to the depolarizing action of acetylcholine (ACh). Early experiments on perfused cat adrenal glands demonstrated that local anesthetics such as tetracaine inhibited the secretory response evoked by ACh or high K^+ by interfering with the inward Ca^{2+} current (Fig. 1; 30). Douglas and Kanno (13) showed that the effect of tetracaine was selective because it inhibited only the Ca^{2+} current, whereas ACh caused depolarization by enhancing both Na^+ and Ca^{2+} inward currents. The clear implication of these experiments is that Na^+ and Ca^{2+} enter the medullary chromaffin cells through two operationally distinct channels. These data also indicate the Ca^{2+} channel as the critical site for Ca^{2+} entry during activation, for though it is true that TTX-sensitive action potentials can be elicited by ACh in cultured adrenal chromaffin cells (5), TTX produces only a small depression of the secretory response to ACh (17, 18). Moreover Na^+ deprivation not only does not block, but in fact enhances spontaneous and ACh-evoked catecholamine secretion (1, 29). Such studies clearly show that it is the activation of the voltage-dependent Ca^{2+} channel that is the critical event associated with enhanced adrenomedullary secretion.

Since the evidence strongly supports the concept that ACh triggers catecholamine secretion by increasing Ca^{2+} entry through the slow Ca^{2+} channel,

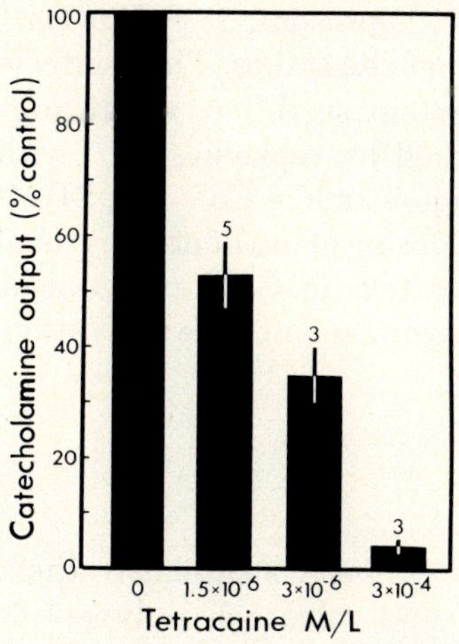

FIG. 1. Inhibitory action of tetracaine on secretory response of perfused cat adrenal gland to ACh. Responses to 2-min exposures to 2×10^{-4} M ACh are expressed as percent of averaged control catecholamine outputs for each individual gland. Control outputs were determined before tetracaine treatment and again after washout of tetracaine (n= 3–5). [From Rubin et al. (30).]

it would be expected that agents that impair Ca^{2+} influx in excitable tissues would block Ca^{2+} uptake and catecholamine release. Indeed D 600, La^{3+}, and Mg^{2+} all block catecholamine secretion in response to ACh stimulation or depolarizing concentrations of high K^+ (26, 29). The specificity of the action of D 600 is indicated by its lack of effect on catecholamine secretion induced by Na^+ deprivation, which does not depend on extracellular Ca^{2+} (1, 2).

A receptor-operated Ca^{2+} channel also apparently exists in adrenal chromaffin cells, since acetylcholine stimulates Ca^{2+}-dependent secretion in the presence of high K^+ or when depolarization is prevented by the absence of extracellular Na^+ (28, 29). The relative contributions of the two channels for Ca^{2+} entry to the physiological mechanism of ACh-induced catecholamine secretion cannot be ascertained, however, since Co^{2+}, Ni^{2+}, and Mg^{2+} (inhibitors of the voltage-sensitive Ca^{2+} channel) also interact with the receptor-operated channels (28).

Pituitary Cells

Neurosecretory cells, such as those found in the neurohypophysis, are also excitable in that stimulation produces a depolarization characteristic of neurons. They release vasopressin and oxytocin by a mechanism that requires extracellular Ca^{2+} (12, 29). Studies from several laboratories have demonstrated that during stimulation Ca^{2+} is transported across the cell membrane of these neurosecretory cells to trigger secretion (35). D 600 completely blocks stimulated release and depresses ^{45}Ca uptake; this inhibition can be reversed by increasing the Ca^{2+} concentration of the medium (35). These data are consistent with the concept that a movement of Ca^{2+} is critical to trigger secretion.

It has also been shown that normal and clonal cells of the adenohypophysis can generate action potentials in response to depolarizing currents or hormonal stimulation to effect the inward movements of both Na^+ and Ca^{2+} (25, 34). The evidence for a Ca^{2+} component is indicated by several findings: *1*) action potentials occur in Na^+-free solution or when TTX is present, *2*) Sr^+ or Ba^{2+} can substitute for Ca^{2+}, and *3*) Mg^{2+} or Co^{2+} abolishes the spikes. These effects on electrical activity parallel those on Ca^{2+}-dependent secretion, where hormone release elicited by high K^+ can be maintained by replacing Ca^{2+} with Sr^{2+} or Ba^{2+} and blocked by inorganic cation antagonists like Co^{2+} (25, 34). It thus seems likely that the adenohypophysis, an endocrine gland of nonneuronal origin, also possesses Ca^{2+} channels that are sensitive to Ca^{2+} antagonists. These channels regulate the inward Ca^{2+} current during membrane activation, which appears responsible for triggering secretion.

Pancreatic β-Cells

The β-cells of the endocrine pancreas possess properties similar to those of pituitary cells. Not only do they synthesize and release a polypeptide (insulin) by a Ca^{2+}-dependent process, but they exhibit electrical activity in response to an appropriate stimulus. Glucose-sensitive electrical activity was first reported by Matthews and associates (11, 22) in pancreatic islets (Fig. 2) and was subsequently corroborated by several other groups (see 14). The

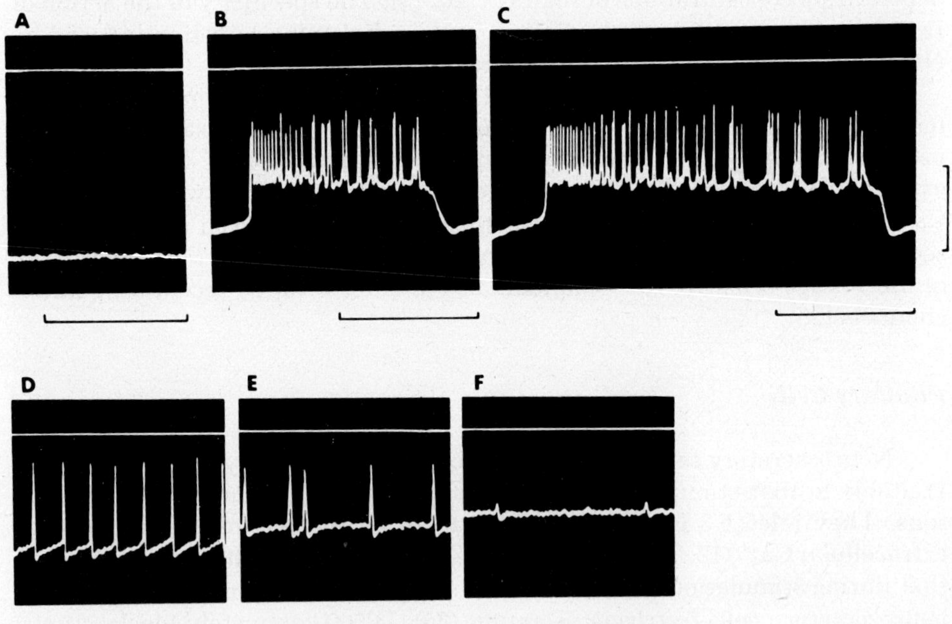

FIG. 2. Glucose-stimulated electrical activity in pancreatic islets and its inhibition by D 600. Sequential panels are derived from a single typical experiment. *A*: 2.8 mM glucose. *B–F*: 11.2 mM glucose. D 600 (5×10^{-5} M) was added immediately after *C* (at 30 min) and continued through remainder of experiment, which lasted 80 min. Subsequent removal of D 600 restored electrical activity (data not shown). Note slow waves with superimposed bursts of action potentials after glucose application (*B* and *C*) and gradual disappearance of slow waves and progressive block of burst activity after application of D 600 (*D–F*). Horizontal bars represent time interval (10 s), and vertical bar indicates 25 mV. [From Matthews and Sakamoto (22).]

spiking activity appears to be generated mainly by Ca^{2+} influx, as revealed by the findings that spike amplitude varies with the external Ca^{2+} concentration, but not with the Na^+ concentration, and that the spikes are blocked by Ca^{2+} deprivation, various divalent cations, and D 600 (Fig. 2), but not by TTX. Moreover the blockade of electrical activity can be reversed by high Ca^{2+} levels. All the Ca^{2+} antagonists also block glucose-induced insulin release (21). The ability of TTX to exert only a slight depression of glucose-induced insulin secretion further underscores the importance of the slow Ca^{2+} channel in regulating activity and also infers that only a small portion of the total Ca^{2+} enters the cell by the fast Na^+ channel. These data, taken with the fact that the Ca^{2+} antagonists depress glucose-stimulated ^{45}Ca uptake into islet tissue, point to a glucose-induced Ca^{2+} entry through a voltage-sensitive Ca^{2+} channel that is directly linked to insulin secretion.

NONEXCITABLE SYSTEMS

One should not infer that in all Ca^{2+}-dependent and secretory systems stimulation results in an enhanced influx of extracellular Ca^{2+}. There is convincing evidence that in certain secretory systems, including the exocrine pancreas, salivary glands, and the adrenal cortex, stimulation brings about a mobilization of Ca^{2+} from cellular stores such as the mitochondria and/or the endoplasmic reticulum (4, 7, 36). Moreover a significant proportion of the total cellular Ca^{2+} is bound to the plasma membrane, so any or all of these sites may be the source of "activator" Ca^{2+} in certain secretory cells, just as they appear to be the sources of activator Ca^{2+} in certain types of muscle. The actions of Ca^{2+} antagonists in such secretory systems would therefore be expected to differ from those in excitable cells (27).

Adrenocortical Cells

Rather than presenting a general survey of the secretory systems that appear to use cellular Ca^{2+}, the discussion here is confined to one such system, the adrenal cortex. My studies have dealt with the role of Ca^{2+} in the complex events associated with adrenocorticotropin (ACTH)-induced activation of steroid production and release from the intact adrenal gland and from isolated cells. A complex system operates to convert cholesterol to corticosteroid during activation of the cortex by ACTH. This pituitary peptide hormone apparently expresses its effects through the adenylate cyclase–cyclic adenosine 3′,5′-monophosphate (cyclic AMP) system and/or the phospholipase-prostaglandin system (31), since ACTH increases both cyclic AMP and prostaglandin synthesis in isolated cat adrenocortical cells, and cyclic AMP and prostaglandins [particularly prostacyclin (PGI_2)] stimulate steroid production and release when administered exogenously (8, 20). Nevertheless the physiological relevance of these putative mediators in the adrenal cortex is still unknown.

One approach used to assess the relative contribution to secretion by extra- and intracellular Ca^{2+} is to examine the effects of extracellular Ca^{2+} deprivation on the secretory response. If the removal of extracellular Ca^{2+} results in depression of the secretory response, it is usually interpreted to mean that extracellular Ca^{2+} is critical for providing the link between the membrane signal and activation of the secretory mechanism. Such experiments, however, by no means present a complete or accurate picture. Thus the removal of Ca^{2+} from the medium incubating isolated adrenocortical cells readily blocks the steroidogenic response to ACTH, whereas the effects of PGI_2 and dibutyryl cyclic AMP only partially depend on the presence of external Ca^{2+} (8, 20). These data might suggest that ACTH uses an external pool of Ca^{2+}, whereas PGI_2 and cyclic AMP mobilize a cellular pool of Ca^{2+}. Indeed the finding that a low ACTH concentration (which has absolutely no effect on steroid output in the absence of external Ca^{2+}) significantly enhances the Ca^{2+}-independent aspect of PGI_2 action confirms that Ca^{2+} mobilization by prostaglandins is performed intracellularly or at least on some pool inaccessible to ACTH (32).

If ACTH were promoting a transmembrane influx of Ca^{2+}, however, then it might be expected that D 600 and tetracaine would block the actions of ACTH. In fact, these agents are weak blockers of ACTH action, even at relatively high concentrations (Fig. 3). D 600 is equally effective in depressing PGI_2-evoked steroid output in the presence and absence of Ca^{2+}, suggesting

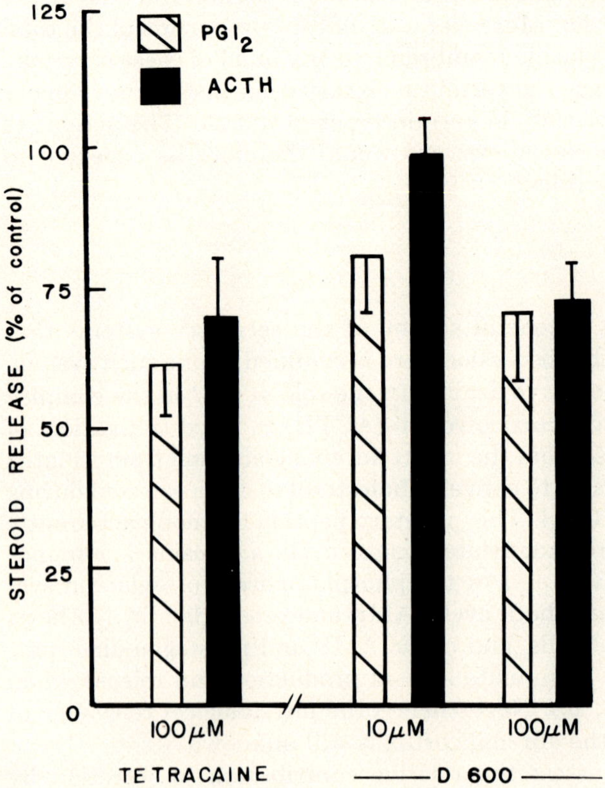

FIG. 3. Comparison of effects of tetracaine and D 600 on steroid output elicited by prostacyclin (PGI_2) and adrenocorticotropin (ACTH). Isolated cat adrenocortical cells were incubated for 60 min with or without tetracaine or D 600 in Ca^{2+}-containing medium in presence or absence of PGI_2 and ACTH. Steroid outputs for tetracaine- and D 600–treated cells are represented as percent of values derived from corresponding untreated cells.

that at these relatively high concentrations it is not blocking transmembrane Ca^{2+} flux but may be exerting a membrane-stablizing action at the surface of the cortical cell or on intracellular membrane systems. The expendability of extracellular Ca^{2+} in this system is also revealed by the weak ability of tetracaine to block ACTH- and PGI_2-induced steroid output, since I have previously shown that is quite effective in blocking catecholamine secretion from medullary chromaffin cells (Fig. 1). The apparently contradictory findings that ACTH action depends on extracellular Ca^{2+} but is only weakly blocked by D 600 may be explained either by a critical pool of Ca^{2+} in a membrane compartment readily exchangeable with the extracellular fluid or by the existence of receptor-linked Ca^{2+} channels in the cortical cell membrane insensitive to channel blockers. Since it is still undecided whether ACTH is capable of enhancing Ca^{2+} influx into the cortical cells (19), it cannot be said whether Ca^{2+} channels exist in adrenocortical cell membranes.

Effects of TMB-8

Since Ca^{2+}-channel blockers are relatively ineffective in inhibiting evoked steroid production and release, attention has been directed to the putative intracellular Ca^{2+} antagonist *N,N*-(diethylamino)octyl 3,4,5-trimethoxybenzoate (TMB-8). Chiou and Malagodi (10) originally reported that this agent was a potent inhibitor of skeletal and smooth muscle contractility and was capable of inhibiting caffeine-induced ^{45}Ca release from skeletal muscle sarcoplasmic reticulum. They suggested that TMB-8 exerted its inhibitor actions by interfering with release of Ca^{2+} from intracellular binding sites. This agent also has interesting actions on secretory cells. Thus TMB-8, in the absence of extracellular Ca^{2+}, inhibits phorbol myristate acetate–induced release of lysozyme from neutrophils (33) and the thrombin-induced platelet release reaction (9). These TMB-8–inhibited release responses could be reactivated by the addition of Ca^{2+}. Similarly TMB-8, in relatively low concentrations (3–30 μM), produces a concentration-dependent inhibition of PGI_2-evoked steroidogenesis in the absence of extracellular Ca^{2+} (Fig. 4). This inhibition is partially reversed by the addition of Ca^{2+}, supporting the hypothesis that the action of TMB-8 is directed toward antagonizing Ca^{2+}. The diminished effects of Ca^{2+} in the presence of the higher TMB-8 concentration (30 μM) lends credence to this thesis, as does the finding that the steroidogenic action of pregnenolone is not at all affected by TMB-8 (Fig. 4). Pregnenolone, a corticoid precursor, promotes a Ca^{2+}-independent activation of steroidogenesis by providing a substrate source for the enzymatic steps distal to the rate-limiting conversion of cholesterol to pregnenolone in the normal steroidogenic sequence. Thus it appears from the evidence in the literature and my own results that when cellular Ca^{2+} is directly involved in cell activation (secretion or contraction), TMB-8 exerts its primary pharmacological action by inhibiting Ca^{2+} mobilization from cellular binding sites, thereby creating a requirement for extracellular Ca^{2+}.

One may argue, however, that TMB-8 exerts its inhibitory actions by

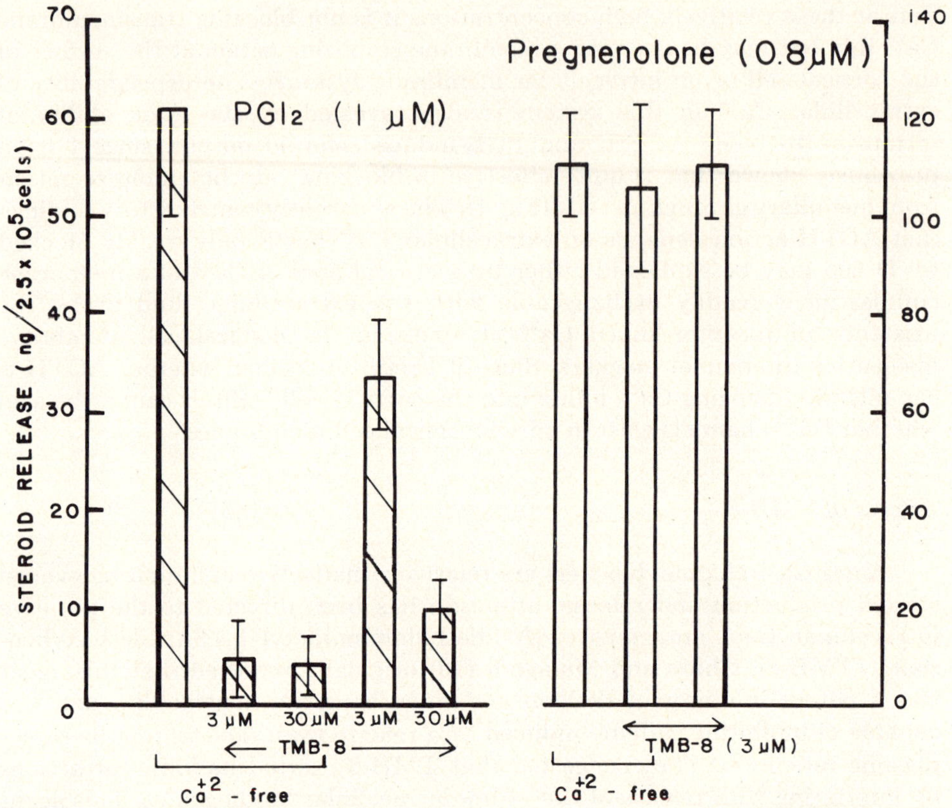

FIG. 4. Depression of Ca^{2+}-independent PGI_2-induced steroid output by TMB-8 and its reversal by Ca^{2+}. Cat adrenocortical cells were incubated for 60 min with PGI_2 in Ca^{2+}-free or Ca^{2+}-containing medium in presence or absence of TMB-8 and analyzed for steroid by competitive protein-binding assay. Lack of effect of TMB-8 on pregnenolone-induced steroid output in presence or absence of Ca^{2+} is also shown.

nonselectively stabilizing Ca^{2+} binding to other functionally important membrane sites, including those on the surface of the cell. Such an action could account for the ability of TMB-8 to block ACTH-induced steroidogenesis, which completely depends on extracellular Ca^{2+} (32). We have already seen, however, that tetracaine, a potent membrane-stabilizing agent, is a weak inhibitor of ACTH action. Moreover TMB-8 does not significantly depress the Ca^{2+}-dependent, ACTH-induced stimulation of adrenal cyclic AMP levels (32). By contrast, La^{3+}, which competes with and displaces Ca^{2+} from membrane sites on the surface of the cell, markedly inhibits the amount of cyclic AMP generated in response to ACTH (15). Thus an internal site for the action of TMB-8 is likely.

The experiments demonstrating a lack of effect of TMB-8 on the elevation of cyclic AMP levels by ACTH not only imply that the inhibitory action of TMB-8 on steroidogenesis cannot be attributed to a suppression of adenylate cyclase, but also imply that TMB-8 blocks Ca^{2+} utilization at a point beyond the synthesis of cyclic AMP. This notion is supported by the finding that the steroidogenic action of cyclic AMP obtained in the absence of external Ca^{2+} is markedly depressed by TMB-8 and can be reversed by the Ca^{2+} ionophore A23187 (Fig. 5). It is well established that A23187 can penetrate intracellularly and release Ca^{2+} from cellular binding sites.

Although there are potential pitfalls associated with ascribing selective actions to a pharmacological agent, it appears as though TMB-8 might be useful in probing and identifying one or another of the cellular pools of Ca^{2+} involved in secretion. Ongoing investigations of the effects of TMB-8 on ^{45}Ca uptake and efflux in subcellular fractions of adrenal glands should help us to more accurately identify the locus of TMB-8 action. The results obtained with TMB-8 must be viewed with caution, however, until further information on its mode of action and its selectivity becomes available.

CONCLUSIONS

The availability of Ca^{2+} antagonists has enabled us to describe in greater detail the nature of the mechanisms by which the primary signal brings about an increase in the free-Ca^{2+} concentration of secretory cells. In the neuron, as well as in neuroendocrine and endocrine glands (which have been designated

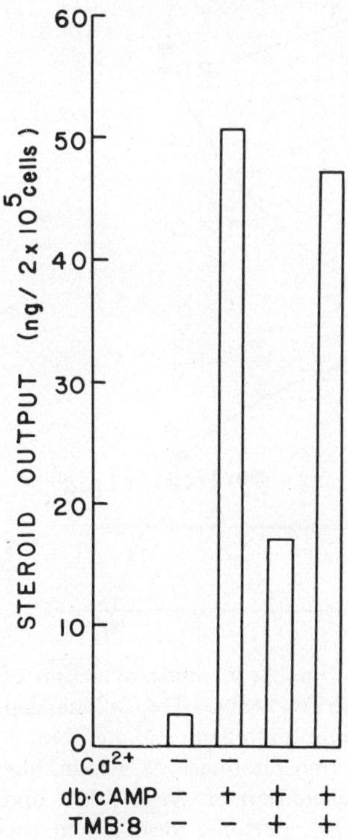

FIG. 5. Depression of dibutyryl cyclic AMP–induced steroid output by TMB-8 and its reversal by A23187. Rat adrenocortical cells were incubated for 90 min with dibutyryl cyclic AMP (1 mM) in Ca^{2+}-free medium in presence or absence of TMB-8 (30 μM) and analyzed for steroid. In certain experiments A23187 (1 μM) was also added with TMB-8.

as excitable because they can respond to high K^+ levels), one can identify voltage-sensitive Ca^{2+} channels analogous to the slow channels in muscle. These channels directly control the Ca^{2+}-gating mechanisms responsible for enhanced Ca^{2+} entry into the cell during activation of the secretory process. Although the pharmacological properties of these channels are qualitatively similar to those in muscle, the Ca^{2+} channels in excitable secretory cells are generally less sensitive to channel blockers (24). Pharmacological dissection of

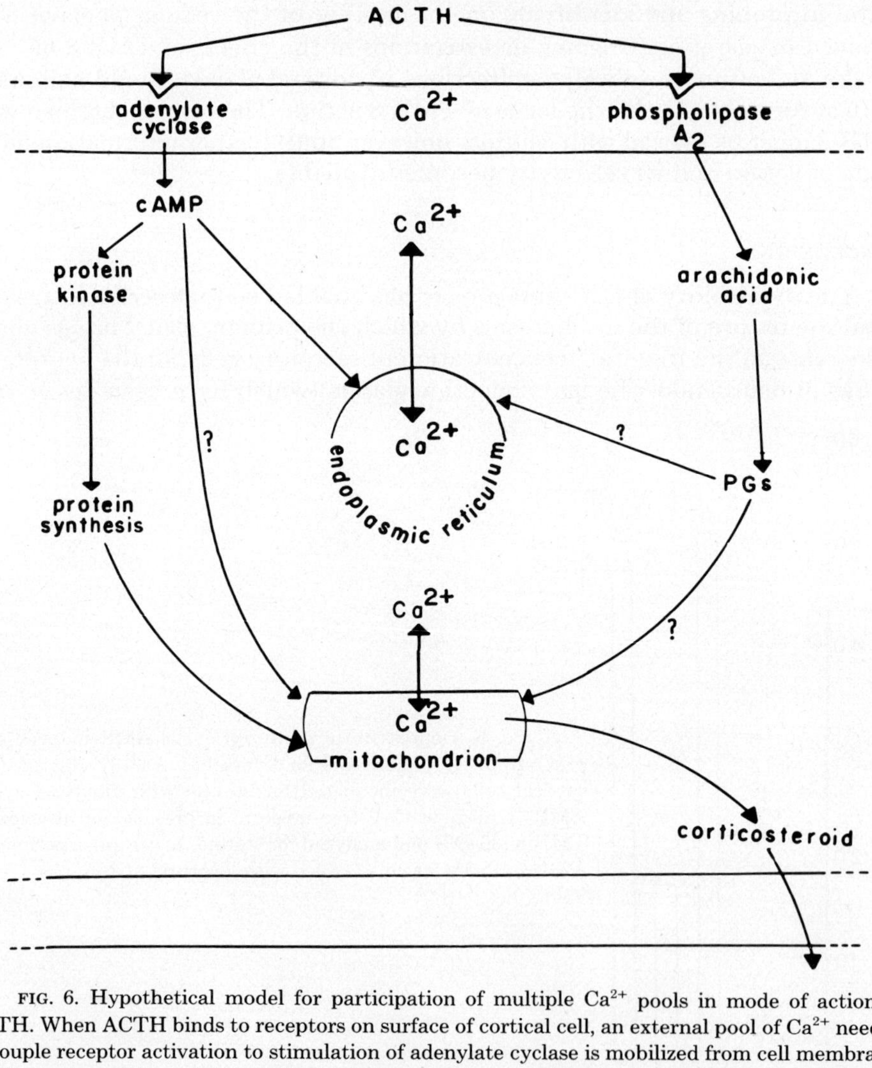

FIG. 6. Hypothetical model for participation of multiple Ca^{2+} pools in mode of action of ACTH. When ACTH binds to receptors on surface of cortical cell, an external pool of Ca^{2+} needed to couple receptor activation to stimulation of adenylate cyclase is mobilized from cell membrane. Calcium released from this superficial compartment also activates phospholipase A_2, which, like adenylate cyclase, is localized to plasma membrane. Resulting formation of cyclic AMP and prostaglandins leads to activation of an internal pool of cell Ca^{2+}, perhaps localized in the endoplasmic reticulum or mitochondria, which is sensitive to TMB-8 and is required for steroid production and release.

these channels clearly shows that the slow inward Ca^{2+} current can function independently of the fast inward Na^+ current. However, we must also consider the view that although in the experimental situation the slow Ca^{2+} channel may operate independently, under physiological conditions activation of the fast Na^+ channel may be a prelude to Ca^{2+} gating (18).

In nonexcitable secretory cells a seemingly more complex system for regulating Ca^{2+} metabolism appears to exist. Thus in the adrenal cortex, evidence reveals the presence of multiple Ca^{2+} pools required for the steroidogenic action of ACTH (Fig. 6). There is apparently a Ca^{2+} pool localized to the surface of the cell (and exchangeable with the extracellular medium) that is required for the stimulation of plasmalemmal adenylate cyclase and phospholipase A_2. This pool can be depleted by Ca^{2+} deprivation but cannot be inactivated by Ca^{2+}-channel blockers. A second Ca^{2+} pool or compartment appears to be localized to an unknown intracellular site that enables cyclic AMP and/or the prostaglandins to express their effects on steroid synthesis and/or release. Additionally receptor-operated channels may exist in both nonexcitable and excitable secretory cells. But despite the apparent complexities of secretory systems such as the adrenal cortex regarding the interactions of multiple Ca^{2+} pools, cyclic nucleotides, and prostaglandins, the continued development and judicious use of selective channel blockers and putative intracellular Ca^{2+} antagonists such as TMB-8 may help us to elucidate the nuances of each individual secretory system.

Some of the work reported in this chapter was supported by National Institutes of Health Grant AM-18066.

REFERENCES

1. AGUIRRE, J., J. FALUTZ, J. E. B. PINTO, AND J. M. TRIFARO. Effects of methoxyverapamil on the stimulation by Ca^{2+}, Sr^{2+}, and Ba^{2+} and on the inhibition by Mg^{2+} of catecholamine release from the adrenal medulla. *Br. J. Pharmacol.* 66: 591–600, 1979.
2. AGUIRRE, J., J. E. B. PINTO, AND J. M. TRIFARO. Calcium movements during the release of catecholamines from the adrenal medulla: effects of methoxyverapamil and external cations. *J. Physiol. London* 269: 371–394, 1977.
3. BAKER, P. F., H. MEVES, AND E. B. RIDGWAY. Effects of manganese and other agents on the calcium uptake that follows depolarization of squid axons. *J. Physiol. London* 231: 511–526, 1973.
4. BERRIDGE, M. J. The interaction of cyclic nucleotides and calcium in the control of cellular activity. *Adv. Cyclic Nucleotide Res.* 6: 1–98, 1975.
5. BIALES, B., M. DICHTER, AND A. TISCHLER. Electrical excitability of cultured adrenal chromaffin cells. *J. Physiol. London* 262: 743–753, 1976.
6. BRINLEY, F. J. Regulation of intracellular calcium in squid axons. *Federation Proc.* 39: 2778–2782, 1980.
7. BUTCHER, F. R., AND J. W. PUTNEY. Regulation of parotid gland function by cyclic nucleotides and calcium. *Adv. Cyclic Nucleotide Res.* 13: 215–249, 1980.
8. CARCHMAN, R. A., J. C. SHEN, S. BILGIN, AND R. P. RUBIN. Diverse effects of Ca^{2+} on the prostacyclin and corticotropin modulation of adenosine 3′: 5″-monophosphate and steroid production in normal cat and mouse tumor cells of the adrenal cortex. *Biochem. Pharmacol.* 29: 2213–2218, 1980.
9. CHARO, I. F., R. D. FEINMAN, AND T. C. DETWILER. Inhibition of platelet secretion by an antagonist of intracellular calcium. *Biochem. Biophys. Res. Commun.* 72: 1462–1467, 1976.
10. CHIOU, C. Y., AND M. H. MALAGODI. Studies on the mechanism of action of a new Ca^{2+} antagonist, 8-(N,N-diethylamino)octyl 3,4,5-trimethoxybenzoate hydrochloride in smooth and skeletal muscles. *Br. J. Pharmacol.* 53: 279–285, 1975.
11. DEAN, P. M., AND E. K. MATTHEWS. Electrical activity in pancreatic islet cells: effect of ions. *J. Physiol. London* 210: 265–275, 1970.
12. DOUGLAS, W. W. Exocytosis and the exocytosis-vesiculation sequence: with special reference to neurohypophysis, chromaffin and mast cells, calcium and calcium ionophores. In: *Secretory Mechanisms of Exocrine Glands*, edited by N. A. Thorn and O. H. Petersen. Copenhagen: Munksgaard, 1974, p. 116–136.
13. DOUGLAS, W. W., AND T. KANNO. The effect of amethocaine on acetylcholine-induced depolarization and catecholamine secretion in the adrenal chromaffin cell. *Br. J. Pharmacol.* 30: 612–619, 1976.
14. GINSBORG, B. L., AND C. R. HOUSE. Stimulus-response coupling in gland cells. *Annu. Rev. Biophys. Bioeng.* 9: 55–80, 1980.
15. HAKSAR, A., D. V. MAUDSLEY, F. G. PERON, AND E.

BEDIGIAN. Lanthanum: inhibition of ACTH-stimulated cyclic AMP and corticosterone synthesis in isolated rat adrenocortical cells. *J. Cell Biol.* 68: 142–153, 1976.
16. KATZ, B., AND R. MILEDI. Tetrodotoxin-resistant electric activity in presynpatic terminals. *J. Physiol. London* 203: 459–487, 1969.
17. KIDOKORO, Y., A. K. RITCHIE, AND S. HAGIWARA. Effect of tetrodotoxin on adrenaline secretion in the perfused rat adrenal medulla. *Nature London* 278: 63–65, 1979.
18. KIRPEKAR, S. M., AND J. C. PRAT. Release of catecholamines from perfused cat adrenal gland by veratridine. *Proc. Natl. Acad. Sci. USA* 76: 2081–2083, 1979.
19. LAYCHOCK, S. G., E. J. LANDON, AND J. G. HARDMAN. The effect of adrenocorticotropin and nucleotides on Ca^{2+} uptake in adrenal cortical microsomal vesicles. *Endocrinology* 103: 2198–2206, 1978.
20. LAYCHOCK, S. G., W. WARNER, AND R. P. RUBIN. Further studies on the mechanisms controlling prostaglandin biosynthesis in the cat adrenal cortex: the role of calcium and cyclic AMP. *Endocrinology* 100: 74–81, 1977.
21. MALAISSE, W. J., A. HERCHUELZ, G. DEVIS, G. SOMERS, A. C. BOSCHERO, J. C. HUTTON, S. KAWAZU, AND A. SENER. Regulation of calcium fluxes and their regulatory roles in pancreatic islets. *Ann. NY Acad. Sci.* 307: 562–582, 1978.
22. MATTHEWS, E. K., AND Y. SAKAMOTO. Electrical characteristics of pancreatic islet cells. *J. Physiol. London* 246: 421–437, 1975.
23. MILEDI, R. Lanthanum ions abolish the "calcium response" of nerve terminals *Nature London* 229: 410–411, 1971.
24. NACHSEN, D. A., AND M. P. BLAUSTEIN. The effects of some organic "calcium antagonists" on calcium influx in presynaptic nerve terminals. *Mol. Pharmacol.* 16: 579–586, 1979.
25. OZAWA, S., AND N. KIMURA. Membrane potential changes caused by thyrotropin-releasing hormone in the clonal GH_3 cell and their relationship to secretion of pituitary hormone. *Proc. Natl. Acad. Sci. USA* 65: 6017–6020, 1979.
26. PINTO, J. E. B., AND J. M. TRIFARO. The different effects of D-600 (methoxyverapamil) on the release of adrenal catecholamines induced by acetylcholine, high potassium, or sodium deprivation. *Br. J. Pharmacol.* 57: 127–132, 1976.
27. PUTNEY, J. W., JR. Stimulus-permeability coupling: role of calcium in the receptor regulation of membrane permeability. *Pharmacol. Rev.* 30: 209–245, 1979.
28. RITCHIE, A. K. Catecholamine secretion in a rat pheochromocytoma cell line: two pathways for calcium entry. *J. Physiol. London* 286: 541–561, 1979.
29. RUBIN, R. P. *Calcium and the Secretory Process.* New York: Plenum, 1974.
30. RUBIN, R. P., M. B. FEINSTEIN, S. D. JAANUS, AND M. PAIMRE. Inhibition of catecholamine secretion and calcium exchange in perfused cat adrenal glands by tetracaine and magnesium. *J. Pharmacol. Exp. Ther.* 155: 463–471, 1967.
31. RUBIN, R. P., AND S. G. LAYCHOCK. Prostaglandins and calcium-membrane interactions in secretory glands. *Ann. NY Acad. Sci.* 307: 377–390, 1978.
32. RUBIN, R. P., J. C. SHEN, AND S. G. LAYCHOCK. Evidence for the mobilization of cellular calcium by prostacyclin in cat adrenocortical cells: the effect of TMB-8. *Cell Calcium* 1: 391–404, 1980.
33. SMITH, R. J., AND S. S. IDEN. Phorbol myristate acetate-induced release of granule enzymes from human neutrophils: inhibition of the calcium antagonist 8-(N,N-diethylamino)octyl 3,4,5-trimethylbenzoate hydrochloride. *Biochem. Biophys. Res. Commun.* 91: 263–271, 1979.
34. TARASKEVICH, P. S., AND W. W. DOUGLAS. Action potentials occur in cells of the normal anterior pituitary gland and are stimulated by the hypophysiotropic peptide thyrotropin-releasing hormone. *Proc. Natl. Acad. Sci. USA* 74: 4064–4067, 1977.
35. THORN, N. A., J. T. RUSSELL, C. TORP-PEDERSEN, AND M. TREIMAN. Calcium and neurosecretion. *Ann. NY Acad. Sci.* 307: 618–639, 1978.
36. WILLIAMS, J. A. Regulation of pancreatic acinar cell function by intracellular calcium. *Am. J. Physiol.* 238 (*Gastrointest. Liver Physiol.* 1): G269–G279, 1980.

12

Roles of Calcium in Mediator Release From Mast Cells

TIMOTHY J. SULLIVAN AND LAURA J. BROWN

Department of Internal Medicine, University of Texas Health Science Center, Dallas, Texas

CALCIUM AND MEDIATOR RELEASE FROM MAST CELLS

Mast cells release inflammatory molecules by two distinct processes. Preformed mediators such as histamine are secreted by an exocytotic process involving fusion of the membranes surrounding cytoplasmic granules with the plasma membrane (23). Mediators such as slow-reacting substance of anaphylaxis (SRS-A), prostaglandin D_2 (PGD_2), and hydroxy derivatives of arachidonic acid are formed at the time of stimulation through cleavage of arachidonate from membrane lipids and subsequent modification by cyclooxygenase or lipoxygenase enzymes (21). These same preformed and newly formed mediators are released after mast cell stimulation by immunoglobulin E (IgE) receptor-mediated signals, lectins, kinins, and a variety of other molecules (23). A substantial body of data indicates that the initial biochemical events after each of these receptor-ligand interactions may differ but that a final common pathway of biochemical and biophysical events exists (21, 23).

Several lines of evidence support the theory that Ca^{2+} plays central roles in regulating mast cell function. Depletion of extracellular Ca^{2+} inhibits the ability of mast cells to respond to antigen, anaphylatoxins, or a variety of other stimuli (4, 9, 15, 23). Mediator release from human basophils absolutely depends on the presence of extracellular Ca^{2+} or Sr^{2+} (9, 12). The Ca^{2+} dependence of mediator release from isolated rat mast cells varies with the stimulus being examined (23) and with the pH of the incubation medium (4). Mediator release in this system becomes dependent on Ca^{2+} if cellular stores of Ca^{2+} are depleted, suggesting an important role for intracellular pools of this cation (1).

Mediator release from rat mast cells is associated with accelerated uptake of extracellular Ca^{2+} (6, 9, 17, 23). The degree of increased Ca^{2+} uptake is directly related to the magnitude of mediator release (4, 6). Accelerated efflux of Ca^{2+} has also been reported during IgE-mediated secretory responses (7). Net Ca^{2+} movements during mast cell mediator release have not been estimated definitively, but net increases in cell-associated Ca^{2+} have been suggested by murexide absorbance data (5).

Experimental introduction of Ca^{2+} into rat mast cells by ionophores (4, 5), fusion with Ca^{2+}-laden liposomes (27), or iontophoresis after cell puncture with micropipettes (8) has been reported to cause the release of inflammatory mediators. Stimulation of mast cells by these maneuvers leads to morphological and biochemical alterations that mimic at least some of the known responses to physiological signals (4, 5, 10, 23, 27). Although the physiological relevance of each of these approaches has been questioned on technical and conceptual grounds (4, 19, 24), reactions that increase cytoplasmic levels of Ca^{2+} appear capable of causing mediator release from mast cells. Recent work in many secretory systems (2, 18) and artificial membrane systems (16) would have predicted this. The release of preformed and newly formed mast cell mediators is apparently provoked by artificially increased levels of cytoplasmic Ca^{2+}, but the extent to which this functional response is mediated by the same biochemical and biophysical reactions activated by physiological stimuli is not known.

Quercetin and local anesthetic agents inhibit Ca^{2+} uptake by stimulated rat mast cells over the same concentration ranges that they inhibit mediator release (3, 9). Little other data are available to assess the functional impact of inhibiting the Ca^{2+} uptake during mediator release. Since mast cells can release substantial stores of mediators in the absence of free extracellular Ca^{2+} (4, 23), these experiments are difficult to interpret. Nevertheless the Ca^{2+}-depletion data, together with these pharmacological modulation data, suggest that movement of Ca^{2+} into stimulated mast cells is an important step in the normal mediator-release process.

As a whole these studies strongly suggest that Ca^{2+} plays an important role in the transduction of signals delivered at the surface of mast cells and basophils to intracellular mediator-release mechanisms.

CALCIUM AND MAST CELL CYCLIC AMP METABOLISM

Studies in several laboratories have demonstrated that cyclic adenosine 3′,5′-monophosphate (cyclic AMP) participates in the activation of mediator release from mast cells and that pharmacological modulation of cyclic AMP levels can cause modulation of the mediator-release process (21, 23–25). Abrupt changes in cyclic AMP levels occur after mast cell stimulation: early increases in cyclic AMP levels followed by rapid falls or simply by decreases in cyclic AMP levels are observed depending on the stimulus (13, 23–25). In general pharmacological agents that increase cyclic AMP in mast cells suppress mediator release. Several agents that decrease cyclic AMP augment mediator release.

The observation that suspension of rat mast cells in a medium containing no added Ca^{2+} is associated with a two- to sevenfold increase in whole-cell cyclic AMP levels (26) and a marked decrease in functional responsiveness (24) raises several possibilities. The increased levels of cyclic AMP associated with low levels of extracellular Ca^{2+} could contribute to the decreased secretory responsiveness noted under these conditions. Alterations in Ca^{2+} levels within mast cells might contribute to the marked changes in cyclic AMP known to occur during mediator release. These data also suggest that under physiological conditions Ca^{2+} levels may regulate cyclic AMP levels. Pharmacological modulation of cyclic AMP levels in resting cells also can modulate mast cell permeability to Ca^{2+}: increased cyclic AMP levels have been reported to decrease Ca^{2+} permeability (4). Antigen-IgE–activated mast cells and basophils display rapid increases in cyclic AMP followed by decreases. During activation increased cyclic AMP levels may increase Ca^{2+} permeability (4). These studies indicate that mast cell cyclic AMP regulatory systems and Ca^{2+} levels are closely linked.

Recent studies of cyclic AMP levels in isolated rat mast cells stimulated by the Ca^{2+} ionophore A23187 have extended our insight into the relationships between Ca^{2+} and cyclic AMP systems (13). The ionophore caused a dose-related noncytotoxic release of mast cell histamine at concentrations ranging from 0.05 to 0.5 μg/ml (Fig. 1). At concentrations of A23187 above 0.5 μg/ml, mast cell viability was decreased. Whole-cell cyclic AMP levels measured 10 min after challenge with A23187 decreased in a dose-related manner over the same concentration range (Fig. 1). The rate of cyclic AMP decrease following A23187 challenge is contrasted with the rate of histamine release in Figure 2. The onsets and periods of maximum rates of mediator release and decreases in cyclic AMP are quite similar. When mast cells were suspended in a medium containing no added Ca^{2+} and then treated with A23187, basal cyclic AMP levels were elevated and did not decrease significantly after ionophore stimulation (Fig. 3). Little mediator release occurred under these conditions (23).

These observations agree with the notion that whole mast cell levels of cyclic AMP are inversely related to the level of Ca^{2+} in critical areas of the cell. Mediator release provoked by A23187 is closely linked to decreased levels

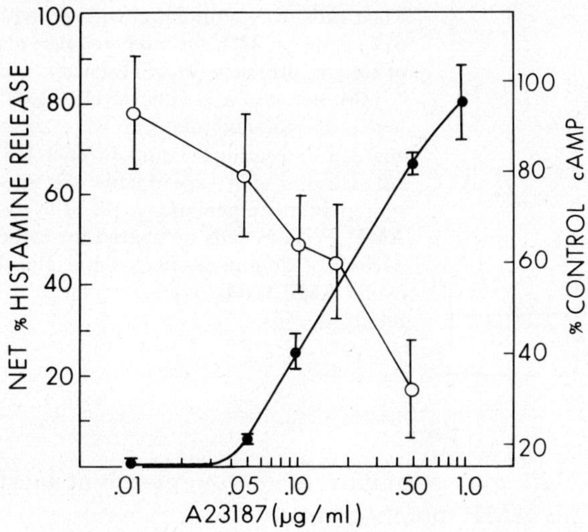

FIG. 1. Effects of Ca^{2+} ionophore A23187 on release of histamine from isolated rat mast cells and on mast cell cyclic AMP levels. Mast cells were challenged with indicated concentrations of A23187 at 37°C for 10 min (13, 26). Release of histamine (●) and total incubation mixture cyclic AMP levels (○) were measured as described elsewhere (13, 26). Data are mean percents ± SE of results of 3 experiments. Spontaneous release of histamine was 4.2 ± 1.3%. This value was subtracted from each experimental result to derive the net histamine release attributable to A23187. Basal cyclic AMP levels averaged 16.8 ± 4.1 pmol/10^6 cells.

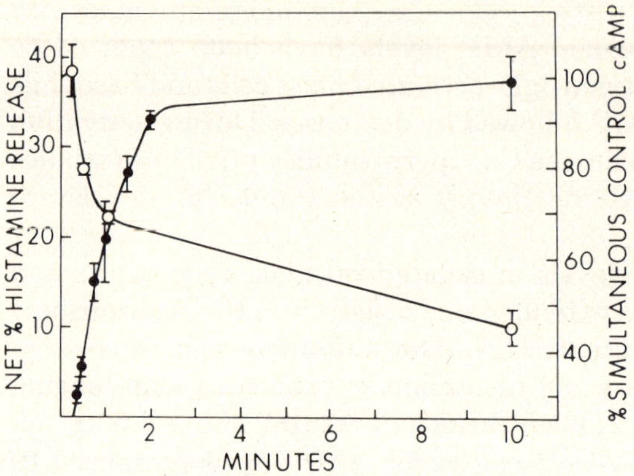

FIG. 2. Kinetic analysis of effects of A23187 on mast cell mediator release and cyclic AMP levels. Mast cells were incubated with A23187 (0.3 μg/ml) or medium alone, as a control, for the indicated periods of time. At the end of incubation, tubes to be assayed for cyclic AMP were plunged into liquid nitrogen and processed for radioimmunoassay as described elsewhere (26). The cyclic AMP data (○) are presented as percents of simultaneous control to compensate for the small rises in cyclic AMP known to occur after agitation (26). Interpretation of results was not significantly different if zero-time control cyclic AMP was used as a reference. Histamine release (●) was measured as described elsewhere (13). Data are mean percents ± SE of results of triplicate measurements in 1 experiment. Very similar results were obtained in 2 additional experiments. Spontaneous release of histamine was 5.2 ± 1.6%. Basal cyclic AMP level averaged 20.3 ± 2.1 pmol/10^6 cells.

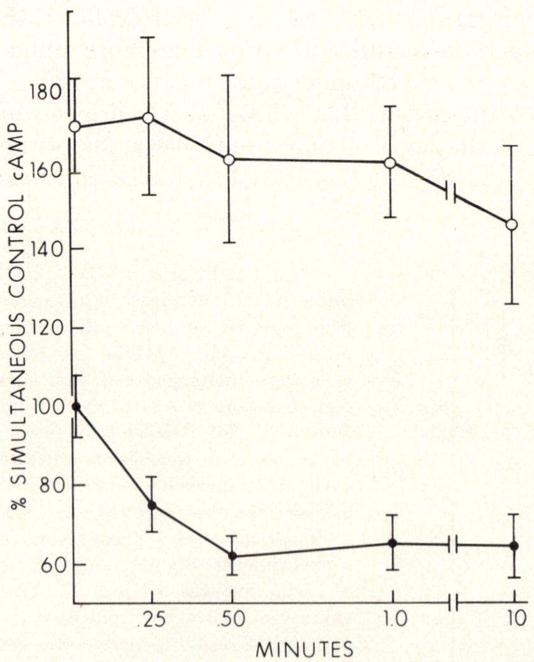

FIG. 3. Effects of A23187 on mast cell cyclic AMP levels in presence or absence of added extracellular Ca^{2+}. Mast cells were stimulated with A23187 (0.3 μg/ml) at 37°C for indicated periods of time in presence (●) or absence (○) of 1 mM added Ca^{2+} (13). Cyclic AMP levels in reaction mixtures were measured after freezing in liquid nitrogen (13, 26). Results of 3 experiments are presented as mean percents ± SE of cyclic AMP levels in cells incubated for same periods of time in medium alone. Basal cyclic AMP levels averaged 14.0 ± 5.3 pmol/10^6 cells.

of cyclic AMP, suggesting that Ca^{2+} may stimulate mediator release at least partly through alterations in cyclic AMP metabolism.

Stimulation of mast cell mediator release by the cationic polymer compound 48/80 also is associated with abruptly decreased levels of cyclic AMP

(Fig. 4; 24). When mast cells were suspended in medium containing no added Ca^{2+} and then challenged with 48/80, a more complex result emerged. Mediator release was reduced by approximately 25% (23). A fall in cyclic AMP did occur, but the basal levels were elevated and the absolute nadir level was not as low as in cells incubated in the presence of 1 mM Ca^{2+}. Thus the fall in cyclic AMP after challenge with 48/80 did not appear to be mediated primarily by an influx of extracellular Ca^{2+}. Although 48/80 might well mobilize Ca^{2+} from intracellular stores (4), the mechanism of the fall in cyclic AMP currently is unknown.

Mediator release induced by cross-linking IgE receptors or by the lectin concanavalin A is associated with an early rise in mast cell cyclic AMP levels (25). After a peak increase 20 s after stimulation, cyclic AMP levels fall rapidly and mediator release begins approximately 30 s after stimulation. To assess the role of a Ca^{2+} influx on these biochemical and functional parameters, responses to concanavalin A in complete medium were compared with responses in medium containing no added Ca^{2+} (Fig. 5). In the absence of added Ca^{2+}, resting cell cyclic AMP levels were 204% (± 24% SE) of the value found in cells incubated in the presence of 1 mM Ca^{2+} (14.2 ± 3.5 pmol/10^6 cells based on three experiments). Stimulation with concanavalin A (100 μg/ml) in the absence of added Ca^{2+} caused the level of cyclic AMP to increase within 20 s to 356% (± 26% SE) of the value recorded in cells incubated with 1 mM Ca^{2+} but no concanavalin A; cyclic AMP levels returned to control (concanavalin A, no added Ca^{2+}) values (230% ± 12% SE) within 1 min (in the same three

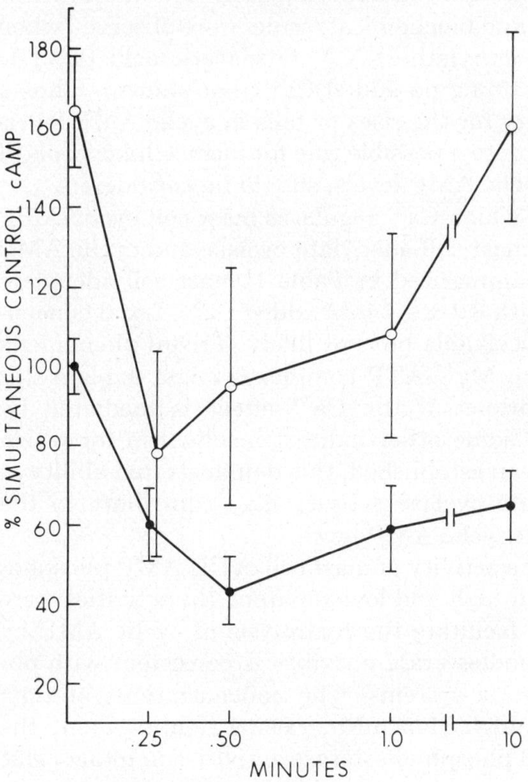

FIG. 4. Effect of compound 48/80 on mast cell cyclic AMP levels in presence or absence of added extracellular Ca^{2+}. Mast cells were incubated with 48/80 (1 μg/ml) or medium alone as a control in presence (●) or absence (○) of 1 mM added Ca^{2+}. Data are mean percents ± SE of results of 3 experiments. Basal cyclic AMP levels averaged 25.2 ± 6.8 pmol/10^6 cells.

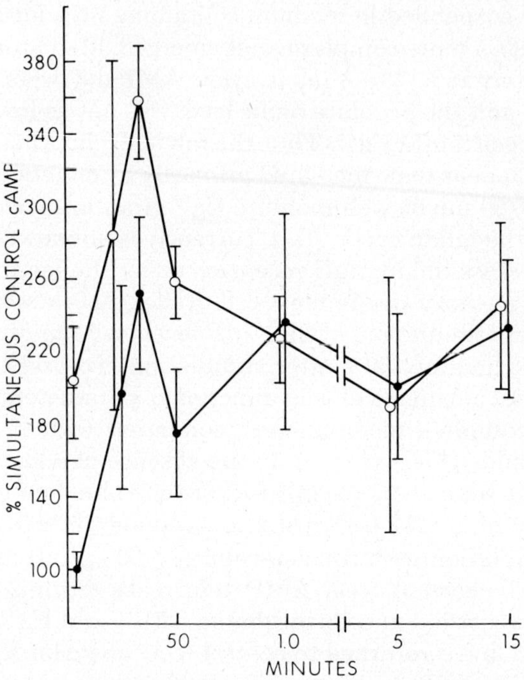

FIG. 5. Effect of concanavalin A on mast cell cyclic AMP levels in presence or absence of added extracellular Ca^{2+}. Mast cells were stimulated with concanavalin A (100 μg/ml) or medium alone in presence (●) or absence (○) of 1 mM added Ca^{2+}. Data are mean percents ± SE of results of 3 experiments.

experiments). Mediator release was reduced approximately 50% under these conditions (23). Similar functional and biochemical results were observed when 2 mM ethylene glycol-bis(β-aminoethylether)-*N,N′*-tetraacetic acid (EGTA) was included in the medium containing no added Ca^{2+} (not shown). Thus a Ca^{2+} influx does not seem necessary for the rises or falls in cyclic AMP levels induced by concanavalin A. Once more a possible role for intracellular pools of Ca^{2+}, at least in the decrease in cyclic AMP levels, should be considered.

To probe the mechanisms by which Ca^{2+} regulates mast cell cyclic AMP levels, we examined its effects on mast cell adenylate cyclase and cyclic AMP phosphodiesterase activities. As summarized in Table 1, mast cell adenylate cyclase activity is inhibited 50% with 8.2 ± 2.7 mM added Ca^{2+}. Local concentrations of this magnitude are conceivable but not likely. Trivial phenomena such as displacement of Mg^{2+} from Mg^{2+}-ATP complexes might explain this effect of Ca^{2+} at high concentrations. If the Ca^{2+} effect is mediated by calmodulin-driven mechanisms or some other indirect mechanism for which optimal conditions have not yet been established, this demonstrated ability of Ca^{2+} to inhibit mast cell adenylate cyclase activity may contribute to the ability of Ca^{2+} to reduce whole-cell cyclic AMP levels.

Calcium modestly increases the activity of mast cell cyclic AMP phosphodiesterase enzymes (Table 2). Both high and low apparent K_m activities were increased. The ability of Mg^{2+} to facilitate the hydrolysis of cyclic AMP by mast cell cyclic nucleotide phosphodiesterase enzymes is consistent with observations in many other eukaryotic systems. The concentrations of Ca^{2+} required to cause an effect were high (1–5 mM), raising doubts about the ability of Ca^{2+} to regulate mast cell phosphodiesterase activities in intact cells.

TABLE 1. *Effect of Ca^{2+} on rat mast cell adenylate cyclase activity*

Concentration of Added Ca^{2+}, mM*	Adenylate Cyclase Activity, $pmol \cdot min^{-1} \cdot 10^{-6}$ cells†
0	4.2 ± 0.2
0.1	3.9 ± 0.2
1.0	3.5 ± 0.2
10	1.0 ± 0.1

* Mast cells (2 × 10^5) were resuspended in 0.10 ml of ice-cold medium containing 50 mM tris(hydroxymethyl)aminomethane (Tris), 2 mM dithiothreitol, 0.2 mM Mg^{2+}, 50 mM theophylline, bovine serum albumin (1 mg/ml), 1 mM ATP, and the indicated concentrations of Ca^{2+} (pH 7.6). Cells were disrupted by sonication and kept at 4°C. † Adenylate cyclase activity was assessed by measuring the appearance of cyclic AMP during 1 min of incubation at 37°C. Reactions were terminated by heating at 100°C for 3 min. Cyclic AMP levels were determined using a radioimmunoassay method (20, 21). Values are means ± SE of results of 3 experiments.

TABLE 2. *Effects of Ca^{2+} and Mg^{2+} on rat mast cell low apparent K_m cyclic AMP phosphodiesterase activity*

Concentrations of Added Ca^{2+} and Mg^{2+}, mM*	Cyclic cAMP Hydrolysis, $pmol \cdot min^{-1} \cdot 10^{-6}$ cells†
0	0.15 ± 0.03
Mg^{2+}, 1	3.22 ± 0.50
Mg^{2+}, 5	4.50 ± 0.71
Ca^{2+}, 1	0.16 ± 0.11
Ca^{2+}, 5	0.20 ± 0.02
Mg^{2+}, 5; Ca^{2+}, 5	6.08 ± 0.78

* Cyclic AMP phosphodiesterase activity was determined at 5 μM and 100 μM cyclic AMP according to methods described elsewhere (22). † Data from 1 experiment are presented as means ± SE of triplicate determinations. Two additional experiments provided very similar results. High apparent K_m phosphodiesterase activity also was modestly augmented by these same concentrations of Ca^{2+}.

If the observed effect of Ca^{2+} is indirect, conditions may not have been optimal to demonstrate effects at lower levels of Ca^{2+}.

Together these studies provide evidence that the level of cyclic AMP in a cell is inversely related to the level of Ca^{2+} and that increased Ca^{2+} levels may provoke mediator release, at least partly, through alterations in cyclic AMP–regulated processes by reducing cyclic AMP levels. Calcium inhibits adenylate cyclase activity and increases cyclic AMP phosphodiesterase activities in mast cell homogenates, but the levels of Ca^{2+} required to achieve these effects in cell homogenates exceed 1 mM. Since the bulk of mast cell cyclic AMP is bound to receptors on intracellular membranes (23), Ca^{2+} may lower cyclic AMP levels by promoting release of cyclic AMP, thereby permitting hydrolysis of molecules previously protected from phosphodiesterase enzymes. The mechanisms that mediate Ca^{2+} effects on mast cell cyclic AMP levels are unclear and warrant detailed study.

CALCIUM AND MAST CELL LIPID METABOLISM

Stimulation of mediator release from rat mast cells is associated with two- to fourfold increases in whole-cell levels of 1,2-diacylglycerol (DAG) (11). This molecule is directly fusogenic (i.e., it causes or promotes membrane fusion) and also appears to be the principal source of arachidonic acid for the generation of newly formed mediators in mast cells (11). Phospholipase C (the enzyme that causes formation of DAG from membrane phospholipids), DAG kinase (the enzyme that causes conversion of DAG to phosphatidic acid), and DAG lipase (the enzyme activity that causes conversion of DAG to monoacylglycerol—an intense fusogen—and to free fatty acid) are apparently the principal determinants of the DAG levels and the degree of activity of the DAG-lipase pathway (11). This pathway is thought to play an important role in the generation of fusogens (and therefore in the release of preformed mediators) and in the liberation of arachidonic acid (and thus in the generation of newly formed mediators).

We have examined the impact of Ca^{2+} on rat mast cell DAG-kinase and DAG-lipase activities. As shown in Table 3, DAG-kinase activity is inhibited 50% with 0.12 ± 0.07 mM added Ca^{2+}. Increases in Ca^{2+} levels in the region of DAG kinases could reach levels sufficient to inhibit the enzyme. Increased DAG levels in stimulated mast cells may result partly from Ca^{2+}-mediated inhibition of DAG-kinase conversion of DAG to phosphatidic acid. This interaction deserves detailed investigation.

The DAG-lipase activity increases with increasing Ca^{2+} concentrations (Table 4). A twofold increase requires a concentration of Ca^{2+} (~ 5 mM). Optimal conditions for stimulation may not have been identified if the effect is mediated by calmodulin or other indirect mechanisms. The influence of Ca^{2+} on DAG lipase deserves careful study.

All the phospholipase C enzymes characterized so far, with the exception of rat liver phospholipase C, have been found to be Ca^{2+} dependent (14). Preliminary studies of the rat mast cell phospholipase C enzymes by Stechschulte (personal communication) and by Sullivan and Brown (unpublished data) indicate that this activity is Ca^{2+} dependent. Thus increased levels of Ca^{2+} would be expected to increase the level of DAG in mast cells by accelerating synthesis of DAG and by inhibiting conversion of DAG to phosphatidic acid by DAG kinase. The DAG-lipase pathway activity would be expected to increase because of increased substrate levels and because of increased DAG-lipase activity in the presence of Ca^{2+}.

TABLE 3. *Effect of Ca^{2+} on rat mast cell DAG-kinase activity*

Concentration of Added Ca^{2+}, mM*	DAG-Kinase Activity, pmol·min^{-1}·10^{-6} cells†
0	1,528 ± 203
2.5	305 ± 49
0.25	733 ± 58
0.025	1,251 ± 87

* The DAG-kinase activity was assessed by measuring the ability of mast cell enzymes to convert 1,2-diolein to phosphatidic acid. Mast cells (2×10^5) were added, suspended in a medium containing 0.1 mM diolein, 2 mM Mg^{2+}, 100 mM Na^+, 1 mM 1,4-piperazine-bis(ethanesulfonic acid) (PIPES), and bovine serum albumin (1 mg/ml) at pH 7.4. All reagents were prewarmed to 37°C, and 10 μCi of γ-$^{32}PO_4$ ATP in 1 mM unlabeled ATP were added. After a 3-min incubation, reactions were stopped and newly formed phosphatidic acid was measured according to methods described elsewhere (11). † Values are means ± SE of results of 3 experiments.

TABLE 4. *Effect of Ca^{2+} on rat mast cell DAG-lipase activity*

Concentration of Added Ca^{2+}, mM*	DAG-Lipase Activity, $pmol \cdot min^{-1} \cdot 10^{-6}$ cells†
0	948 ± 58
0.1	1,213 ± 102
1.0	1,687 ± 131
5.0	1,868 ± 183

* The DAG-lipase activity was determined by methods described elsewhere (11). † Values are means ± SE of results of 1 experiment. Two additional experiments provided very similar results.

SUMMARY AND CONCLUSIONS

Several independent lines of evidence indicate that increased levels of Ca^{2+} in mast cells can trigger the release of inflammatory mediators. Although less well documented, Ca^{2+} levels probably do rise in mast cells stimulated by physiologically relevant signals. Calcium levels in the extracellular and intracellular compartments appear to exert control over cyclic AMP metabolism: for example, increased Ca^{2+} levels decrease cyclic AMP, perhaps through liberation of cyclic AMP from membrane binding sites permitting hydrolysis, adenylate cyclase inhibition, and cyclic AMP–phosphodiesterase activation. These regulatory systems appear closely linked in mast cells.

Studies in homogenates of mast cells suggest that increased levels of Ca^{2+}, by mechanisms not yet fully defined, increase the level of DAG and the passage of DAG down the DAG-lipase pathway. The full spectrum of mast cell reactions regulated by Ca^{2+} and cyclic AMP is just beginning to emerge. Much work remains to be done to clarify the roles of Ca^{2+} in mast cell function, but the available information indicates that Ca^{2+} is likely to be a pivotal molecule in the mediator-release process.

We are attracted to the hypothesis that stimulation of mast cell surface receptors leads to increased levels of Ca^{2+} in critical sites, resulting in accelerated conversion of phospholipid to DAG. Inhibition of enzymes that convert DAG to phospholipids, and increased DAG-lipase activity, at least partly mediated by increased levels of Ca^{2+}, could lead to membrane fusion and liberation of metabolites of arachidonic acid. The availability of a testable hypothesis and the growing number of Ca^{2+} antagonists should permit rapid progress in our pursuit of understanding of mast cell mediator release. Potent new pharmacological methods to regulate this process may emerge from resolution of the roles of Ca^{2+} in this system.

REFERENCES

1. COCHRANE, D. E., AND W. W. DOUGLAS. Calcium-induced extrusion of secretory granules (exocytosis) in mast cells exposed to 48/80 or the ionophores A23187 and X537A. *Proc. Natl. Acad. Sci. USA* 71: 408–412, 1974.
2. DOUGLAS, W. W. Exocytosis and the exocytosis-vesiculation sequence: with special reference to neurohypophysis, chromaffin and mast cells, calcium and calcium ionophores. In: *Secretory Mechanisms of Exocrine Glands*, edited by N. A. Thorn and O. H. Petersen. Copenhagen: Munksgaard, 1974, p. 116–136.
3. FEWTRELL, C. M. S., AND B. D. GOMPERTS. Quercetin: a novel inhibitor of Ca^{2+} influx and exocytosis in rat peritoneal mast cells. *Biochim. Biophys. Acta* 469: 52–60, 1977.

4. FOREMAN, J. C. Calcium and secretion from mast cells and leukocytes. In: *Biochemistry of the Acute Allergic Reaction*, edited by E. L. Becker and K. F. Austen. Santa Barbara, CA: Kroc Found., in press.
5. FOREMAN, J. C., J. L. MONGAR, AND B. D. GOMPERTS. Calcium ionophores and movement of calcium ions following the physiological stimulus to a secretory process. *Nature London* 245: 249–251, 1973.
6. ISHIZAKA, T., J. C. FOREMAN, A. R. STERK, AND K. ISHAZAKA. Induction of calcium flux across the rat mast cell membrane by bridging IgE receptors. *Proc. Natl. Acad. Sci. USA* 76: 5858–5862, 1979.
7. KALINER, M., AND K. F. AUSTEN. Cyclic AMP, ATP, and reversed anaphylactic histamine release from rat mast cells. *J. Immunol.* 112: 664–674, 1974.
8. KANNO, T., D. E. COCHRANE, AND W. W. DOUGLAS. Exocytosis (secretory granule extension) induced by injection of calcium into mast cells. *Can. J. Physiol. Pharmacol.* 51: 1001–1004, 1973.
9. KAZIMIERCZAK, W., AND B. DIAMANT. Mechanisms of histamine release in anaphylactic and anaphylactoid reactions. *Prog. Allergy* 24: 295–365, 1978.
10. KENNERLY, D. A., C. SECOSAN, C. W. PARKER, AND T. J. SULLIVAN. Modulation of stimulated phospholipid metabolism in mast cells by pharmacologic agents that increase cyclic 3′,5′-adenosine monophosphate levels. *J. Immunol.* 123: 1519–1524, 1979.
11. KENNERLY, D. A., T. J. SULLIVAN, P. SYLWESTER, AND C. W. PARKER. Diacylglycerol metabolism in mast cells: a potential role in membrane fusion and arachidonic acid release. *J. Exp. Med.* 150: 109–1044, 1979.
12. LICHTENSTEIN, L. M. The immediate allergic response: in vitro separation of antigen activation, decay and histamine release. *J. Immunol.* 107: 1122–1130, 1971.
13. MARQUARDT, D. L., C. W. PARKER, AND T. J. SULLIVAN. Potentiation of mast cell mediator release by adenosine. *J. Immunol.* 120: 871–878, 1978.
14. MATSUZAWA, Y., AND K. Y. HOSTETLER. Properties of phospholipase C isolated from rat liver lysosomes. *J. Biol. Chem.* 255: 646–652, 1980.
15. MONGAR, J. L., AND H. O. SCHILD. The effect of calcium and pH on the anaphylactic reaction. *J. Physiol. London* 140: 272–284, 1958.
16. PAPAHADJOPOULOS, D., G. POSTE, AND W. J. VAIL. Studies on membrane fusion with natural and model membranes. In: *Methods in Membrane Biology*, edited by Edward V. Korn. New York: Plenum, 1979, vol. 10, p. 1–122.
17. RANADIVE, N. S., AND D. NUSRAT. Movement of calcium ions and release of histamine from rat mast cells. *Int. Arch. Allergy Appl. Immunol.* 61: 9–18, 1980.
18. RUBIN, R. P. *Calcium and the Secretory Process*. New York: Plenum, 1974.
19. SIRAGANIAN, R. P., A. KULCZYCKI, G. MENDOZA, AND H. METZGER. Ionophore A-23187 induced histamine release from rat mast cells and rat basophil leukemia (RBL-1) cells. *J. Immunol.* 115: 1599–1602, 1975.
20. STEINER, A. L., D. M. KIPNIS, R. UTIGER, AND C. PARKER. Radioimmunoassay for the measurement of adenosine 3′5′-cyclic phosphate. *Proc. Natl. Acad. Sci. USA* 64: 367–373, 1969.
21. SULLIVAN, T. J., AND A. KULCZYCKI, JR. Immediate hypersensitivity responses. In: *Clinical Immunology*, edited by C. W. Parker. Philadelphia: Saunders, 1980, p. 115–142.
22. SULLIVAN, T. J., AND C. W. PARKER. Cyclic AMP phosphodiesterase activation by the histamine releasing agent compound 48-80. *Biochem. Biophys. Res. Commun.* 55: 1334–1339, 1973.
23. SULLIVAN, T. J., AND C. W. PARKER. Pharmacologic modulation of inflammatory mediator release by rat mast cells. *Am. J. Pathol.* 85: 437–463, 1976.
24. SULLIVAN, T. J., K. L. PARKER, S. A. EISEN, AND C. W. PARKER. Modulation of cyclic AMP in purified rat mast cells. II. Studies on the relationship between intracellular cyclic AMP concentrations and histamine release. *J. Immunol.* 114: 1480–1485, 1975.
25. SULLIVAN, T. J., K. L. PARKER, A, KULCZYCKI, AND C. W. PARKER. Modulation of cyclic AMP in purified rat mast cells. III. Studies on the effects of anti-IgE and concanavalin A on cyclic AMP concentrations during histamine release. *J. Immunol.* 117: 713–716, 1976.
26. SULLIVAN, T. J., K. L. PARKER, W. STENSON, AND C. W. PARKER. Modulation of cyclic AMP in purified rat mast cells. I. Responses to pharmacologic, metabolic, and physical stimuli. *J. Immunol.* 114: 1473–1479, 1975.
27. THEOHARIDES, T. C., AND W. W. DOUGLAS. Secretion in mast cells induced by calcium entrapped within phospholipid vesicles. *Science* 202: 1143–1145, 1978.

13

Calcium Antagonists and Calcium-Gating Mechanisms in the Exocrine Glands

JAMES W. PUTNEY, JR.

Department of Pharmacology, Medical College of Virginia, Virginia Commonwealth University, Richmond, Virginia

THIS CHAPTER DEALS WITH THE role of Ca^{2+} and the action of Ca^{2+} antagonists in the exocrine glands. It focuses mainly on recent research on the rat parotid and lacrimal glands studied as model exocrine systems in my laboratory.

Douglas and Poisner (4) first demonstrated that Ca^{2+} was required for maximum salivary secretion. Accordingly (as is subsequently discussed) appropriately chosen Ca^{2+} antagonists block the responses of cells of the exocrine glands to neurotransmitters or drugs mimicking the actions of neurotransmitters. To understand the actions of the Ca^{2+} antagonists and their relation to Ca^{2+} mechanisms, three somewhat related questions are addressed. First, what is the role of Ca^{2+} in regulating cellular responses in the exocrine glands? Second, what kinds of substances act as Ca^{2+} antagonists in these cells? Third, can the biochemical properties of these antagonists provide information about the molecular mechanism of Ca^{2+} channels in exocrine (and perhaps other) cells?

CALCIUM AND EXOCRINE CELLULAR FUNCTION

Recall that exocrine glands perform two rather distinct secretory functions. In the rat parotid gland, the secretion of macromolecules by exocytosis is mediated by cyclic adenosine 3′,5′-monophosphate (cyclic AMP; 3, 26). In the lacrimal gland and exocrine pancreas, this function is apparently mediated by Ca^{2+} (21, 29).

Neurotransmitters and hormones also activate the secretion of water by the exocrine glands. The mechanisms involved are quite distinct from those involved in exocytosis. The cellular events necessary to activate transepithelial flux of water necessarily involve alterations in membrane permeability and transport of osmotically important cations such as Na^+, K^+, Cl^-, or HCO_3^- (1). In the salivary glands, lacrimal gland, and exocrine pancreas, Ca^{2+} is apparently the second messenger for this response (15). Accordingly, by analogy with the role of Ca^{2+} in excitation-contraction coupling and stimulus-secretion coupling, the term *stimulus-permeability coupling* has been used to describe the events linking receptor activation to altered ion fluxes in these cells (15).

The effects of receptor activation on membrane permeability to K^+ in exocrine tissues can be assessed in experiments such as those summarized in Figure 1. Unidirectional efflux of $^{86}Rb^+$, a marker for K^+ flux, was monitored with thin slices of rat lacrimal gland incubated in vitro (19). In the presence of Ca^{2+}, the addition of carbachol to the Ringer's solution from 20 to 40 min causes a precipitous and sustained increase in the efflux of $^{86}Rb^+$, which probably indicates an increased membrane permeability to K^+. The response is completely blocked by atropine. When the experiment is done in media that lacks added Ca^{2+} and to which 10^{-4} M ethylene glycol-bis(β-aminoethylether)-*N,N′*-tetraacetic acid (EGTA) has been added, the initial increase in K^+ permeability persists (with some attenuation), but the response is transient; the rate coefficient returns to control levels in 6–8 min. Restoration of the elevated efflux rate on reintroduction of Ca^{2+} to the medium illustrates that the effect of Ca^{2+} omission is reversible. This protocol—introduction of agonist in the absence of Ca^{2+}, followed by restoration of Ca^{2+} to physiological levels—serves as a convenient assay of Ca^{2+}-dependent and Ca^{2+}-independent phases of the $^{86}Rb^+$-efflux response.

There is considerable evidence that the Ca^{2+}-dependent phase of the K^+-permeability response represents a receptor activation of surface membrane Ca^{2+}-influx sites or Ca^{2+} channels. Presumably the elevation in intracellular Ca^{2+} somehow activates surface membrane channels for K^+ (15). Thus the divalent cationophore A23187 mimics the effects of receptor activation in causing the second, but not the first phase of $^{86}Rb^+$ efflux (6, 14, 18). Addition-

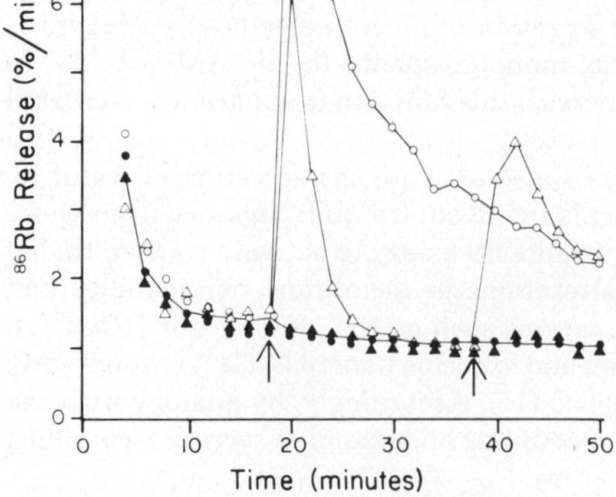

FIG. 1. Stimulation of $^{86}Rb^+$ efflux from rat lacrimal slices by carbachol and Ca^{2+}. Slices were loaded with $^{86}Rb^+$ (19), and unidirectional efflux was monitored in Ringer's solution containing 3.0 mM $CaCl_2$, unless stated otherwise. ● Control; ○ 10^{-5} M carbachol from 18 to 50 min; ▲ 10^{-4} M atropine from 12 to 50 min, 10^{-5} M carbachol from 18 to 50 min; △ no added Ca^{2+}, but 10^{-4} M EGTA added from 0 to 50 min, 10^{-5} M carbachol from 18 to 50 min, and 3.1 mM Ca^{2+} for 38–50 min. Arrows indicate 18 min and 38 min after initiation of washout. Each point is mean from 4 experiments. [From Putney et al. (19).]

ally agents that stimulate K^+ efflux also augment Ca^{2+} influx as measured with ^{45}Ca (10, 13, 20).

In the parotid gland, the Ca^{2+}-dependent K^+ efflux can be activated through muscarinic, α-adrenergic, or substance P receptors, and in the lacrimal gland by muscarinic and α-adrenergic receptors (15). When agonists acting on different receptors were tested in combination, they did not summate in causing the second phase of $^{86}Rb^+$ efflux despite the fact that the external Ca^{2+} concentration was submaximal (6, 14). The conclusion, therefore, is that these different receptor mechanisms can activate the same Ca^{2+}-influx sites (15).

ACTIONS OF CALCIUM ANTAGONISTS

The effects of a number of putative Ca^{2+} antagonists on the receptor-activated $^{86}Rb^+$ efflux from parotid slices were determined (6). Addition of carbachol and Ca^{2+} sequentially to the Ringer's solution (see Fig. 1) permitted the effects of these substances on the transient (Ca^{2+}-independent) and sustained (Ca^{2+}-dependent) phases of the responses to be assayed separately. The results (with carbachol used as the agonist) are summarized in Figure 2. Magnesium, at a concentration of 10 mM, had no statistically detectable effect in these studies, although subsequently it was found that in rather high concentrations it could act as a Ca^{2+} antagonist. The actions of the remaining substances fit one of two general patterns. Lanthanum, cobalt, nickel, and neomycin showed some selectivity in antagonizing the Ca^{2+}-dependent phase of the response, whereas procaine, tetracaine, and methoxyverapamil (D 600)

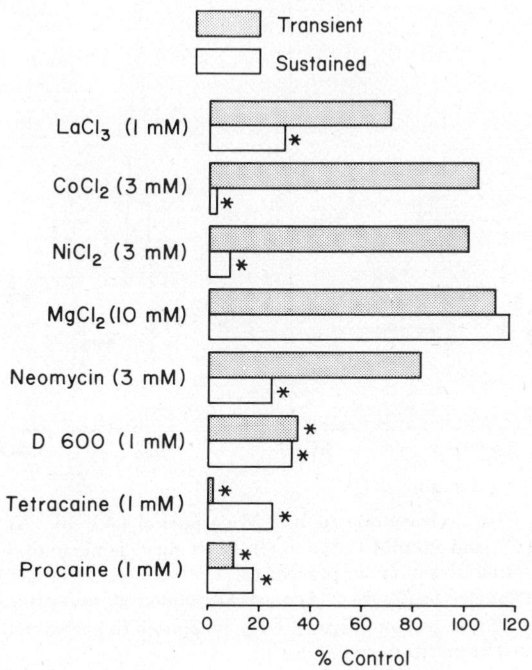

FIG. 2. Effect of putative Ca^{2+} antagonists on $^{86}Rb^+$-efflux response to carbachol. Protocol was identical to that for experiments indicated by open triangles in Figure 1, with 10^{-5} M carbachol producing the transient phase and 10^{-5} M carbachol + 1.0 mM Ca^{2+} producing the sustained phase. Concentration for each antagonist is given in millimolars. Data are expressed as percent response relative to paired controls. Means of 3–6 paired experiments are given. * Statistically significant inhibition. [Adapted from Marier et al. (6).]

blocked both phases of the response to a similar degree. The action of the first group is most easily explained by assuming that these strongly cationic agents antagonize the inward movement of Ca^{2+} at the level of the plasma membrane. This is consistent with demonstrated actions of these substances in a number of systems (17, 28) and has been directly demonstrated for Co^{2+} by ^{45}Ca-flux techniques (20).

The actions of D 600 and the local anesthetics are not so readily determined. The concentration of D 600 required for antagonism is substantially higher than that required for inhibition of Ca^{2+} channels in excitable tissues. When agonists acting on other receptors were tested, it was found that responses to substance P were not blocked by these agents (Fig. 3; 2). As all three receptors presumably regulate the same Ca^{2+} channels, the antagonism of muscarinic and α-adrenergic responses must occur at some step preceding Ca^{2+} mobilization. Considerable evidence suggests that these compounds inhibit agonist binding to muscarinic and α-adrenergic receptors but not to peptide receptors in a variety of tissues (16).

MOLECULAR MECHANISMS FOR CALCIUM CHANNELS AND CALCIUM ANTAGONISM

The simplest explanation for the mechanism of Ca^{2+} antagonism by cationic substances such as Co^{2+} is that they compete with Ca^{2+} for interaction

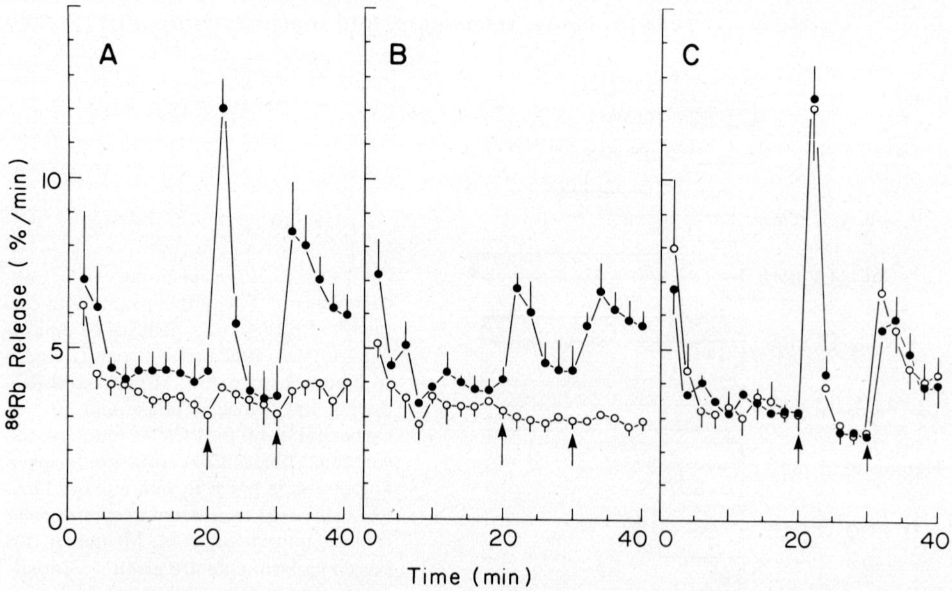

FIG. 3. Effect of 1.0 mM procaine on $^{86}Rb^{+}$ release due to 10^{-5} M carbachol (*A*), 10^{-5} M phenylephrine (*B*), or 10^{-7} M substance P (*C*), and 1.0 mM Ca^{2+} (*A–C*). Each curve is mean of 3 experiments in which 1.0 mM procaine was either absent (●) or present (○). Protocol was identical to that for experiments indicated by open triangles in Figure 1. Agonist was added 20 min after start of washout (first arrow) and Ca^{2+} after 30 min (second arrow). Only responses to carbachol and phenylephrine were affected by procaine. [From Marier et al. (6).]

with some membrane molecule that functions as a Ca^{2+} channel. To propose a more detailed mechanism, a hypothesis about the molecular nature of the Ca^{2+} channels is necessary.

Michell (7, 8) has proposed that in many tissues, receptor-activated turnover of phospholipids may be intimately involved in the Ca^{2+}-gating mechanism. The evidence for this conjecture is that *1*) the reaction is ubiquitous for membrane receptors that gate Ca^{2+}; *2*) whereas most tissue responses are Ca^{2+} dependent, the phospholipid effect is not; and *3*) whereas most tissue responses are mimicked by the divalent cationophore A23187, the phospholipid effect is not. In the parotid gland, activation of muscarinic, α-adrenergic, or substance P receptors stimulates turnover of phosphatidylinositol (5). The reaction is independent of the presence of Ca^{2+} and is not activated by A23187 (5, 12, 23).

A net increase in tissue levels of phosphatidic acid is generally associated with the phosphatidylinositol turnover (9). Noting that phosphatidic acid can function as a calcium ionophore in a Pressman chamber (27), Salmon and Honeyman (24, 25) proposed that the phosphatidic acid formed on receptor activation may mediate cellular Ca^{2+} influx. To support this contention, they demonstrated that phosphatidic acid could mimic the contractile effects of acetylcholine on smooth muscle cells (24, 25).

In the parotid gland, muscarinic-receptor activation increases tissue levels of phosphatidic acid (22). When tested for effects on $^{86}Rb^+$ efflux, phosphatidic acid stimulated the efflux with a pattern similar to that for ionophores; that is, a sustained but not a transient phase was obtained (22). The response is unaffected by receptor-blocking concentrations of phentolamine or atropine. It is reasonable to suggest, therefore, that phosphatidic acid formed on receptor activation mediates the inward Ca^{2+} movement responsible for the Ca^{2+}-dependent phase of the $^{86}Rb^+$-efflux response (22).

Quantitative studies of the interactions of Ca^{2+} antagonists with receptor-activated Ca^{2+} channels and with phosphatidic acid support this hypothesis (22). Concentration-effect relationships for Ca^{2+} with submaximal methacholine were determined in the presence or absence of fixed concentrations of La^{3+}, Tm^{3+}, Co^{2+}, Ni^{2+}, Mg^{2+}, and neomycin. From the rightward shift in the log concentration-effect curve, an antagonist-dissociation constant (K_a), presumably relating to binding to the Ca^{2+} channel, can be estimated (22).

The relative ability of these same substances to inhibit Ca^{2+} binding to phosphatidic acid was also estimated (22). When Ringer's solution containing ^{45}Ca was extracted with chloroform, radioactivity partitioned into the organic phase only if phosphatidic acid was added to the system. Increasing concentrations of La^{3+}, Tm^{3+}, Co^{2+}, Ni^{2+}, Mg^{2+}, neomycin, and nonradioactive Ca^{2+} progressively inhibited this partitioning activity such that IC_{50} values (concentrations inhibiting the partitioning of ^{45}Ca by 50%) could be determined. Procaine, tetracaine, and D 600 did not inhibit partitioning. Presumably the rank order of IC_{50} values would be the same as for the inhibition of Ca^{2+} binding to the divalent negative head group of phosphatidic acid (22).

For both the antagonism of Ca^{2+} channels and the inhibition of Ca^{2+} binding to phosphatidic acid the rank order of potency was the same: La^{3+} >

Tm^{3+} > neomycin > Co^{2+} > Ni^{2+} > Ca^{2+} ≫ Mg^{2+} (22). The correlation of the two parameters is illustrated in Figure 4.

Taken with previous observations, these data suggest that these cationic Ca^{2+} antagonists block receptor-activated Ca^{2+} channels by competing with Ca^{2+} for binding to phosphatidic acid. Accordingly binding to phosphatidic acid (newly synthesized after receptor activation) is proposed as a necessary and sufficient event for surface membrane Ca^{2+} gating in the exocrine glands. Whether calcium phosphatidic acid can traverse the plasmalemma unaided by other specialized membrane constituents is unknown; observations suggesting that increasing membrane phosphatidic acid content does not always increase membrane permeability to Ca^{2+} (9, 11) are relevant to this point. It is hoped that continued research in this area will more clearly define the role of phosphatidic acid in Ca^{2+} gating and in the actions of certain Ca^{2+} antagonists.

SUMMARY

In the parotid gland, occupation of muscarinic, α-adrenergic, or substance P receptors activates surface membrane Ca^{2+} gates. A number of cationic substances antagonize Ca^{2+} entry through this pathway. Local anesthetics and organic Ca^{2+} antagonists such as D 600 inhibit only at high concentrations and then by receptor blockade. The mechanism of Ca^{2+} gating may involve Ca^{2+} binding to phosphatidic acid newly synthesized on receptor activation. Agents

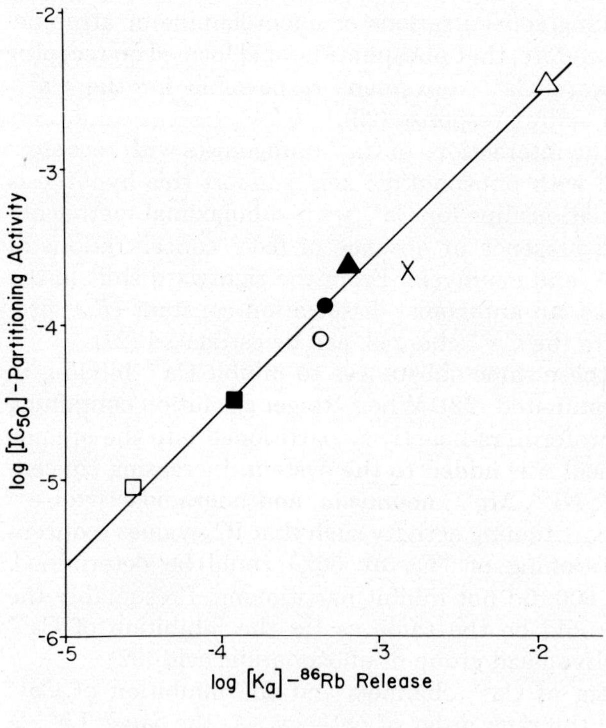

FIG. 4. Relationship between inhibition of ^{45}Ca-partitioning activity of phosphatidic acid and inhibition of receptor-activated, Ca^{2+}-dependent $^{86}Rb^{+}$ release by various agents. Ordinate values are concentrations of agent necessary to inhibit ^{45}Ca transport by phosphatidic acid into $CHCl_3$ by 50%. Abscissa values are apparent dissociation constants (K_a) for antagonism of Ca^{2+} gating in parotid slices. Agents were La^{3+} (□), Tm^{3+} (■), neomycin (○), Co^{2+} (●), Ni^{2+} (▲), Mg^{2+} (△), Ca^{2+} (×; K value as agonist estimated). Line was fitted by least-squares analysis. Calcium was omitted, however, because K value, as agonist, is uncertain. Slope (using logarithms) was 0.998, and correlation coefficient was 0.993. [From Putney et al. (22).]

that act as Ca^{2+} antagonists may do so by inhibiting the binding of Ca^{2+} to phosphatidic acid.

Work from my laboratory described here was primarily supported by National Institutes of Health Grants DE-05764 and EY-03533.

REFERENCES

1. BERRIDGE, M. J., AND J. L. OSCHMAN. *Transporting Epithelia.* New York: Academic, 1972.
2. BUTCHER, F. R. Calcium and cyclic nucleotides in the regulation of secretion from the rat parotid by autonomic agonists. *Adv. Cyclic Nucleotide Res.* 9: 707–721, 1978.
3. BUTCHER, F. R., AND J. W. PUTNEY, JR. Role of calcium and cyclic nucleotides in exocrine glands. *Adv. Cyclic Nucleotide Res.* 13: 215–249, 1980.
4. DOUGLAS, W. W., AND A. M. POISNER. The influence of calcium on the secretory response of the submaxillary gland to acetylcholine or to noradrenaline. *J. Physiol. London* 165: 528–541, 1963.
5. JONES, L. M., AND R. H. MICHELL. The relationship of calcium to receptor-controlled stimulation of phosphatidylinositol turnover. Effects of acetylcholine, adrenaline, calcium ions, cinchocaine and a bivalent cation ionophore on rat parotid-gland fragments. *Biochem. J.* 148: 479–485, 1975.
6. MARIER, S. H., J. W. PUTNEY, JR., AND C. M. VAN DE WALLE. Control of calcium channels by membrane receptors in the rat parotid gland. *J. Physiol. London* 279: 141–151, 1978.
7. MICHELL, R. H. Inositol phospholipids and cell surface receptor function. *Biochim. Biophys. Acta* 415: 81–147, 1975.
8. MICHELL, R. H. Inositol phospholipids in membrane function. *Trends Biochem. Sci.* 4: 128–131, 1979.
9. MICHELL, R. H., S. S. JAFFERJI, AND L. M. JONES. The possible involvement of phosphatidylinositol breakdown in the mechanism of stimulus-response coupling at receptors which control cell-surface calcium gates. *Adv. Exp. Biol. Med.* 83: 447–464, 1977.
10. MILLER, B. E., AND D. L. NELSON. Calcium fluxes in isolated acinar cells from rat parotid: the effect of adrenergic and cholinergic stimulation. *J. Biol. Chem.* 252: 3629–3636, 1977.
11. MONGAR, J. L., AND P. SVEC. The effect of phospholipids on anaphylactic histamine release. *Br. J. Pharmacol.* 46: 741–752, 1972.
12. ORON, Y., M. LOWE, AND Z. SELINGER. Incorporation of inorganic [^{32}P] phosphate into rat parotid phosphatidylinositol. Induction through activation of alpha adrenergic and cholinergic receptors and relation to K^+ release. *Mol. Pharmacol.* 11: 79–86, 1975.
13. PAROD, R. J., B. A. LESLIE, AND J. W. PUTNEY, JR. Muscarinic and alpha-adrenergic stimulation of Na and Ca uptake by dispersed lacrimal cells. *Am. J. Physiol.* 239 (*Gastrointest. Liver Physiol.* 2): G99–G105, 1980.
14. PAROD, R. J., AND J. W. PUTNEY, JR. Role of calcium in the receptor mediated control of potassium permeability in the rat lacrimal gland. *J. Physiol. London* 281: 371–381, 1978.
15. PUTNEY, J. W., JR. Stimulus-permeability coupling: role of calcium in the receptor regulation of membrane permeability. *Pharmacol. Rev.* 30: 209–245, 1978.
16. PUTNEY, J. W., JR. Calcium in the regulation of membrane permeability in the exocrine glands. In: *Calcium in Normal and Pathological Biological Systems*, edited by L. J. Anghileri. Boca Raton, FL: CRC, in press.
17. PUTNEY, J. W., JR., AND A. ASKARI. Modification of membrane function by drugs. In: *Physiology of Membrane Disorders*, edited by T. E. Andreoli, J. F. Hoffman, and D. D. Fanestil. New York: Plenum, 1978, p. 417–445.
18. PUTNEY, J. W., JR., B. A. LESLIE, AND S. H. MARIER. Calcium and the control of potassium efflux in the sublingual gland. *Am. J. Physiol.* 235 (*Cell Physiol.* 4): C128–C135, 1978.
19. PUTNEY, J. W., JR., R. J. PAROD, AND S. H. MARIER. Control by calcium of exocytosis and membrane permeability to potassium in the rat lacrimal gland. *Life Sci.* 20: 1905–1912, 1977.
20. PUTNEY, J. W., JR., C. M. VAN DE WALLE, AND B. A. LESLIE. Receptor control of calcium influx in parotid acinar cells. *Mol. Pharmacol.* 14: 1046–1053, 1978.
21. PUTNEY, J. W., JR., C. M. VAN DE WALLE, AND B. A. LESLIE. Stimulus-secretion coupling in the rat lacrimal gland. *Am. J. Physiol.* 235 (*Cell Physiol.* 4): C188–C198, 1978.
22. PUTNEY, J. W., JR., S. J. WEISS, C. M. VAN DE WALLE, AND R. A. HADDAS. Is phosphatidic acid a calcium ionophore under neurohumoral control? *Nature London* 284: 345–347, 1980.
23. ROSSIGNOL, B., G. HERMAN, A. M. CHAMBAUT, AND G. KERYER. The calcium ionophore A-23187 as a probe for studying the role of Ca^{2+} ions in the mediation of carbachol effects on rat salivary glands: protein secretion and metabolism of phospholipids and glycogen. *FEBS Lett.* 43: 241–246, 1974.
24. SALMON, D. M., AND T. W. HONEYMAN. Increased phosphatidate accumulation during single contractions of isolated smooth muscle cells. *Biochem. Soc. Trans.* 7: 986–988, 1979.
25. SALMON, D. M., AND T. W. HONEYMAN. Proposed mechanism of cholinergic action in smooth muscle. *Nature London* 284: 344–345, 1980.
26. SCHRAMM, M., AND Z. SELINGER. The functions of cyclic AMP and calcium as alternative second messengers in parotid gland and pancreas. *J. Cyclic Nucleotide Res.* 1: 181–192, 1975.
27. TYSON, C. A., H. V. ZANDE, AND D. E. GREEN. Phospholipids as ionophores. *J. Biol. Chem.* 251: 1326–1332, 1976.
28. WEISS, G. B. Cellular pharmacology of lanthanum. *Annu. Rev. Pharmacol.* 14: 343–354, 1974.
29. WILLIAMS, J. A. Regulation of pancreatic acinar cell function by intracellular calcium. *Am. J. Physiol.* 238 (*Gastrointest. Liver Physiol.* 1): G269–G279, 1980.

14

Calcium Antagonists and the Secretion of Insulin and Thyrotropin

NORMAN FLEISCHER, ULRICH K. SCHUBART, ADRIENNE FLECKMAN, AND JACK ERLICHMAN

Department of Medicine, Diabetes Research and Training Center, Albert Einstein College of Medicine, Bronx, New York

IT IS WELL ESTABLISHED THAT Ca^{2+} plays a fundamental role in coupling certain extracellular signals to the appropriate physiological response of a cell. During the early 1960's Douglas and Rubin showed that Ca^{2+} entry into the chromaffin cell was a critical event in the stimulus-secretion coupling of catecholamine release (14, 15). Depolarization of the plasma membrane, whether induced by neurotransmitters or elevated concentrations of extracellular K^+, promoted an increased rate of cellular Ca^{2+} entry and stimulated catecholamine release. It was this entry of Ca^{2+} rather than the depolarization of the membrane that was responsible for the enhanced hormone release, since omission of Ca^{2+} from the incubation media resulted in cessation of secretion without affecting the ability of acetylcholine to depolarize the cell. Studies of the role of Ca^{2+} in hormone secretion from a variety of other secretory tissues extended the hypothesis that Ca^{2+} was the stimulus-secretion coupling factor to many if not all the hormones stored in and released from secretory granules (54). This applies to tissues that are innervated by secretomotor neurons and to neuroendocrine and endocrine glands such as the anterior pituitary and the pancreatic islet. The Ca^{2+} that is involved in stimulus-secretion coupling may be extracellular, as originally proposed by Douglas and Rubin, or it may be released from intracellular membranes or organelles.

The mechanism by which cytosolic Ca^{2+} mediates the physiological response of the cell is still unclear. Recent studies have shown that the intracellular Ca^{2+}-binding protein calmodulin (8, 48, 70) may regulate a number of Ca^{2+}-dependent cellular processes, including *1*) regulation of cyclic nucleotide levels through calmodulin's action on adenylate cyclase and cyclic nucleotide phosphodiesterase, *2*) control of intracellular Ca^{2+} levels by stimulating Ca^{2+}-dependent ATPase activity in cells such as erythrocytes and fat cells, *3*) energy production through action on phosphorylase kinase, *4*) smooth muscle contraction through activation of calmodulin-dependent myosin light-chain kinase, and *5*) Ca^{2+}-mediated neurotransmitter and polypeptide hormone secretion. Thus calmodulin may mediate the action of increased cystolic-free Ca^{2+} in the diverse physiological responses of a variety of cells.

Agents that prevent the penetration of Ca^{2+} into intracellular compartments have been valuable for studying the role of Ca^{2+} in excitation-contraction and stimulus-secretion coupling (22). Drugs such as verapamil and methoxyverapamil (D 600) that are relatively specific for the voltage-dependent slow Ca^{2+} channel have been used to demonstrate the dependence on Ca^{2+} uptake of hormone secretion from normal pituitaries (20), isolated pancreatic islets (13), and hormone-secreting tumor cells (18, 58). Calcium antagonists of the ω-(*N*,*N*′-diethylamino)alkyl 3,4,5-trimethoxybenzoate (TMB) series, which are thought to act on intracellular Ca^{2+} (9), have been shown to inhibit secretion from platelets (7) and insulinoma cells (18). Recent studies have shown that trifluoperazine, a phenothiazine that blocks calmodulin action (40, 68), inhibits Ca^{2+}-dependent hormone release from normal islets (55, 61) and insulinoma cells (55, 56). Trifluoperazine also inhibits Ca^{2+} uptake by inside-out plasma membrane phagocytic vesicles from rabbit lung macrophages (41). Thus the phenothiazines may comprise a potentially useful class of Ca^{2+} antagonists for studying Ca^{2+} action at the molecular level.

This chapter focuses on some of our studies on the effects of Ca^{2+} antagonists on insulin and thyrotropin (TSH) release as a basis for determining how Ca^{2+} mediates polypeptide hormone secretion.

INSULIN SECRETION

Insulin secretion is governed by a complex interplay of a variety of agents that act on the pancreatic β-cell. There are at least three separate mechanisms for the regulation of insulin secretion. The major physiological stimulus is an elevation in the extracellular concentration of glucose (1, 2). Changes in the concentration of leucine or certain other amino acids also stimulate insulin release independent of glucose action. In contrast glucagon and β-adrenergic agonists, through their stimulation of cyclic adenosine 3′,5′-monophosphate (cyclic AMP) production, potentiate glucose action in the normal islet (50). Other polypeptide hormones and biogenic amines can also influence the β-cell response to glucose. Somatostatin (43, 65), α-adrenergic agonists (71), and serotonin (21, 39) inhibit insulin secretion. Thus many factors are important for the regulation of the physiological response of the β-cell.

Isolated Islets

Although the precise mechanism by which glucose stimulates insulin secretion is still not completely understood, the requirement for extracellular Ca^{2+} suggests a key regulatory role for Ca^{2+} in stimulus-secretion coupling in the β-cell. Glucose action appears to be related to changes in Ca^{2+} fluxes across the plasma membrane. When isolated pancreatic islets are exposed to elevated glucose levels, an increased influx of Ca^{2+} is seen (25, 45). Recent data suggest that glucose initially inhibits Ca^{2+} efflux from the islet; this is soon followed by an increased Ca^{2+} uptake resulting in an accumulation of intracellular free Ca^{2+} (46). It is believed that this intracellular free Ca^{2+} somehow activates the secretory process.

Glucose stimulation of insulin secretion can be prevented by elimination of extracellular Ca^{2+} or by addition of agents that block Ca^{2+} currents in excitable cells. For example, the Ca^{2+}-channel blocker D 600 can inhibit glucose-induced insulin in a dose-dependent manner (Fig. 1). Verapamil has been shown to prevent both the first and second phases of insulin secretion, and this inhibition can be partially overcome by increasing the extracellular concentrations of Ca^{2+} (13). In addition drugs such as D 600 inhibit the increased electrical activity provoked by glucose in pancreatic islets (47).

Recent studies from this and other laboratories have shown that the antipsychotic drug trifluoperazine inhibits glucose-induced insulin release from isolated rat islets (Fig 1; 55, 61). The existence of calmodulin in islet tissue (61) and the observation that trifluoperazine inhibits Ca^{2+}-dependent insulin secretion have led to the proposal that calmodulin may be involved in insulin release (2, 55, 61). The available evidence for calmodulin involvement in insulin secretion is circumstantial, however. No direct interaction of trifluoperazine with calmodulin in the intact cell has been demonstrated. More definitive studies are needed before one can conclude that calmodulin is indeed the intracellular mediator of Ca^{2+}-dependent insulin secretion. Nevertheless the proposal that calmodulin mediates glucose action in the islet by regulating Ca^{2+}-calmodulin–dependent processes is intriguing and will probably be the focus of many studies.

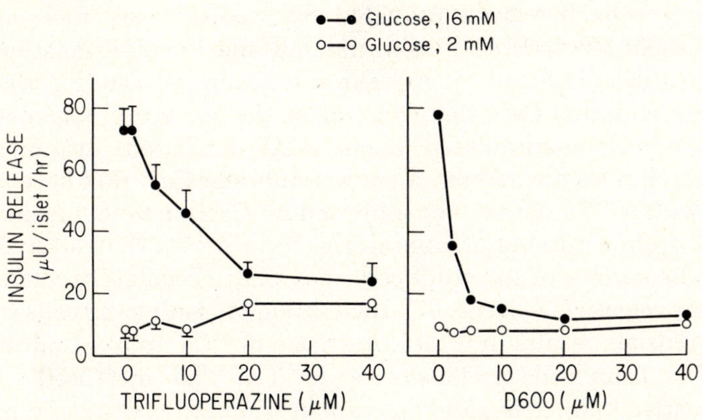

FIG. 1. Effect of trifluoperazine and D 600 on glucose-induced insulin release from isolated rat islets. Rat islets, isolated by collagenase digestion, were preincubated in Krebs-Ringer bicarbonate solution containing 2 mM glucose. After 1 h buffer was removed and fresh buffer was added containing either 2 mM or 16 mM glucose and the indicated concentration of drug. After 1 h buffer was removed and assayed for insulin content by radioimmunoassay (n = 6).

Hamster Insulinoma Cells

The limited amount of islet tissue available for biochemical studies has prompted the development of alternative model systems for the study of insulin release. One such system is the Syrian hamster insulinoma, which was originally described by Kirkman (30, 35). This tumor produces normal hamster insulin, is easily grown as a solid tumor in animals, and is available in gram quantities. The tumor is composed of a relatively homogeneous population of insulin-producing cells. Single-cell suspensions prepared by collagenase digestion of the tumor secrete insulin in response to several stimuli in a manner similar to normal islets. Insulin release in response to glucagon stimulation is the result of elevations in cellular levels of cyclic AMP (57). Alternatively it can be induced by promoting a transmembrane Ca^{2+} current (58). External Ca^{2+} is required for insulin release regardless of the stimulant, and insulin release is proportional to the external Ca^{2+} concentration. However, glucose does not induce a Ca^{2+} current in these cells like it does in normal islet cells. As a result insulin release is not affected by glucose (58). The mechanism of this defect is unknown, but it is possible to bypass this block in glucose action. An increased ^{45}Ca influx into these cells can be achieved by incubating them with depolarizing concentrations of K^+, ouabain, or certain organic mercurials (18, 58). This enhanced uptake is accompanied by an increase in insulin release, making the insulinoma a suitable model for studying Ca^{2+}-dependent hormone secretion. Furthermore the availability of large quantities of tissue makes it possible to study the mechanism of Ca^{2+} action at the biochemical level.

Calcium-channel blockers

As for normal islet cells, insight into the role of Ca^{2+} in insulin secretion in the tumor cells has been provided by studies using Ca^{2+} antagonists. Calcium-channel blockers such as verapamil and D 600 dramatically inhibit both depolarization-induced ^{45}Ca uptake and insulin release (18, 58), indicating that a transmembrane Ca^{2+} flux can trigger the secretory response in insulinoma cells. Glucagon-stimulated (cyclic AMP–mediated) insulin release, on the other hand, does not require a transmembrane Ca^{2+} flux because it is not associated with a ^{45}Ca uptake nor inhibited by Ca^{2+}-channel blockers (58).

Studies from a number of laboratories have shown that sulfhydryl groups present on the surface of secretory cells may play a regulatory role in polypeptide hormone release (18, 32, 66, 67). Incubation of insulinoma cells with certain organic mercurials results in a rapid increase in ^{45}Ca uptake and in increased isotope efflux from cells prelabeled with ^{45}Ca. This increased Ca^{2+} flux is associated with a four- to fivefold increase in stimulation of insulin release (18). Unlike depolarization-induced Ca^{2+} uptake and insulin release, however, sulfhydryl reagent–induced ^{45}Ca uptake and insulin release are not blocked by Ca^{2+} antagonists. Thus their effect on secretion may be the result of mobilizing internal Ca^{2+} as well as promoting transmembrane Ca^{2+} uptake through some other pathway than the slow Ca^{2+} channel.

Trifluoperazine

Since trifluoperazine inhibited glucose-stimulated insulin secretion from isolated islets, it was of interest to determine if it also blocked insulin release from insulinoma cells. Trifluoperazine specifically inhibited depolarization-induced insulin release without affecting basal or glucagon-stimulated release (Table 1). Thus the specificity of trifluoperazine in inhibiting insulin release was apparently the same as that of the Ca^{2+}-channel blockers. Inhibition of K^+-stimulated insulin release was dose related: half-maximal inhibition occurred with 2.5 μM trifluoperazine and complete blockade of K^+-stimulated release with 20 μM trifluoperazine (Fig. 2).

The mechanism by which trifluoperazine inhibits Ca^{2+}-dependent insulin release is still unknown. Since trifluoperazine inhibits calmodulin action in vitro, it may act on a calmodulin-dependent process in the insulinoma cell, thus inhibiting Ca^{2+} action. Alternatively trifluoperazine could block Ca^{2+} entry in a manner similar to the Ca^{2+}-channel blockers. Preliminary experiments measuring the effects of trifluoperazine on Ca^{2+} uptake by the insulinoma cells showed a biphasic mode of action on ^{45}Ca uptake. Increasing concentrations of trifluoperazine between 1 μM and 10 μM resulted in a partial reduction of ^{45}Ca uptake induced by 40 mM K^+. Concentrations of trifluoperazine above 10 μM progressively increased net ^{45}Ca uptake beyond that seen with 40 mM K^+ alone. In contrast, insulin release remained completely inhibited with concentrations of trifluoperazine above 10 μM. These data suggest that trifluoperazine affects Ca^{2+} uptake by the insulinoma cells in a complex way. Entry of ^{45}Ca is inhibited by concentrations of trifluoperazine below 10 μM, whereas with concentrations above 10 μM, Ca^{2+} uptake is enhanced. Since insulin release is inhibited at all concentrations of trifluoperazine, it seems likely that trifluoperazine may be blocking Ca^{2+} action within the cell. Studies to determine the locus and mechanism of phenothiazine action on insulinoma cells are currently being performed in our laboratory.

Protein phosphorylation and insulin release

Studies from our laboratory have shown that hamster insulinoma cells possess a Ca^{2+}-calmodulin–dependent protein kinase activity (55). Cytosol prepared from single-cell suspensions of insulinoma cells contains several

TABLE 1. *Effect of trifluoperazine on insulin release by insulinoma cells*

Secretagogue	n	Insulin Release, % of control ± SE	
		0 μM TFP	20 μM TFP
Control	14	100	94 ± 3
Glucagon, 3 μM	14	195 ± 8	192 ± 18
Ouabain, 1 μM	12	184 ± 7	99 ± 4
Potassium, 40 mM	12	191 ± 11	101 ± 6

TFP, trifluoperazine. [Data from Schubart et al. (55).]

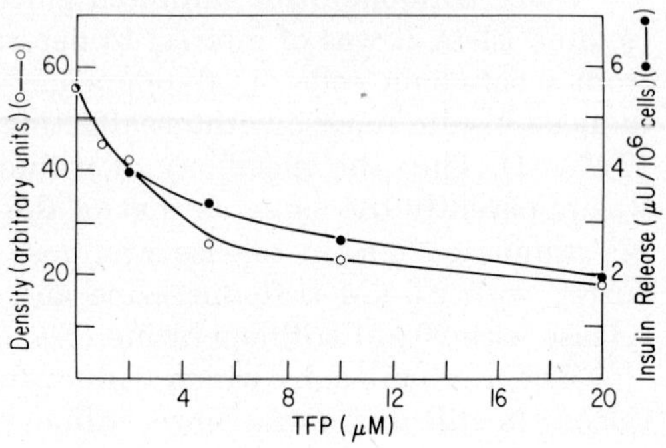

FIG. 2. Effect of trifluoperazine (TFP) on K^+-induced insulin release and phosphorylation of cellular protein P 60, mol wt 60,000, in hamster insulinoma cells. For measuring insulin release (●), insulinoma cells were incubated for 20 min at 37°C in media containing 50 mM K^+ and TFP as indicated. Concentration of insulin released into incubation media was determined by radioimmunoassay. Phosphorylation of P 60 (○) was assessed using ^{32}P-labeled cells incubated for 10 min at 37°C in media containing 50 mM K^+. Cellular phosphoproteins were resolved by sodium dodecyl sulfate–polyacrylamide slab gel electrophoresis and autoradiography, and the degree of ^{32}P-labeling of P 60 was estimated by densitometry. [From Schubart et al. (56).]

proteins that are phosphorylated in vitro in a Ca^{2+}-dependent manner. Phosphorylation of these proteins is enhanced by addition of Ca^{2+} or purified calmodulin and inhibited by ethylene glycol-bis(β-aminoethylether)-*N,N′*-tetraacetic acid (EGTA) or trifluoperazine. In more recent studies (56) we have investigated the role of Ca^{2+} on protein phosphorylation in intact insulinoma cells. Depolarizing concentrations of K^+ promote phosphorylation of a specific protein (P 60), of mol wt 60,000, in the intact cell (Fig. 3). Phosphorylation of P 60 is Ca^{2+} dependent, since omission of extracellular Ca^{2+} or addition of the Ca^{2+}-channel blocker D 600 prevents the K^+-stimulated phosphorylation of this protein. Similarly K^+-stimulated phosphorylation of P 60 could be prevented by addition of trifluoperazine (Fig. 3). The intriguing question is whether phosphorylation of P 60 is involved in Ca^{2+} influx–mediated insulin release. The concentration of trifluoperazine required to inhibit K^+-stimulated phosphorylation of P 60 is strikingly similar to the concentration required to inhibit K^+-stimulated insulin release (Fig. 2), supporting this proposal. Furthermore the time course of K^+-induced phosphorylation of P 60 parallels the time course of insulin release (Fig. 3). These results are consistent with the notion that Ca^{2+}-dependent phosphorylation of P 60 may help regulate Ca^{2+} influx–mediated insulin release from insulinoma cells. Calcium-dependent protein phosphorylation has been implicated as a possible intermediate step in secretory processes in several other tissues, including serotonin release from platelets (4, 31, 44), histamine release from mast cells (59), and neurotransmitter release from synaptosomes (11, 36, 49). It is not known if the kinase responsible for phosphorylation of P 60 is the same calmodulin-dependent protein kinase demonstrated in the cytosol of insulinoma cells. Our data, however, support the hypothesis that an increase in cytosolic free Ca^{2+} activates a Ca^{2+}-calmodulin–dependent protein kinase resulting in increased protein phosphorylation as an intermediate event in the activation of the secretory apparatus for insulin (56).

PITUITARY THYROTROPIN SECRETION

The secretion of the various anterior pituitary hormones is partly regulated by specific releasing factors (hormones) from the hypothalamus that are delivered to the anterior pituitary through the hypothalamic-hypophyseal portal blood supply. A number of these factors have been identified and characterized, including TSH-releasing hormone (TRH), which stimulates TSH and prolactin secretion; luteinizing hormone–releasing factor, which stimulates secretion of luteinizing hormone and follicle stimulating hormone; and somatostatin, which inhibits TSH and growth hormone secretion (65). It is expected that other factors affecting pituitary functions will be isolated. These factors are important not only because of their pituitary action, but also because they apparently represent a class of modified neurotransmitters widely distributed throughout the central nervous system (17). Peptides such as somatostatin are also present at extraneural sites (43).

Thyrotropin-releasing hormone is a potent stimulator of TSH secretion in vitro (65) and in vivo (24). Secretion of TSH in animals can be inhibited by the infusion of anti-TRH antisera, establishing the importance of endogenous TRH for TSH secretion. The sequence of events by which TRH stimulates TSH release is not completely understood. Both cyclic AMP and Ca^{2+} have been proposed as mediators of TRH action (38, 63, 64). Studies during the past several years have yielded conflicting results, however, and no clear understanding of the mechanism of TRH action has emerged.

Calcium and Thyrotropin-Releasing Hormone Action

It has long been recognized that TRH action in vitro requires extracellular Ca^{2+} (63, 64) and that Ca^{2+} may serve as a second messenger to couple TRH action and hormone secretion. Removal of extracellular Ca^{2+} by washing

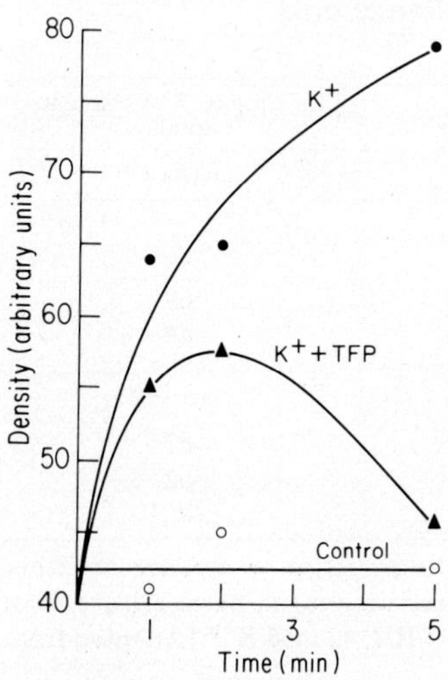

FIG. 3. Time course for phosphorylation of cellular protein P 60, mol wt 60,000, induced by 50 mM K^+ in intact insulinoma cells. ^{32}P-labeled cells were incubated in buffer containing 4 mM K^+ (control), 50 mM K^+ (K^+), or 50 mM K^+ + 40 μM trifluoperazine (K^+ + TFP). Phosphorylation of P 60 was assessed as in Figure 2. [From Schubart et al. (56).]

pituitary tissue in Ca^{2+}-free media before stimulation with TRH results in a decreased effectiveness of TRH to induce TSH release. Prolonged washing of pituitary tissue, however, may alter the distribution of intracellular Ca^{2+} by causing Ca^{2+} to efflux from internal sites. In this regard such experiments do not prove that extracellular Ca^{2+} is responsible for mediating TRH action. Thus, although it is generally accepted that Ca^{2+} plays a critical role in TRH action, it is unclear which pools of Ca^{2+} are affected by TRH. To explore the mechanism by which TRH stimulates TSH release, we have used several Ca^{2+} antagonists as probes for studying the interrelationship between TRH and Ca^{2+}.

Calcium-channel blockers

Studies from several laboratories have shown that TRH induces Ca^{2+}-dependent action potentials in cultured pituitary cells (16, 34, 52, 62). Calcium efflux from cells prelabeled with ^{45}Ca has also been found to occur in response to TRH (27). Such studies show that Ca^{2+} movement across the plasma membrane occurs when cells are stimulated with TRH. Since TRH does not cause a measurable increase in ^{45}Ca uptake by the pituitary, its action does not appear to be identical to membrane depolarization (19).

To determine if transmembrane Ca^{2+} currents were occurring in response to TRH, we compared the effects of D 600 and phenytoin, agents that block depolarization-induced ^{45}Ca uptake in excitable tissues (22, 51, 53, 60) on TRH- and K^+-induced TSH release (23). Phenytoin and D 600 inhibited both TRH- and K^+-induced TSH release (Table 2); D 600 was slightly more effective in inhibiting K^+-induced release than in inhibiting TRH action. Phenytoin, however, showed marked preference for preventing K^+-stimulated release (Table 2). In addition to inhibiting hormone secretion, these drugs also blocked

TABLE 2. *Effect of D 600 and phenytoin on TSH release and ^{45}Ca uptake by rat pituitaries in vitro*

Concentration, μM	TSH Release, % of stimulated release		^{45}Ca Uptake, % of stimulated uptake
	14 nM TRH	40 mM K^+	40 mM K^+
D 600			
1	100	80	79
10	61	57	69
100	33	9	43
Phenytoin			
20	63	20	67
80	55	19	52
200	53	18	38

Experiments were conducted as paired incubations in the presence or absence of either D 600 or phenytoin. Stimulated release was calculated by subtracting mean basal release from TSH release observed in the presence of a secretagogue (14 nM TRH, 40 mM K^+). [Adapted from Fleckman et al. (23).]

K^+-induced ^{45}Ca uptake (Table 2). In contrast, verapamil, a less potent analogue of D 600, blocked K^+-induced TSH release without affecting TRH action (20). These findings are consistent with the idea that TRH alters membrane Ca^{2+} flux in a manner similar but not identical to membrane depolarization and suggest that TRH action may require Ca^{2+} movement through a voltage-dependent Ca^{2+} channel without necessarily affecting net Ca^{2+} uptake. However, the observations that phenytoin inhibited depolarization-induced hormone release more effectively than TRH-stimulated secretion and that D 600 inhibited TRH action whereas verapamil did not indicate that these drugs may inhibit TRH- and depolarization-induced TSH release by a mechanism unrelated to the voltage-dependent Ca^{2+} channel. Thus the specificity of action of these Ca^{2+} antagonists must be determined before definite conclusions are made about the site affected by these drugs.

Sodium-channel activation

Because D 600 and phenytoin have been reported to block Na^+ channels (26, 42, 53), it was of interest to determine if TSH secretion could be regulated by Na^+-channel activation. To explore this possibility, pituitaries were incubated in the presence of veratridine, a specific Na^+-channel activator. Veratridine stimulated TSH release, and its effect was blocked by tetrodotoxin (TTX), an inhibitor of Na^+-channel activation (6, 33). In contrast, when TSH secretion was induced with TRH or 40 mM K^+, TTX did not inhibit TSH release (23). Thus, although TSH secretion can be elicited by activation of fast Na^+ channels, the inability of TTX to block TRH action makes it unlikely that TRH-induced TSH release involves Ca^{2+} currents through the Na^+ channel.

Trifluoperazine

The effects of trifluoperazine as a Ca^{2+} antagonist on insulin release (see above) prompted us to examine if it would also inhibit TSH release from the pituitary. As shown in Figure 4, trifluoperazine inhibited both TRH- and K^+-

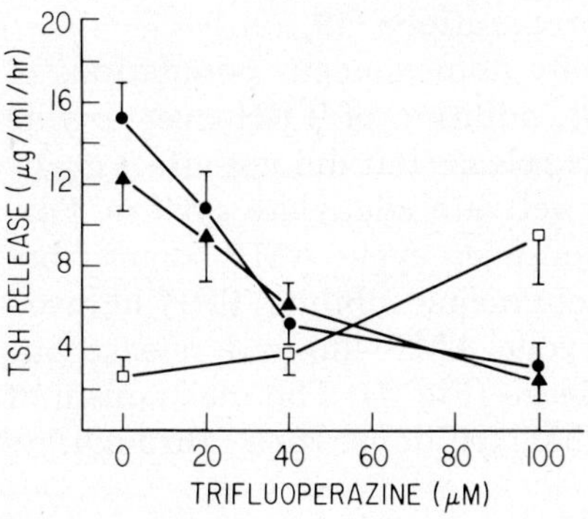

FIG. 4. Effect of trifluoperazine on TSH release induced by TRH, 40 mM K^+, and 8-bromo cyclic AMP. Pituitaries were incubated with varying concentrations of trifluoperazine in presence or absence of a secretagogue. Effect of trifluoperazine on TSH release was determined by subtracting TSH released in the presence of trifluoperazine alone from TSH released in the presence of both trifluoperazine and a secretagogue. ● 14 nM TRH; ▲ 40 mM K^+; □ 2.5 mM 8-bromo cyclic AMP. [From Fleckman et al. (23).]

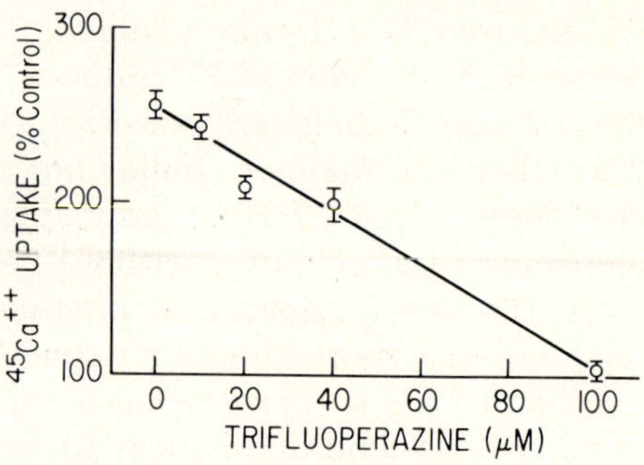

FIG. 5. Effect of trifluoperazine on ^{45}Ca uptake induced by 40 mM K^+. ^{45}Ca uptake by pituitaries was determined as described in Ref. 23. Data are expressed as percent of ^{45}Ca uptake by pituitaries incubated in 40 mM K^+ relative to pituitaries incubated in 4 mM K^+. [From Fleckman et al. (23).]

induced TSH release. The dose dependency for inhibiting TSH release was similar for both TRH and K^+. However, higher concentrations of trifluoperazine were required to inhibit TSH release from the pituitary than insulin release from insulinoma cells. When trifluoperazine was tested for its ability to inhibit depolarization-induced ^{45}Ca uptake, it was found that inhibition of K^+-stimulated ^{45}Ca uptake was proportional to the concentration of trifluoperazine and that 100 μM trifluoperazine completely blocked uptake (Fig. 5). Although the precise mechanism by which trifluoperazine inhibits TSH release is not known, our finding that it inhibits ^{45}Ca uptake at the same concentrations at which it inhibits TSH release suggests that inhibition of Ca^{2+} uptake rather than inhibition of Ca^{2+} action is the mode by which trifluoperazine inhibits depolarization-induced TSH release. The possibility remains, however, that trifluoperazine also inhibits Ca^{2+} action in the pituitary.

Cyclic Nucleotides and Thyrotropin-Releasing Hormone Action

Specific receptors for TRH have been identified on TRH-responsive cells (29, 37). It was proposed that interaction of TRH with these receptors activates adenylate cyclase, thereby increasing intracellular levels of cyclic AMP and thus initiating the secretory process for TSH release (38). Indeed analogues of cyclic AMP and inhibitors of cyclic nucleotide phosphodiesterase can promote TSH release from the pituitary (69). Although small increases in intracellular levels of cyclic AMP in response to TRH have been seen in intact pituitary cells by some investigators (3, 5), others have not been able to confirm this observation (10, 28). To date, no direct stimulation of adenylate cyclase by TRH has been demonstrated in broken cell preparations (19, 28).

In studies in our laboratory with relatively homogeneous populations of TRH-responsive thyrotropic tumor cells (19), addition of TRH over a wide concentration and time range stimulated TSH release but did not affect cyclic AMP production. Additionally TRH did not activate adenylate cyclase. Further support for the segregation of TRH action from cyclic AMP comes from studies with trifluoperazine. Although trifluoperazine inhibits TRH-induced TSH release, it does not inhibit 8-bromo cyclic AMP-induced release but instead potentiates cyclic AMP-mediated release (Fig. 4). The mechanism of this action is unclear, but it is possible that trifluoperazine, through its interaction with calmodulin, inhibits the calmodulin-dependent cyclic nucleotide phosphodiesterase, thereby potentiating cyclic AMP–mediated TSH secretion. The failure of trifluoperazine to inhibit cyclic AMP action while inhibiting TRH-stimulated release adds to the evidence that cyclic AMP is not the mediator of TRH action.

CONCLUSION

Calcium antagonists have been used to study the role of Ca^{2+} in polypeptide hormone secretion. They have facilitated the demonstration that secretion of a number of hormones in response to certain stimuli is the consequence of an increase in net Ca^{2+} uptake into the cell, and they have helped to distinguish between Ca^{2+}- and cyclic AMP–mediated events.

In our laboratory we have used a variety of Ca^{2+} antagonists to help elucidate the role of Ca^{2+} in insulin and TSH secretion. Insulin secretion from the normal islet cell is a consequence of glucose-induced increase in net Ca^{2+} uptake in the cell. The Ca^{2+}-channel blockers and the Ca^{2+} antagonists whose actions are thought to inhibit intracellular Ca^{2+} block glucose-stimulated insulin release from isolated islets. These agents also block depolarization-induced insulin release from insulinoma cells. Parallel results are seen when protein phosphorylation in the intact insulinoma cell is studied, suggesting a link between Ca^{2+}-dependent protein phosphorylation and insulin secretion. Although Ca^{2+} and cyclic AMP–mediated insulin release from insulinoma cells are under separate control, it is possible that protein phosphorylation might represent analogous mechanisms for the two pathways.

In the pituitary it is likely that TRH action is the consequence of effects on Ca^{2+} movement either from cellular stores or from the extracellular space. Although an increased ^{45}Ca uptake is not seen, the observation that certain Ca^{2+}-channel blockers inhibit TRH-induced TSH release supports (but does not prove) the notion that TRH affects the Ca^{2+} channel.

Trifluoperazine and related phenothiazines constitute a class of Ca^{2+} antagonists that are thought to act on intracellular Ca^{2+}. We studied these agents because they are known calmodulin inhibitors and found that they block hormone secretion with the same specificity as D 600 and verapamil. If these drugs do indeed block Ca^{2+} action, they may provide new insights into the intracellular action of Ca^{2+}. Since phenothiazines can also block Ca^{2+} uptake in certain tissues, however, one must be cautious about drawing conclusions regarding the site of phenothiazine action.

Because Ca^{2+}-channel blockers such as verapamil and phenothiazines such as trifluoperazine are used clinically, it is important to determine their effects on polypeptide secretion. Based on their effects in isolated tissues, it might be anticipated that these drugs would inhibit polypeptide secretion in humans. Indeed verapamil has been shown to inhibit insulin secretion in humans and to affect insulin secretion in patients with insulinomas (12). The phenothiazines have also been reported to cause hyperglycemia in patients, but this has not been a common problem. Since the concentrations of Ca^{2+} antagonists required to inhibit hormone secretion are higher than the concentrations required for clinical therapy, it may be that the failure to detect substantial effects on secretion in vivo is the result of differences in tissue sensitivity. Nevertheless it is important to evaluate Ca^{2+} antagonists proposed for clinical use for their effects on endocrine function.

The research reported in this chapter was supported by National Institutes of Health Grants AM-20541, AM-25326, and AM-7004.

REFERENCES

1. ASHCROFT, S. J. H. The control of insulin release by sugars. In: *Ciba Foundation Symposium 41*, edited by R. Porter and D. W. Fitzsimons. Amsterdam: Elsevier, 1976, p. 117–139.
2. ASHCROFT, S. J. H. Glucoreceptor mechanisms and the control of insulin release and biosynthesis. *Diabetologia* 18: 5–15, 1980.
3. BARNES, G. D., B. L. BROWN, T. G. GARD, D. ATKINSON, AND R. P. EKINS. Effect of TRH and dopamine on cyclic AMP levels in enriched mammotroph and thyrotroph cells. *Mol. Cell. Endocrinol.* 12: 273–284, 1978.
4. BENNETT, W. F., J. S. BELVILLE, AND G. LYNCH. A study of protein phosphorylation in shape change and Ca^{++} dependent serotonin release by blood platelets. *Cell* 18: 1015–1023, 1979.
5. BOWERS, C. Y. Studies on the role of cyclic AMP in the release of anterior pituitary hormones. *Ann. NY Acad. Sci.* 185: 263–290, 1971.
6. CATTERALL, W. A. Neurotoxins that act on voltage-sensitive sodium channels in excitable membranes. *Annu. Rev. Pharmacol. Toxicol.* 20: 15–43, 1980.
7. CHARO, I. F., R. D. FEINMAN, AND T. C. DETWILER. Inhibition of platelet secretion by an antagonist of intracellular calcium. *Biochem. Biophys. Res. Commun.* 72: 1462–1467, 1976.
8. CHEUNG, W. Y. Calmodulin plays a pivotal role in cellular regulation. *Science* 207: 19–27, 1980.
9. CHIOU, C. Y., AND M. H. MALAGODI. Studies on the mechanism of action of a new Ca^{2+} antagonist, 8-(N,N-diethylamino)octyl 3,4,5-trimethoxybenzoate hydrochloride in smooth and skeletal muscles. *Br. J. Pharmacol.* 53: 279–285, 1975.
10. DANNIES, P. S., AND A. H. TASHJIAN. Action of cholera toxin on hormone synthesis and release in GH cells: evidence that adenosine 3′5′-monophosphate does not mediate the decrease in growth hormone synthesis caused by thyrotropin-releasing hormone. *Endocrinology* 106: 1532–1536, 1980.
11. DELORENZO, R. J., S. D. FREEDMAN, W. B. YOHE, AND S. C. MAURER. Stimulation of Ca^{2+}-dependent neurotransmitter release and presynaptic nerve terminal protein phosphorylation by calmodulin and a calmodulin-like protein isolated from synaptic vesicles. *Proc. Natl. Acad. Sci. USA* 76: 1838–1842, 1979.
12. DEMARINIS, L., AND A. BARBARINO. Calcium antagonists and hormone release. I. Effects of verapamil on insulin release in normal subjects and patients with islet-cell tumor. *Metabolism* 29: 599–604, 1980.
13. DEVIS, G., G. SOMERS, E. VAN OBBERGHEN, AND W. J. MALAISSE. Calcium antagonists and islet function. I. Inhibition of insulin release by verapamil. *Diabetes* 24: 547–551, 1975.
14. DOUGLAS, W. W. Stimulus-secretion coupling: the concept and clues from chromaffin and other cells. *Br. J. Pharmacol.* 34: 451–474, 1968.
15. DOUGLAS, W. W., AND R. P. RUBIN. The mechanism of catecholamine release from the adrenal medulla and the role of calcium in stimulus-secretion coupling. *J. Physiol. London* 167: 288–310, 1963.
16. DUFY, B., J. D. VINCENT, H. FLEURY, P. DUPASQUIER, D. GOURDJI, AND A. TIXIER-VIDAL. Membrane effects of thyrotropin-releasing hormone and estrogen shown by intracellular recording from pituitary cells. *Science* 204: 509–511, 1979.
17. ELDE, R. Localization of hypophysiotropic peptides and other biologically active peptides within the brain. *Annu. Rev. Physiol.* 41: 587–602, 1979.
18. ERLICHMAN, J., U. K. SCHUBART, AND N. FLEISCHER. Sulfhydryl regeant-induced insulin release and $^{45}Ca^{2+}$ fluxes in Syrian hamster insulinoma cells. *Endocrinology* 105: 1183–1190, 1979.
19. ETO, S., AND N. FLEISCHER. Regulation of thyrotropin (TSH) release and production in monolayer cultures of transplantable TSH-producing mouse tumors. *Endocrinology* 98: 114–122, 1976.
20. ETO, S., J. M. WOOD, M. HUTCHINS, AND N. FLEISCHER. Pituitary $^{45}Ca^{2+}$ uptake and release of ACTH, GH and TSH: effects of verapamil. *Am. J. Physiol.* 226: 1315–1320, 1974.
21. FELDMAN, J. M., AND H. E. LEBOVITZ. Serotonin inhibition of *in vitro* insulin release from golden hamster pancreas. *Endocrinology* 86: 66–70, 1970.
22. FLECKENSTEIN, A. Specific pharmacology of calcium in myocardium, cardiac pacemakers, and vascular smooth muscle. *Annu. Rev. Pharmacol. Toxicol.* 17: 149–166, 1977.
23. FLECKMAN, A., J. ERLICHMAN, U. K. SCHUBART, AND N. FLEISCHER. Effect of trifluoperazine, D600, and phenytoin on depolarization- and TRH-induced TSH release from rat pituitary tissue. *Endocrinology* in press.
24. FLEISCHER, N., AND R. GUILLEMIN. Hypothalamic hypophysiotropic hormones. In: *Peptide Hormones*, edited by J. A. Parsons. London: Macmillan, 1976, p. 317–335.
25. FRANKEL, B. J., J. A. KROMHOUT, W. IMAGAWA, H. D. LANDAHL, AND G. M. GRODSKY. Glucose-stimulated ^{45}Ca uptake in isolated rat islets. *Diabetes* 27: 365–369, 1978.
26. GALPER, J. D., AND W. A. CATTERALL. Inhibition of sodium channels by D600. *Mol. Pharmacol.* 15: 174–178, 1979.
27. GERSHENGORN, M. C. Thyrotropin-releasing hormone stimulation of prolactin release: evidence for a membrane potential-independent, Ca^{2+}-dependent mechansim of action. *J. Biol. Chem.* 255: 1801–1803, 1980.
28. GERSHENGORN, M. C., M. J. REBECCHI, E. GERAS, AND C. O. AREVALO. Thyrotropin-releasing hormone (TRH) action in mouse thyrotropic tumor cells in culture: evidence against a role for adenosine 3′5′ monophosphate

as a mediator of TRH-stimulated thyrotropin release. *Endocrinology* 107: 665–670, 1980.

29. GRANT, G., W. VALE, AND R. GUILLEMIN. Interaction of thyrotropin-releasing factor with membrane receptors of pituitary cells. *Biochem. Biophys. Res. Commun.* 46: 28–34, 1972.
30. GRILLO, T. A. I., A. J. WHITTY, H. KIRKMAN, P. P. FOA, AND S. D. KOBERNICK. Biological properties of a transplantable islet-cell tumor of the golden hamster. I. Histology and histochemistry. *Diabetes* 16: 409–414, 1967.
31. HASLAM, R. J., AND J. A. LYNHAM. Relationship between phosphorylation of blood platelet proteins and secretion of platelet granule constituents. I. Effects of different aggregating agents. *Biochem. Biophys. Res. Commun.* 77: 714–722, 1977.
32. HELLMAN, B., J. SEHLIN, M. SODERBERG, AND I. -B. TALJEDAL. The mechanism of action of chloromercuribenzene-p-sulphonic acid as insulin secretagogue: fluxes of calcium, sodium and rubidium in islets exposed to mercurial and membrane-active antagonists. *J. Physiol. London* 252: 701–712, 1975.
33. KAO, C. Y. Tetrodotoxin, sagitoxin and their significance in the study of excitation phenomena. *Pharmacol. Rev.* 18: 997–1049, 1966.
34. KIDOKORO, Y. Spontaneous calcium action potentials in a clonal pituitary cell line and their relationship to prolactin secretion. *Nature London* 258: 741–742, 1975.
35. KIRKMAN, H. A. A preliminary report concerning tumors observed in Syrian hamsters. *Stanford Med. Bull.* 20: 163, 1962.
36. KRUEGER, B. K., J. FORN, AND P. GREENGARD. Depolarization-induced phosphorylation of specific proteins mediated by calcium ion influx in rat brain synaptosomes. *J. Biol. Chem.* 252: 2764–2773, 1977.
37. LABRIE, F., N. BARDEN, G. POIRIER, AND A. DELEAN. Binding of thyrotropin-releasing hormone to plasma membranes of bovine anterior pituitary gland. *Proc. Natl. Acad. Sci. USA* 69: 283–287, 1972.
38. LABRIE, F., L. LAGACE, L. FERLAND, M. BEAULIEU, J. MASSICOTTE, AND N. RAYMOND. New aspects of the control of pituitary secretion. *Ann. Clin. Res.* 10: 109–119, 1978.
39. LERNMARK, A. The significance of 5-hydroxytryptamine for insulin secretion. *Horm. Metab. Res.* 3: 305–309, 1971.
40. LEVIN, R. M., AND B. WEISS. Binding of trifluoperazine to the calcium-dependent activator of cyclic nucleotide phosphodiesterase. *Mol. Pharmacol.* 13: 690–697, 1977.
41. LEW, P. D., AND T. P. STOSSEL. Calcium transport by macrophage plasma membranes. *J. Biol. Chem.* 255: 5841–5846, 1980.
42. LIPICKY, R. J., D. L. GILBERT, AND I. M. STILLMAN. Diphenylhydatoin inhibition of sodium conductance in squid giant axon. *Proc. Natl. Acad. Sci. USA* 69: 1758–1760, 1972.
43. LUFT, R., S. EFENDIC, AND T. HOKFELT. Somatostatin—both hormone and neurotransmitter. *Diabetologia* 14: 1–13, 1978.
44. LYONS, R. M., N. STANFORD, AND P. W. MAJERUS. Thrombin-induced protein phosphorylation in human platelets. *J. Clin. Invest.* 56: 924–936, 1975.
45. MALAISSE, W. J. Calcium fluxes and insulin release in pancreatic islets. *Biochem. Soc. Trans.* 5: 872–875, 1977.
46. MALAISSE, W. J., A. HERCHUELZ, G. DEVIS, G. SOMERS, A. C. BOSCHERO, J. C. HUTTON, S. KAWAZU, A. SENER, I. J. ATWATER, G. DUNCAN, B. RIBALET, AND E. ROJAS. Regulation of calcium fluxes and their regulatory roles in pancreatic islets. *Ann. NY Acad. Sci.* 307: 562–581, 1978.
47. MATTHEWS, E. K., AND Y. SAKAMOTO. Electrical characteristics of pancreatic islet cells. *J. Physiol. London* 246: 421–437, 1975.
48. MEANS, A. R., AND J. R. DEDMAN. Calmodulin—an intracellular calcium receptor. *Nature London* 285: 73–77, 1980.
49. MICHAELSON, D. M., AND S. AVISSAR. Ca^{2+}-dependent protein phosphorylation of purely cholinergic Torpedo synaptosomes. *J. Biol. Chem.* 254: 12542–12546, 1979.
50. MONTAGUE, W. Cyclic nucleotide, calcium and insulin secretion. In: *Cyclic 3'5' Nucleotides: Mechanisms of Action*, edited by H. Cramer and J. Schultz. New York: Wiley, 1977, p. 133–146.
51. NAWRATH, H., R. E. TEN EICK, T. F. McDONALD, AND W. TRAUTWEIN. On the mechanism underlying the action of D 600 on slow inward current and tension in mammalian myocardium. *Circ. Res.* 40: 408–414, 1977.
52. OZAWA, S., AND N. KIMURA. Membrane potential changes caused by thyrotropin-releasing hormone in the clonal GH_3 cell and their relationship to secretion of pituitary hormone. *Proc. Natl. Acad. Sci. USA* 76: 6017–6021, 1979.
53. PERRY, J. G., L. McKINNEY, AND P. DE WEER. The cellular mode of action of the anti-epileptic drug 5,5'-diphenylhydantoin. *Nature London* 272: 271–273, 1978.
54. RUBIN, R. P. The role of calcium in the release of neurotransmitter substances. *Pharmacol. Rev.* 22: 389–428, 1970.
55. SCHUBART, U. K., J. ERLICHMAN, AND N. FLEISCHER. The role of calmodulin in the regulation of protein phosphorylation and insulin release in hamster insulinoma cells. *J. Biol. Chem.* 255: 4120–4124, 1980.
56. SCHUBART, U. K., N. FLEISCHER, AND J. ERLICHMAN. Ca^{2+}-dependent protein phosphorylation and insulin release in intact hamster insulinoma cells. *J. Biol. Chem.* 255: 11063–11066, 1980.
57. SCHUBART, U. K., S. SHAPIRO, N. FLEISCHER, AND O. M. ROSEN. Cyclic adenosine 3'5'-monophosphate-mediated insulin secretion and ribosomal protein phosphorylation in a hamster islet cell tumor. *J. Biol. Chem.* 252: 92–101, 1977.
58. SHAPIRO, S., Y. KANEKO, S. G. BAUM, AND N. FLEISCHER. The role of calcium in insulin release from hamster insulinoma cells. *Endocrinology* 101: 485–493, 1977.
59. SIEGHART, W., T. C. THEOHARIDES, S. L. ALPER, W. W. DOUGLAS, AND P. GREENGARD. Calcium-dependent protein phosphorylation during secretion by exocytosis in the mast cell. *Nature London* 275: 329–331, 1978.
60. SOHN, R. S., AND J. A. FERRENDELLI. Inhibition of Ca^{2+} transport into rat brain synaptosomes by diphenylhydantoin. *J. Pharmacol. Exp. Ther.* 185: 272–275, 1973.
61. SUGDEN, M. C., M. R. CHRISTIE, AND S. J. H. ASHCROFT. Presence and possible role of calcium-dependent regulator (calmodulin) in rat islets of Langerhans. *FEBS Lett.* 105: 95–100, 1979.
62. TARASKEVICH, P. S., AND W. W. DOUGLAS. Action potentials occur in cells of the normal anterior pituitary gland and are stimulated by the hypophysiotropic peptide thyrotropin-releasing hormone. *Proc. Natl. Acad. Sci. USA* 74: 4064–4067, 1977.
63. TASHJIAN, A. H., M. E. LOMEDICO, AND D. MAINA. Role of calcium in the thyrotropin-releasing hormone-stimulated release of prolactin from pituitary cells in culture. *Biochem. Biophys. Res. Commun.* 81: 798–806, 1978.
64. VALE, W., R. BURGUS, AND R. GUILLEMIN. Presence of calcium ions as a requisite for the *in vitro* stimulation of TSH release by hypothalamic TRF. *Experientia* 23: 853–855, 1967.

65. VALE, W., C. RIVIER, AND M. BROWN. Regulatory peptides of the hypothalamus. *Annu. Rev. Physiol.* 39: 475–527, 1977.
66. WATKINS, D. T., AND S. DALY. Presence of sulfhydryl groups in glucagon secretion granule membranes. *Endocrinology* 103: 24–28, 1978.
67. WATKINS, D. T., AND M. MOORE. Effect of sulfhydryl-binding reagents on insulin release from isolated secretion granules. *Endocrinology* 95: 485–491, 1974.
68. WEISS, B., AND R. M. LEVIN. Mechanism for selectively inhibiting the activation of cyclic nucleotide phosphodiesterase and adenylate cyclase by antipsychotic agents. In: *Adv. Cyclic Nucleotide Res.* 9: 285–303, 1978.
69. WILBER, J. F., G. T. PEAKE, AND R. D. UTIGER. Thyrotropin release *in vitro*: stimulation by cyclic 3′,5′-adenosine monophosphate. *Endocrinology* 84: 758–760, 1969.
70. WOLFF, D. J., AND C. O. BROSTROM. Properties and functions of the calcium-dependent regulator protein. *Adv. Cyclic Nucleotide Res.* 11: 27–87, 1979.
71. WOODS, S., AND D. PORTE. Neural control of the endocrine pancreas. *Physiol. Rev.* 54: 596–619, 1974.

15

Calcium-Channel Blockers: Possible Mechanisms of Protective Effects on the Ischemic Myocardium

A. SCHWARTZ, G. GRUPP, R. W. MILLARD, I. L. GRUPP, D. A. LATHROP, M. A. MATLIB, P. L. VAGHY, AND J. R. VALLE

Departments of Pharmacology and Cell Biophysics, Physiology, Medicine, and Pediatrics, University of Cincinnati College of Medicine, Cincinnati, Ohio

AMONG THE DRUGS ADVOCATED to reduce morbidity and mortality from acute myocardial infarction, those referred to collectively as *Ca^{2+} antagonists* or *Ca^{2+}-channel blockers* offer a unique combination of actions. The most prominent Ca^{2+} antagonists are verapamil, nifedipine, and diltiazem. Nisoldipine and lidoflazine are also under investigation (Fig. 1). These agents differ widely in chemical structure and include both lipophilic and hydrophilic moieties. Their cardiac and vascular actions appear to be mediated at least partly by the inhibition of Ca^{2+} influx, which accompanies a voltage-dependent process. The diminution of Ca^{2+} influx, and possibly other actions as well, can result in depressed myocardial contractile force, decreased heart rate, shortened action potential duration, and coronary vasodilation. This combination of effects, when considered in the face of impending myocardial infarction or further damage to an ischemic or infarcted area, predicts a therapeutic modality that concomitantly diminishes myocardial metabolic demand, decreases systemic blood pressure, and improves myocardial blood flow to margins of the ischemic myocardium and within ischemic regions via improved coronary collateral blood flow. Calcium-channel blockers apparently exert this unique combination of actions.

There is evidence that some of the Ca^{2+}-channel blockers may penetrate the cell membrane and accumulate intracellularly. It is reasonable that at least two possible regulating sites exist for the slow channel, one on the external

DILTIAZEM

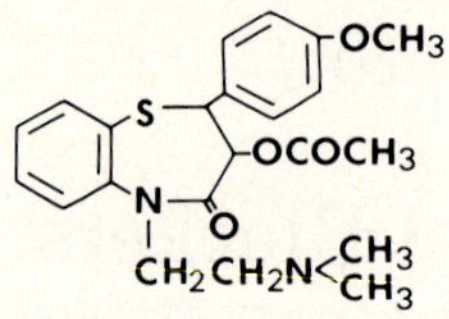

LIDOFLAZINE

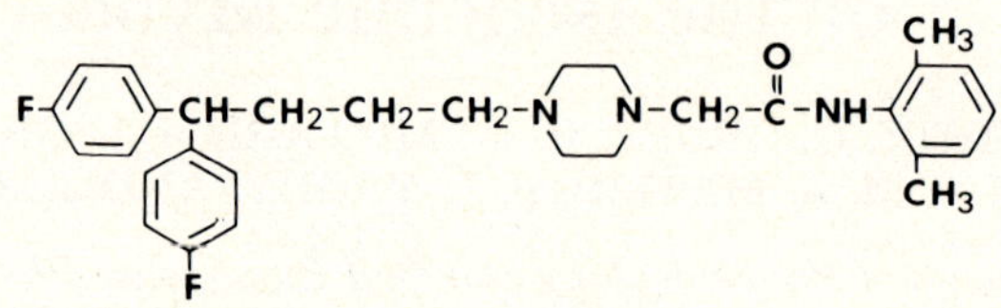

NIFEDIPINE

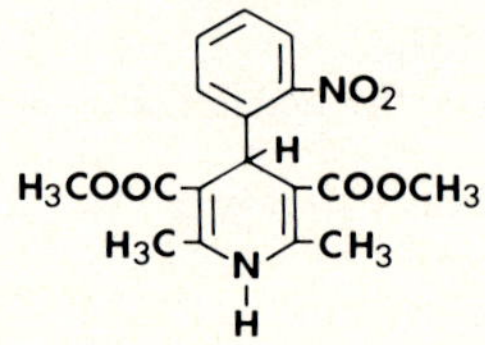

NISOLDIPINE

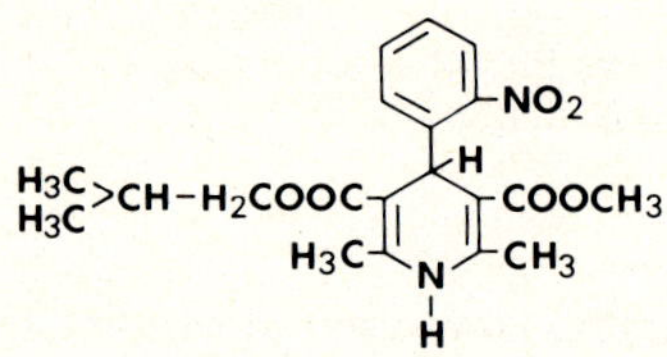

VERAPAMIL

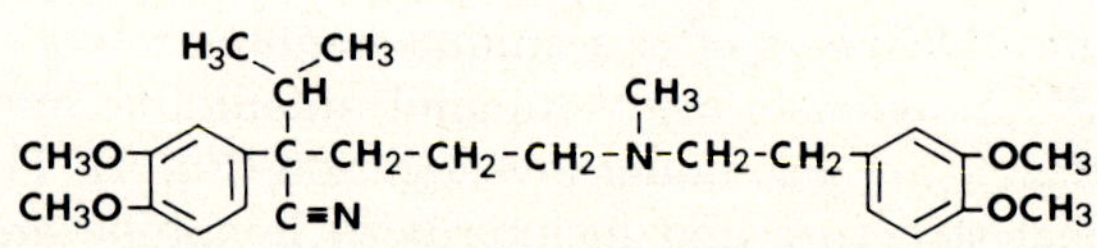

FIG. 1. Structural formulas of certain Ca^{2+}-channel blockers.

sarcolemmal surface and the other perhaps within the membrane or on its internal surface (Fig. 2). A Ca^{2+} blocker might exert its action through insertion into the sarcolemma with its hydrophilic chain directed outward and its lipophilic end directed inward. These agents might also exert significant effects on intracellular sites.

The major focus of this chapter is on the effects of Ca^{2+}-channel blockers in acute myocardial infarction. Special attention is given to the possible mechanisms by which these agents might exert effects beneficial to the jeopardized ischemic myocardium. The discussion is divided into sections presenting brief overviews with exemplary figures. The sections include vascular actions, electrophysiological effects, myocardial mechanics, subcellular actions, actions in ischemia, and finally, a summary of potential mechanisms.

VASCULAR EFFECTS OF CALCIUM-CHANNEL BLOCKERS

All the diverse chemical agents that are loosely classified as Ca^{2+} antagonists are potent relaxers of vascular smooth muscle and act probably by inhibiting some Ca^{2+}-dependent event. For example, in porcine coronary artery strips in vitro, nifedipine ($> 10^{-9}$ M), verapamil ($> 10^{-7}$ M), and diltiazem ($> 10^{-7}$ M) exert relaxant actions (7). Coronary flow in dogs is increased by nifedipine (11) and diltiazem (8). Coronary collateral flow to the ischemic myocardium is increased by verapamil (13) and diltiazem (8). Interestingly these agents appear to have different effects depending on whether the vessel is activated by a receptor-mediated pathway or by a voltage-dependent (e.g., KCl) site or sites (34).

The vascular effects of diltiazem and nisoldipine on coronary blood flow can be demonstrated in dogs with separately controlled and perfused circumflex and left anterior descending coronary arteries (Fig. 3). Dogs were anesthetized with pentobarbital sodium (30 mg/kg iv), anticoagulated with heparin sodium (5,000 U USP/kg iv), and artificially respired. The chest was opened in the left 5th intercostal space, and a pericardial cradle was constructed to support the heart. The left carotid artery was connected to the left circumflex coronary artery, and the femoral artery to the left descending coronary artery. Coronary flow in both arteries was kept constant and independent by in-line adjustable finger pumps. Mean coronary perfusion pressures were made equal to the systemic systolic arterial blood pressure. Control coronary flows ranged between 40 and 80 ml/min. Contractile force of the left ventricle was recorded from a strain gauge sutured to the anterior wall close to the apex of the heart. Heart rate, electrocardiogram, aortic pressure, coronary pressures, coronary flows, contractile force, left ventricular pressure (LVP), and the dP/d*t* of the LVP were recorded on a Grass P7 polygraph. The flows were measured by Biotronex electromagnetic flowmeters. This permitted us to change flow and/or pressure in one or both of the perfused areas. We could stop antegrade flow and collect retrograde blood to assess collateral flow during complete coronary

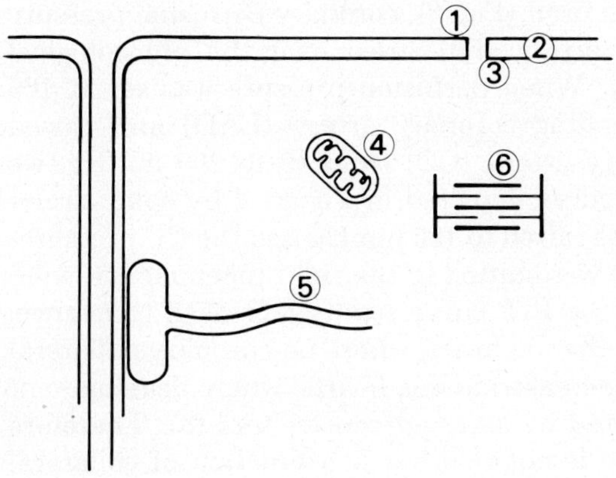

FIG. 2. Potential sites of action of Ca^{2+} antagonist. Site 1, exterior of sarcolemma; site 2, inside sarcolemma (possibly via calmodulin occupancy and/or phospholipids); site 3, on inner surface of sarcolemma; site 4, mitochondrial membrane; site 5, sarcoplasmic reticulum; site 6, troponin, tropomyosin, actin, and myosin.

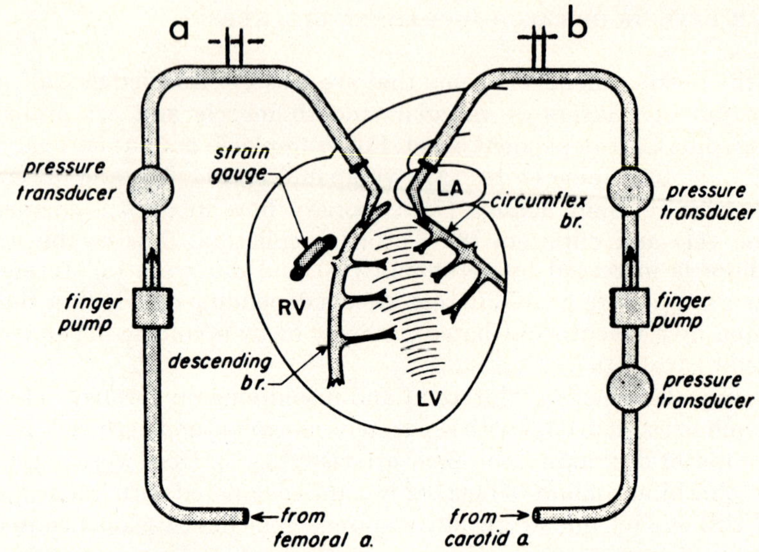

FIG. 3. Schematic perfusion of 2 coronary arteries in dog heart. Left anterior descending coronary artery (descending br.) perfused from left femoral artery (flow is adjusted through a finger pump). Left circumflex artery (circumflex br.) perfused from carotid artery. Valves a and b allow collection of backflow during descending (a) or circumflex (b) occlusion. RV and LV, right and left ventricles, respectively; LA, left atrium.

occlusion of one or both vessels (see Fig. 3). Pressure-flow curves were obtained at different perfusion pressure levels in the adjacent areas. Drugs were infused intravenously or directly into either one or both of the cannulated coronary arteries.

Pressure-Flow Relationships in the Circumflex and Left Anterior Descending Coronary Arteries

In the double-perfusion system used (Fig. 3), coronary perfusion pressure can be fixed at any level in one artery and varied over the physiological pressure range in the other artery. When perfusion pressure was set at 100 mmHg in the left anterior descending coronary artery (LAD) and varied stepwise in the circumflex coronary artery (CC) from 40 to 200 mmHg, we obtained the pressure-flow (P/F) curve depicted in Figure 4 by solid circles and a solid line. If LAD pressure was raised to 195 mmHg and the CC pressure-flow curve repeated, the P/F curve was shifted to the right (open circles, solid line in Fig. 4). Rightward shifts of the P/F curve signify increased resistance. This rightward shift was absent in the pig heart, where no coronary collateral arterial blood vessels exist; it was increased in dog hearts, where development of collateral blood flow is augmented by ameroid constrictors (8). Therefore the apparent increase in resistance is not real but is a function of collateral circulation. In the dog, when the perfusion pressure is higher in one coronary artery, blood flows from that area to other regions principally supplied by a second coronary artery. To describe this collateral connection, we use the term *leak flow*. The leak flow can be estimated from the distance between two P/F curves. It is about 14 ml/min in this experiment at 120 mmHg CC pressure. During intravenous infusion of diltiazem (4 $\mu g \cdot kg^{-1} \cdot min^{-1}$) for 10 min, the CC pressure-flow curve at 100 mmHg LAD pressure shifted to the left (solid circles, broken line in Fig. 4). At 120 mmHg CC pressure, diltiazem increased leak flow by 77%. When the P/F curves were repeated in the presence of diltiazem, with the LAD perfusion pressure at 195 mmHg (open circles, broken line in Fig. 4), the leak flow was about 23 ml/min or about 64% higher than in the control. Intravenous infusion of diltiazem (4 $\mu g \cdot kg^{-1} \cdot min^{-1}$) for 10 min dilated the coronary vessels maximally, since an additional injection of adenosine did not produce further coronary dilation and occlusion of the coronary vessels no longer produced reactive hyperemia. We concluded that diltiazem produces maximal dilation of the resistance vessels in the coronary vascular bed. The increase in leak flow, however, must have been produced partly by a dilation of the collateral vessels.

Intravenous infusion of nisoldipine (0.2 $\mu g \cdot kg^{-1} \cdot min^{-1}$) for 10 min caused identical effects on coronary blood flow and coronary leak flow, producing maximal vasodilation (tested with adenosine and induction of reactive hyperemia).

Coronary Arterial Occlusion Pressures and Backflow of Blood From Completely Occluded Coronary Arteries

In the dog heart preparation illustrated in Figure 3, separate perfusion of the two coronary arteries permits collateral blood flow to be estimated. For instance, when the CC pressure and flow are monitored with pressure and blood flow transducers, lowering the perfusion pressure in the LAD causes an increase of blood flow in the CC. For dogs the increment in flow, without

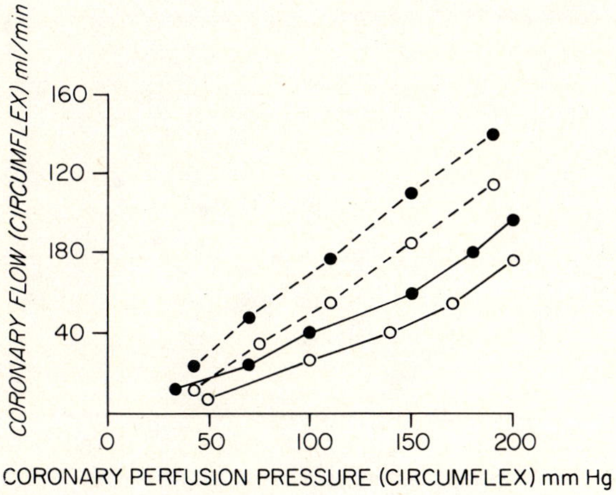

FIG. 4. Pressure-flow curves in CC at 100 mmHg LAD pressure (●) and 195 mmHg LAD pressure (○) during control (solid lines) and intravenous infusion of diltiazem (4 $\mu g \cdot kg^{-1} \cdot min^{-1}$) (broken lines).

experimentally induced collaterals, is about 5 ml/min. As a consequence, the pressure upon complete occlusion of a coronary artery, that is *zero flow pressure*, is actually never at zero (2), since there is always blood flowing in the occluded area that comes through collaterals from areas of the heart that are still perfused (10). This can be verified by occluding one coronary artery and collecting the flow from the occluded area, that is, the *backflow*. This flow is also about 5 ml/min and is entirely arterial in origin, as verified with measurements of blood Po_2 and Pco_2. When the existing collaterals are experimentally augmented by ameroid constrictors, the pressure during complete occlusion of a coronary artery can be close to the systemic pressure and in fact shows pulsation; the backflow then can exceed 50 ml/min.

In separate experiments we completely occluded the inflow into the LAD and varied perfusion pressure in the CC stepwise from 50 to 220 mmHg. Occlusion pressure in the LAD was lower at lower CC pressure. In Figure 5*A* the control observations are shown as solid circles. Infusion of diltiazem (0.5–4 $\mu g \cdot kg^{-1} \cdot min^{-1}$) lowered LAD occlusion pressures, but these pressures still depend on CC perfusion pressure. One interpretation of the data in Figure 5*A* is that diltiazem is reaching the occluded area and is dilating the arterial resistance vessels. The lowest occlusion pressure during infusion of diltiazem (4 $\mu g \cdot kg^{-1} \cdot min^{-1}$) was 8.5 mmHg, whereas during control observations the

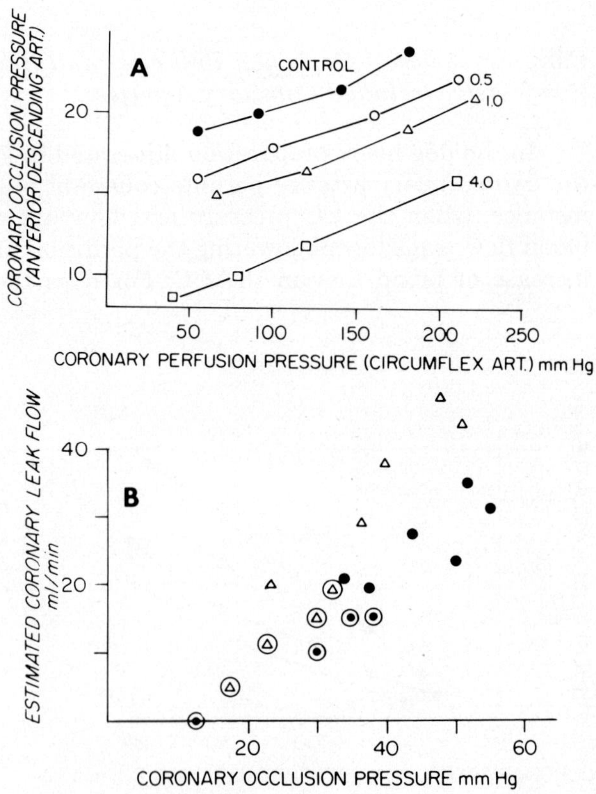

FIG. 5. *A*: dependence of LAD coronary occlusion pressure on CC perfusion in dog heart. ● Control; ○ diltiazem 0.5 at $\mu g \cdot kg^{-1} \cdot min^{-1}$ iv; △ diltiazem at 1.0 $\mu g \cdot kg^{-1} \cdot min^{-1}$ iv; □ diltiazem at 4.0 $\mu g \cdot kg^{-1}$ min^{-1} iv. *B*: relationship of estimated leak flow to coronary occlusion pressure. For LAD: ● control; ⊙ during infusion of diltiazem (4 $\mu g \cdot kg^{-1} \cdot min^{-1}$). For CC: △ control; ◬ during infusion of diltiazem (4 $\mu g \cdot kg^{-1} \cdot min^{-1}$).

lowest occlusion pressure was 19 mmHg. The same experiment with nisoldipine yielded similar results.

Complete occlusion of one coronary artery also affected the artery that continued to be perfused, causing a drop in perfusion pressure in the artery (flow remained constant because of the finger pump). Since we determined P/F relationships in the two cannulated coronary arteries with and without drugs, we could calculate how much flow would have balanced the drop in perfusion pressure. That flow is called *estimated leak flow* because presumably this amount of flow leaked from the perfused arterial bed to the occluded arterial bed. When leak flow was plotted against the occlusion pressure, we found that at high occlusion pressures leak flows also were high and vice versa (Fig. 5*B*). Diltiazem and nisoldipine (data shown only for diltiazem) did not disturb this relationship but moved it to very low occlusion pressures and very low estimated leak flows (Fig. 5*B*). We conclude that diltiazem and nisoldipine dilated all arterial vessels, collaterals, and major thoroughfare vessels and also the arteriolar resistance vessels.

Thus diltiazem and nisoldipine are as effective as adenosine in producing maximal dilation of coronary arteries. The concentrations of diltiazem and nisoldipine required for maximal dilation upon infusion in the LAD were about 10^{-7} M and 10^{-8} M, respectively.

Regional Flow Distribution

To delineate the effects of Ca^{2+} antagonists on regional myocardial blood flow distribution, experiments were done in instrumented dogs. Myocardial blood flow was measured using 15-μm, radionuclide-labeled microspheres. Both endocardial and epicardial flows were increased by all Ca^{2+}-channel blockers. The effects of diltiazem and nisoldipine are compared in Figure. 6. Although both agents increase blood flow to all myocardial regions, they have strikingly different effects on the specific regions. For example, diltiazem increased epicardial blood flow slightly more than endocardial blood flow, resulting in a small reduction of the transmural flow distribution ratio from 1.57 to 1.20, indicating a continuing predominance of endocardial perfusion. In contrast, nisoldipine increased blood flow to the epicardial layer considerably more than to the endocardium, significantly reducing the flow distribution ratio from 1.28 to 0.67. This cannot be interpreted as subendocardial ischemia, but it does indicate markedly different coronary vascular reserves in different myocardial layers as revealed by the action of these representative Ca^{2+}-channel blockers. The effects of the agents in this experiment were not due to any significant actions on heart rate or blood pressure.

ELECTROPHYSIOLOGICAL EFFECTS OF CALCIUM-CHANNEL BLOCKERS

As discussed in the introduction, the Ca^{2+}-channel blockers can interfere with myocardial excitation-contraction coupling, they may depress automat-

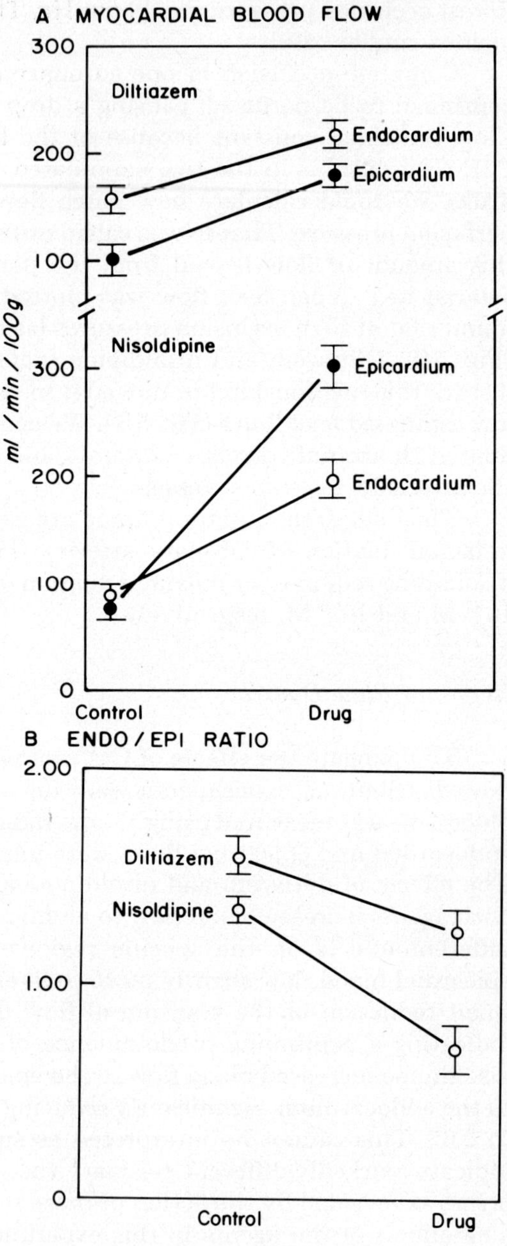

FIG. 6. *A:* effects of intravenous infusion of diltiazem (20 $\mu g \cdot kg^{-1} \cdot min^{-1}$; *top*) and nisoldipine (2 $\mu g \cdot kg^{-1} \cdot min^{-1}$; *bottom*) on myocardial blood flow in dog heart, measured with 15-μm, radionuclide-labeled microspheres. *B*: endocardial and epicardial flow ratios. Values are means ± SE.

icity of spontaneous cardiac tissue, and they can decrease coronary vascular tone (7). The effects of these compounds are reversed by increasing extracellular Ca^{2+} and by addition of catecholamines. The mechanism of action of this diverse group of compounds is not fully understood, but it seems universally accepted that they somehow interfere with transmembrane Ca^{2+} flux. For cardiac tissue it is generally believed that they block entry of Ca^{2+} into the cell through the slow channels, which are activated after the initiation of each cardiac depolarization. Since these compounds were categorized as a distinct pharmacological group in the late 1960's, it has been recognized that some of them (e.g., nifedipine) have potent effects on coronary resistance but little effect on normal atrioventricular (AV) conduction (1, 27). Others (e.g, verapamil), however, are not such potent vasodilators, but they significantly slow AV conduction (37). These findings suggest that the effects on contractility and slow conduction produced by Ca^{2+} antagonists may not be as interdependent as believed. Perhaps the negative inotropic effects are not entirely mediated by blockade of the slow inward current. Alternatively, perhaps the effects on AV conduction are not totally due to abolishing initiation of slow-channel action potentials, as presently thought.

In these studies we investigated the effects of three drugs—verapamil, diltiazem, and nisoldipine—on isolated canine Purkinje fiber tension development and transmembrane electrical activity. We were interested in whether each drug had concurrent effects on contractile and electrical parameters. If not, perhaps the differences might help to explain the apparently paradoxical differences in the order of specificity these three drugs have on various areas of the cardiovascular system (e.g., coronary resistance and AV conduction). Free-running Purkinje fibers obtained from adult canine hearts were used. The fibers, less than 1 mm in diameter and 7 mm in length, were dissected from their ventricular muscle connections and mounted in a modified Hodgkin-Horowicz muscle chamber that had a total volume of 0.5 ml. The fibers were superfused with Tyrode's solution heated to 37°C and gassed with 100% oxygen. The Tyrode's solution consisted of (in mmol/l) Na^+, 153.0; K^+, 5.4; Cl^-, 162.9; Ca^{2+}, 2.7; Mg^{2+}, 1.05; *N*-2-hydroxyethylpiperazine-*N*′-2-ethanesulfonic acid (HEPES) buffer, 5.04; and glucose, 11.0. Transmembrane potentials were recorded using conventional microelectrode techniques. Tension was simultaneously recorded by connecting one end of the fiber to an RCA 5734 mechanoelectronic transducer tube mounted on a micromanipulator. Cumulative dose-response relations were established for each drug over the range of 10^{-10}–10^{-5} M. Each drug produced the expected excitation-contraction uncoupling (Fig. 7). The transmembrane voltage and tension recordings on the *left* of Figure 7 are control recordings; those on the *right* show the effect of a 30-min exposure to 3×10^{-6} M diltiazem, 10^{-6} M verapamil, and 10^{-6} M nisoldipine. In addition to abolishing the contractile response, each drug also decreased the plateau amplitude, the action potential duration at 50% repolarization (plateau duration), and the maximum rising velocity of the action potential upstroke (Fig. 7). These are configurational changes that would be expected after blockade of the slow inward current (17). However, the effects of each of these drugs on final repolarization and total action potential duration were different. Nisoldipine and diltiazem shortened the total duration of the action potential slightly, whereas verapamil greatly lengthened it. This interesting differential effect is illustrated both in Figure 7 and in Figure 8*C*. The difference in the ability to wash out the effects of verapamil on the 50% and 90% repolarization times suggests that these represent different specific effects. For

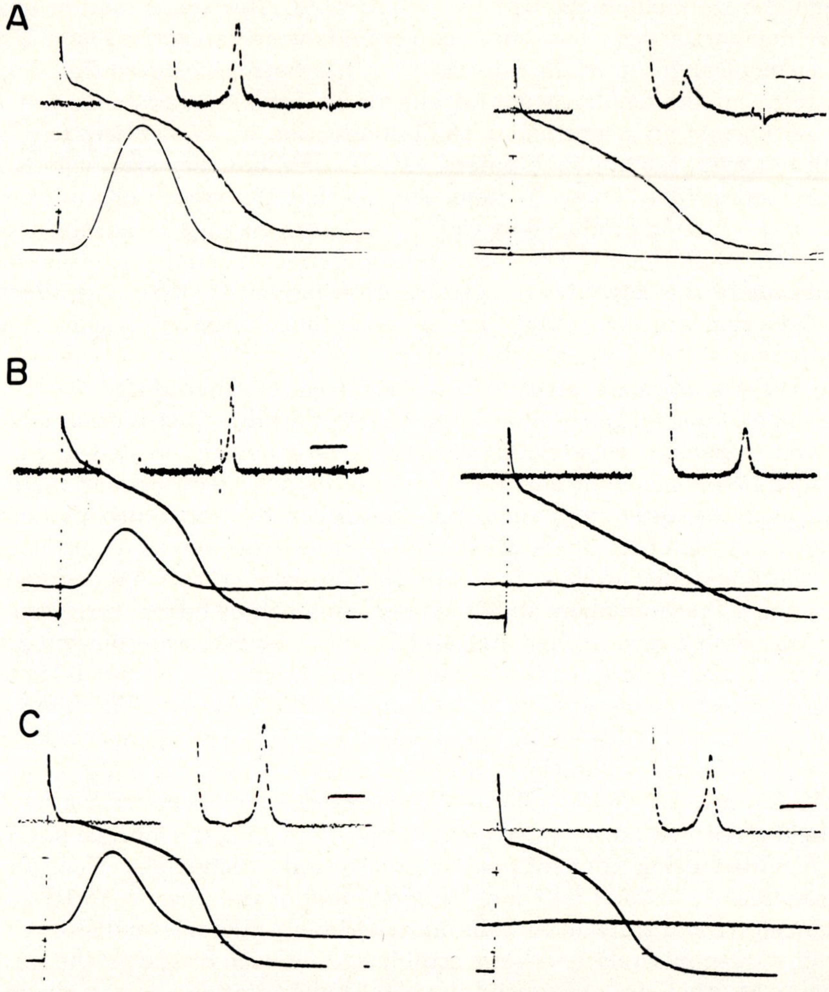

FIG. 7. Simultaneous transmembrane potential and tension recordings from canine Purkinje fibers showing effects of 3×10^{-6} diltiazem (*A*), 10^{-6} M verapamil (*B*), and 10^{-6} M nisoldipine (*C*). Each panel contains 3 traces: *top* is 1st derivative of voltage with respect to time, *middle* is transmembrane voltage, and *bottom* is tension development. Recordings on *left* were made during control, those on *right* were made 30 min after administration of drug. Calibrations are for derivative 200 $V \cdot s^{-1}$/div, 1 ms/div; for transmembrane voltage 20 mV/div, 50 ms/div; and for tension 10 mg/div, 50 ms/div. Calibration marks appearing on voltage recordings are 1 div wide and 5 div high.

each drug, tension development is first decreased at doses that have no apparent effect on the plateau duration (APD_{50}; Fig. 8). If plateau shortening is used as an index of blockade of the slow inward current, this finding would suggest that the negative inotropic action does not occur subsequent to elimination of transmembrane Ca^{2+} entry during the action potential. Rather it might be suggested that these drugs depress the Ca^{2+}-releasing mechanism

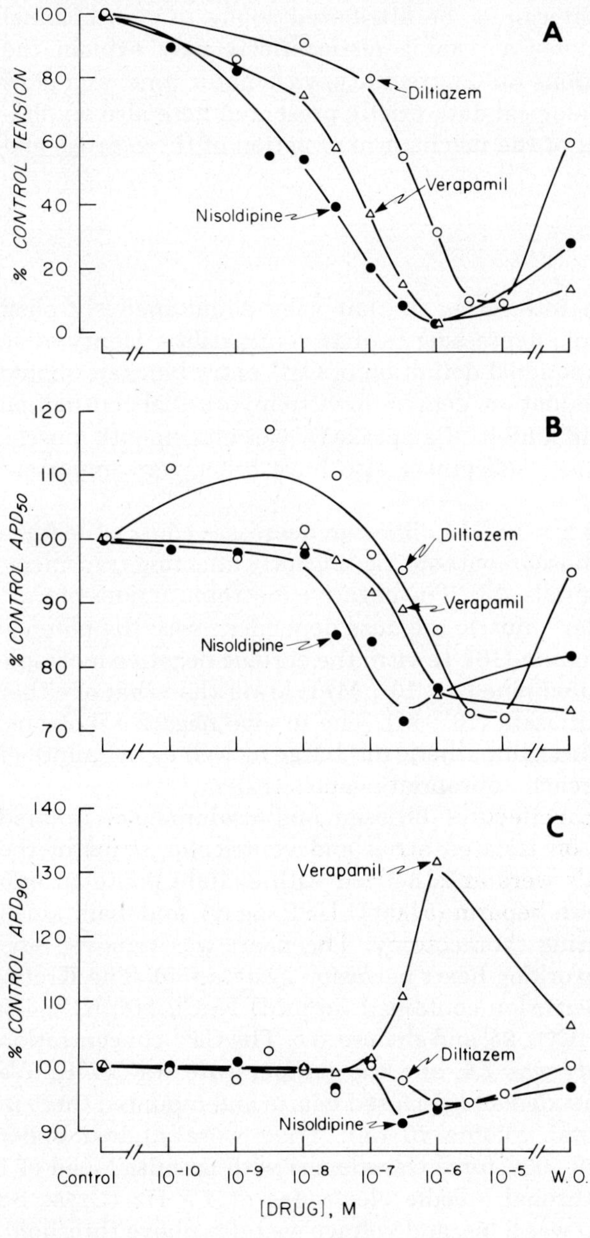

FIG. 8. Dose-response relations obtained by successive 30-min exposures to doses (10^{-10} M–10^{-5} M) of diltiazem (○), verapamil (△), and nisoldipine (●). Washout (W.O.) measurements were obtained 30 min after returning to normal Tyrode's solution after highest dose of each drug. Points shown on graphs are means of values obtained from 3 to 10 canine cardiac Purkinje fibers. Effects of each drug on peak tension development (*A*) and on action potential duration (APD) at 50% of repolarization (*B*) and 90% of repolarization (*C*) are expressed as percent of control measurement made in each fiber.

inside the cell. The caveat, however, is that several currents control the action potential plateau duration (20) and that such a conclusion cannot be made until dose-response relations are established for blockade of the slow inward current under voltage-clamp conditions. These data are presented to illustrate that these three Ca^{2+} antagonists have different effects on the electrical activity of the cardiac membrane that cannot be attributed solely to slow-channel inhibition. These different non–Ca^{2+} antagonistic effects may explain the paradoxical differences in actions on contractility, vasomotor tone, and slow conduction. The electrophysiological data briefly presented here also emphasize the lack of understanding of the mechanism of action of these drugs and the need for further study.

MECHANICAL EFFECTS

Calcium antagonists are interesting in that they simultaneously elicit smooth muscle relaxation while depressing cardiac contractility. Henry et al. (12) has proposed that the functional definition of Ca^{2+} entry blockers should include agents that cause vasodilation, depress in vitro myocardial contraction at concentrations $>10^{-6}$ M, and inhibit ^{45}Ca uptake by vascular smooth muscle. It is entirely possible that these substances also have other very important effects.

It has been reported that 2×10^{-6} M diltiazem decreases contractile force of isolated cardiac papillary muscles without significantly affecting transmembrane resting and action potentials (23). The negative inotropic actions of Ca^{2+} antagonists on isolated papillary muscle are dose dependent with the potency nifedipine > verapamil > diltiazem (16). In vitro the cardiac negative inotropic threshold concentration for nifedipine (2×10^{-7} M) is lower than that of either verapamil (5×10^{-7} M) or diltiazem (10^{-7} M). The in vivo negative inotropic effects may be masked by reflex sympathetic discharge as well as by failure of therapeutic plasma levels to reach appropriate values (12).

We studied the mechanical effects of diltiazem and nisoldipine on isolated work-performing hearts and on isolated atrial and ventricular strips of the guinea pig heart. The animals were anesthetized with sodium pentobarbital (30 mg/kg), anticoagulated with heparin (5,000 U USP/kg iv), and maintained with artificial respiration during thoracotomy. The heart was removed and immediately suspended in a working heart perfusion system (35). The Krebs-Henseleit solution used for perfusion contained (in mM) NaCl, 118; KCl, 4.7; $MgSO_4$, 1.2; KH_2PO_4, 1.2; $NaHCO_3$, 25; and glucose, 5.5. The Ca^{2+} concentration was either 0.9 or 2.5 mM, pH was 7.4, and the temperature was 37°C. We dissected strips from the suspended and perfused hearts and mounted them in the same bath chamber (total volume 70 ml). They were studied under isometric conditions (Grass FT. 03C force transducer) with an initial load of 1 g, stimulation was applied through needle electrodes at 1.5 Hz (Grass S8 stimulator), stimulus duration was 2 ms, and voltage was 20% above threshold. Contractile force and its first derivative dF/dt (Grass 7 P20C differentiator) were recorded for both strips. The preparations were equilibrated for 90–120 min. All dose-response curves were obtained on a cumulative basis.

To study the mounted heart under work conditions, we allowed the perfusion fluid (venous return) to enter through the left atrium into the ventricle. It was pumped by the ventricle through the aortic resistance and the coronary system. A strain-gauge arch was sutured to the right ventricle to record isometric wall tension from the ventricle, and a catheter was introduced into the left ventricle to record left intraventricular pressure and its derivative dP/d*t*. Heart rate, electrocardiogram, coronary flow, and temperature of the perfusion fluid were recorded, and venous return was monitored with a Biotronex electromagnetic flowmeter. Cardiac output, cardiac work, and coronary resistance were calculated from the recorded parameters. Because changes in the concentrations of agonists and antagonists in the perfusion may also alter ion concentration, temperature, or flow, we infused all agents with Harvard infusion pumps at rates of 0.04–0.91 ml/min (added to venous return of about 26–34 ml/min). In so doing we were able to deliver exact concentrations of the agents, to start and stop infusions at precise times, and to add additional drugs at any time.

We first determined the effects of diltiazem and nisoldipine on contractile force in isolated atria and papillary muscles to estimate the approximate dose to be administered to isolated work-performing hearts. The results are shown in Tables 1 and 2. The mean effective dose (ED_{50}) of diltiazem that produced negative inotropic effects in the left atrial strips was 2×10^{-6} M, whereas that of nisoldipine was 3.4×10^{-7} M (Table 1). We therefore decided to use 10^{-6} M diltiazem and 10^{-7} M nisoldipine in the isolated work-performing guinea pig preparation to test the effects of the two agents on myocardial performance at Ca^{2+} concentrations of 0.9 and 2.5 mM (Table 2). We selected 0.9 mM because it is considered the physiological Ca^{2+} concentration (3), whereas 2.5 mM is the Ca^{2+} concentration of the Krebs-Henseleit solution. At 2.5 mM Ca^{2+}, nisoldipine produced an insignificant increase in heart rate, arterial resistance, dP/d*t* (and therefore cardiac work), and the staircase response, but did maximally dilate the coronary arteries (based on comparison with adenosine effects). Diltiazem also insignificantly altered heart rate, arterial resistance, dP/d*t*, cardiac work, and the staircase response. Coronary flow, however, increased maximally. The effects at the 0.9-mM Ca^{2+} concentration were reduced but showed a similar pattern. Note that neither agent at the concentration given had any pronounced effect on AV conduction of the isolated heart.

INTRACELLULAR EFFECTS AND INORGANIC PHOSPHATE

We have reported (22) a highly significant protective effect of diltiazem on experimentally induced myocardial infarction. We attributed this beneficial

TABLE 1. *Isolated muscle strips of guinea pig heart, ED_{50} of negative inotropic effects*

	Diltiazem, M	Nisoldipine, M
Left atrium	2×10^{-6}	3.4×10^{-7}
Right ventricular papillary muscle	3.5×10^{-6}	1.2×10^{-7}

$n = 4$.

TABLE 2. *Mechanical effects in isolated work-performing guinea pig hearts*

	$[Ca^{2+}]$, mM					
	Control		10^{-6} M Diltiazem		10^{-7} M Nisoldipine	
	0.9	2.5	0.9	2.5	0.9	2.5
Heart rate, beats/min	160 ± 18	215 ± 20	150 ± 20	190 ± 18	176 ± 24	242 ± 22
Arterial pressure, mmHg*	57/27	61/31	50/20	55/25	60/30	64/42
Left ventricle, dP/dt_{max}, mmHg/s	840 ± 16	1,090 ± 28	720 ± 18	980 ± 15	920 ± 28	1,240 ± 47
Coronary flow, ml/min	8.4 ± 0.5	9.3 ± 0.2	12.5 ± 0.6	14.8 ± 0.9	13.8 ± 0.4	14.9 ± 0.7
Cardiac work, mmHg · ml·min^{-1}	1,385 ± 71	1,518 ± 65	1,155 ± 88	1,320 ± 72	1,485 ± 74	1,749 ± 102
Staircase response† mmHg·s^{-1}·(100 beats)$^{-1}$	98 ± 28	155 ± 24	74 ± 31	120 ± 25	94 ± 38	140 ± 13

* Systolic/diastolic. † Calculated as increase in dP/dt per 100 beats. Values are means ± SE; $n = 5$.

effect partly to an inhibition of reperfusion-induced damage to mitochondria that we thought was strictly an in vivo action of diltiazem (Fig. 9). More recent experiments, however, have revealed some intriguing new effects of the Ca^{2+} blockers that may involve intracellular sites.

During myocardial ischemia, the level of intracellular P_i rises (9). It is well known that P_i can produce mitochondrial swelling, which in turn reduces oxidative phosphorylation. During studies of in vitro simulation of ischemic conditions with cardiac mitochondria, Vaghy et al. (34) showed that both verapamil and diltiazem effectively prevent the loss of mitochondrial oxidative phosphorylation that we think occurs secondary to the increase in free P_i.

In the experiments described here, mitochondria were isolated from the rabbit heart in isolation medium containing 180 mM KCl, 10 mM ethylene glycol-bis(β-aminoethylether)-*N,N'*-tetraacetic acid (EGTA), and 0.5% bovine serum albumin (Sigma fraction V); pH was adjusted to 7.4 by adding tris

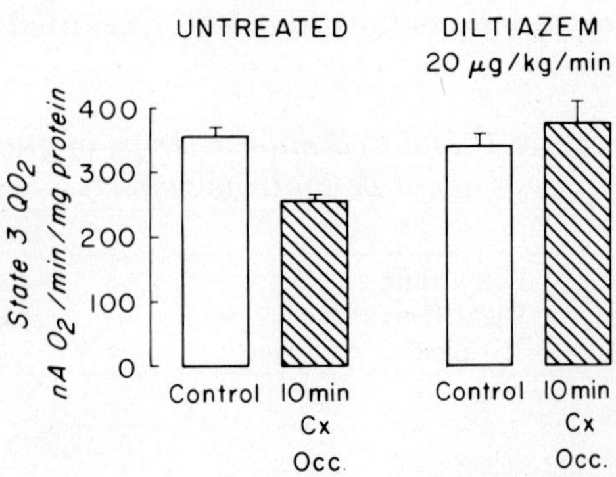

FIG. 9. Effect of diltiazem infusion on mitochondria isolated from ischemic myocardium. Dogs were treated with diltiazem (20 μg·kg^{-1}·min^{-1}) 20 min before occlusion and during the 10 min of occlusion of LAD. Mitochondria were isolated from control and ischemic areas of heart. Respiration was determined at 30°C in 250 mM sucrose, 10 mM HEPES, 2.6 mM P_i (pH 7.0) with glutamate + malate as substrates.

(hydroxymethyl)aminomethane (Tris) base at 4°C. The ventricular tissue was minced with a pair of scissors and then homogenized with an electrically driven teflon pestle in a glass homogenizing vessel. The homogenate was centrifuged at 500 *g* for 10 min in a Beckman J-21C centrifuge fitted with a JA-20 rotor. The supernatant was transferred through two layers of cheesecloth into clean centrifuge tubes and centrifuged at 8,000 *g* for 10 min. The pellet was rinsed, resuspended, and centrifuged at 8,000 *g* for 10 min. This washing procedure was repeated and the final mitochondrial pellet was resuspended in a small volume of isolation medium to make a mitochondrial protein concentration of 20–30 mg/ml. The protein concentration was determined by the method of Lowry et al. (18). The oxygen consumption of mitochondria was determined polarographically with the aid of an Oxygraph (Gilson Medical Electronics) fitted with a Clark type of oxygen electrode. In our initital studies, we used a conventional sucrose-containing assay medium to assay oxygen consumption of isolated mitochondria. In these conditions only very high concentrations of diltiazem or verapamil had an effect, and this was limited to inhibition of state 3 respiration (Fig. 10). Because sucrose-containing solutions are certainly unphysiological, we attempted to simulate conditions that are thought to exist in the cell. Thus studies were carried out in a 150-mM KCl medium containing 1 mM pyruvate and different concentrations of potassium phosphate buffer at 37°C. The state 3 respiration and adenosine diphosphate:oxygen (ADP:O) ratios were determined according to Estabrook (6). Osmotic volume changes of mitochondria were measured by determining the change in the light absorbance at 640 nm with the aid of an Aminco DW 2a spectrophotometer at 37°C.

Elevation of the P_i concentration from 2 to 5 mM resulted in a marked drop in the state 3 rate of respiration (state 3 Qo_2). Both verapamil and diltiazem prevented the decrease in the state 3 Qo_2 induced by 5 mM phosphate in a dose-dependent manner (Fig. 11). In this respect verapamil was found to be more potent than diltiazem. Photometric determination of the osmotic volume changes of mitochondria indicate that 2 mM P_i had only a moderate effect, but 5 mM P_i induced a considerable decrease in the light absorbance; that is, there was a swelling of mitochondria. Although the changes produced

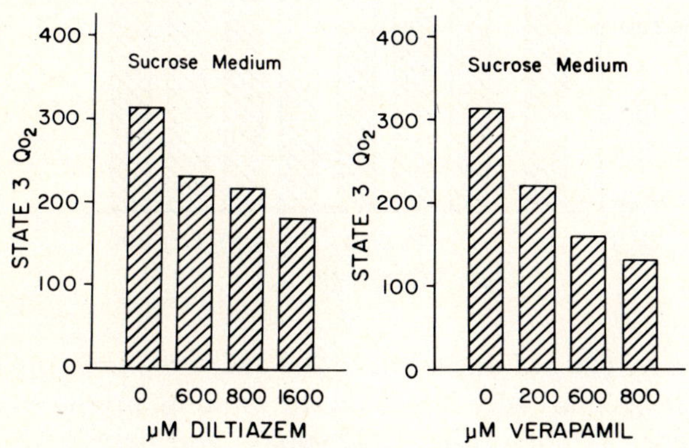

FIG. 10. Effects of diltiazem and verapamil on state 3 Qo_2 of mitochondria in sucrose medium. Assay medium contained 250 mM sucrose, 1 mM pyruvate, and 2 and 5 mM potassium phosphate buffer (pH 7.0 at 37°C).

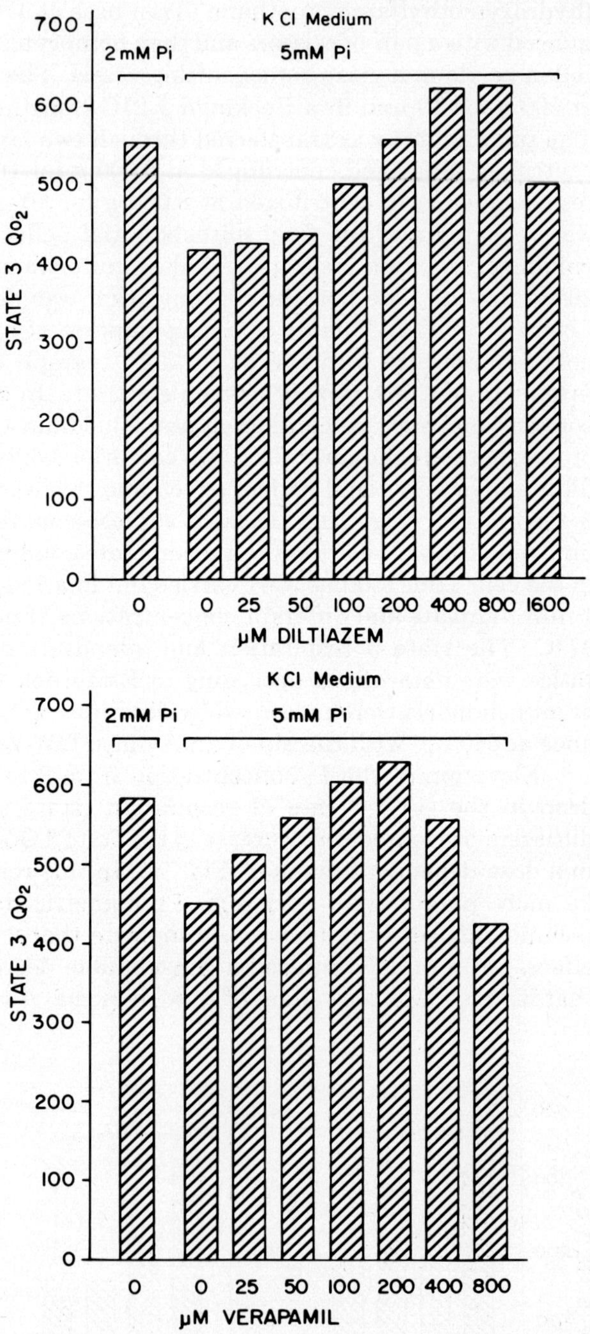

FIG. 11. Effects of diltiazem and verapamil on state 3 Q_{O_2} of mitochondria in medium containing 150 mM KCl, 1 mM pyruvate, and 2 and 5 mM P_i (pH 7.0 at 37°C).

by 2 mM P_i were not influenced by 400 μM diltiazem or 200 μM verapamil, the same concentrations of these drugs markedly inhibited the swelling produced by 5 mM P_i even in the presence of adenosine triphosphate and palmitoyl-coenzyme A (palmitoyl-CoA). Under this condition both 50–200 μM diltiazem and 25–200 μM verapamil exerted a dose-dependent protection of mitochondria against the swelling effects of P_i (Fig. 13). Parallel measurements of oxygen consumption under the same conditions indicate that these concentrations of diltiazem and verapamil stimulated oxygen consumption in a dose-dependent manner.

The ability of these Ca^{2+} antagonists to inhibit the slow Ca^{2+} channel in the sarcolemma is well established (7). On the other hand, the suggestion that intracellular sites are involved in the pharmacological action of these drugs (5) is less well accepted. The profound intracellular accumulation of verapamil, recently demonstrated by Lullman et al. (19), indicate that direct effects of these drugs on subcellular organelles may occur and may have important therapeutic significance. The prevention of the ischemia-induced mitochon-

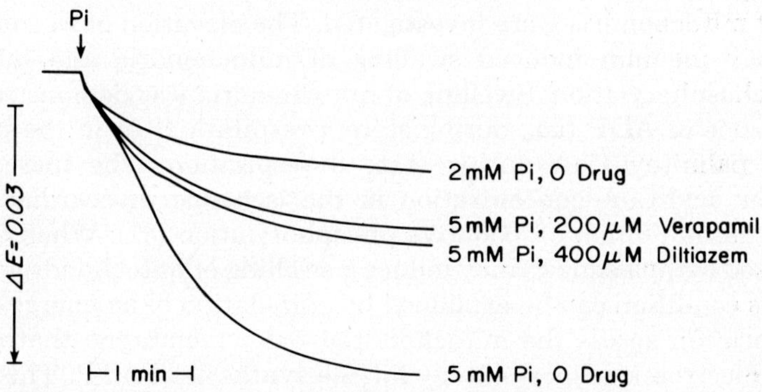

FIG. 12. Effects of verapamil and diltiazem on P_i-induced swelling of mitochondria. Assay medium contained 150 mM KCl and 1 mM pyruvate (pH 7.0 at 37°C). Mitochondrial protein (2 mg) was added to 3 ml medium and changes in light absorbance (extinction, ΔE) were recorded at 640 nm. P_i was added 2 min later.

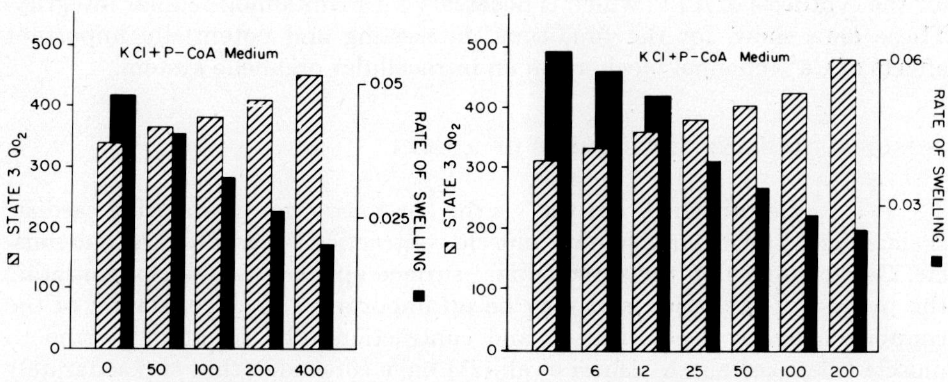

FIG. 13. Effects of diltiazem and verapamil on P_i-induced state 3 Qo_2 and swelling of mitochondria. Assay medium contained 150 mM KCl, 1 mM pyruvate, 2 mM ADP, and 10 μM palmitoyl-CoA (pH 7.0 at 37°C). Mitochondria (1 mg/1.5 ml) were preincubated for 2 min in absence of P_i and then 5 mM P_i was added.

drial damage in the myocardium by verapamil (24–26) and diltiazem (22, 30) has recently been demonstrated. Whether a direct effect of these drugs on mitochondria is involved in the demonstrated protective effects is not known. We have investigated this possibility by studying the influence of verapamil and diltiazem on the function of isolated heart mitochondria. For these studies the KCl assay medium was chosen because the K^+ concentration of this medium is much more similar to that of the cystosol than the conventionally used sucrose medium, and the experiments were carried out at 37°C rather than at the lower temperatures traditionally used. When the KCl assay medium contained only 2 mM P_i, which corresponds to the P_i concentration of the normoxic myocardium (36), a rather high OPR was measured (more than 1,600 nmol $ATP \cdot min^{-1} \cdot mg\ protein^{-1}$). Since the rapid and considerable increase in the intracellular P_i concentration is one of the early characteristics of the ischemic myocardial metabolic changes, the effects of increases in P_i on the function of mitochondria were investigated. The elevation of P_i concentration in the assay medium induced swelling of mitochondria and inhibition of oxidative phosphorylation. Swelling of mitochondria was demonstrated either in the absence of ADP (i.e., during state 4 respiration) or in the presence of ADP and palmitoyl-CoA during state 3 respiration. The increase in the intracellular acyl-CoA concentration in the ischemic myocardial tissue can induce partial inhibition of oxidative phosphorylation (31). When this occurs, the increased cytoplasmic P_i may induce a swelling of mitochondria. The effect of P_i in this condition can be explained by stimulation of an energy-dependent ion translocation across the mitochondrial inner membrane that diverts the use of the electron transport energy for the synthesis of ATP. This can result in an inhibition of oxidative phosphorylation. As previously discussed, we found that verapamil and diltiazem were effective in preventing the P_i-induced swelling and the resulting inhibition of oxidative phosphorylation. Thus under the influence of these drugs, the mitochondria can use the available oxygen more economically even in the presence of unfavorable metabolic conditions for the synthesis of ATP, which is necessary for maintaining cellular integrity. These data show, for the first time, interesting and potentially important effects of Ca^{2+}-channel blockers on an intracellular organelle system.

POSSIBLE SITES AND MECHANISMS OF ACTION

It is well established that Ca^{2+} is the key messenger in initiating cardiac, skeletal, and particularly smooth muscle contraction. Regarding cardiac muscle, Ca^{2+} channels on the sarcolemmal surface appear essential for initiating the process of excitation and may be an important site of regulation of the coupling between depolarization and contraction. The situation in smooth muscle is less certain. Meisheri et al. (21) have concluded that two separately controlled Ca^{2+} channels exist in smooth muscle plasmalemma. Triggle and Swamy (32) indicate that both receptor-operated and potential-dependent

Ca^{2+} channels exist in the sarcolemmal membrane and that some Ca^{2+} antagonists such as nifedipine and verapamil may act at intracellular sites as well as at the cell membrane surface. In considering potential intracellular sites of action for the Ca^{2+} antagonists, it is intriguing to implicate certain intracellular calcium receptors, particularly a 16,000-dalton protein called calmodulin (4). Recently Hidaka et al. (14, 15) suggested that chlorpromazine, prenylamine, and other drugs may significantly affect smooth muscle relaxation by interacting with calmodulin. Recent evidence from our laboratory (29) demonstrates that one new Ca^{2+} blocker, a butylaminoindene, inhibits calmodulin activation of adenylate cyclase and phosphodiesterase.

As we have suggested, it is possible that some Ca^{2+}-channel blockers accumulate within the cell. Payet et al. (28) suggest that both verapamil and its methoxy derivative, D 600, probably penetrate the sarcolemmal membrane to exert an action inside the cell. This is based on the observation that verapamil has a high affinity for lipids and that its distribution space in muscle is greater than the water space. Verapamil may concentrate in beating myocardial tissue in vitro to a level 30 times that found in the extracellular space (19).

CONCLUSION

We have shown that diltiazem has a powerful protective effect on experimentally induced myocardial ischemia, preventing the damage to jeopardized cardiac tissue adjacent to the occlusive site. This certainly is due to multiple actions unique to the class of drugs called Ca^{2+} antagonists or Ca^{2+}-channel blockers. Their effects include potent coronary vasodilation, inhibition of certain currents that describe the action potential, in some cases a slight negative inotropic and chronotropic influence, a drop in afterload, and a very interesting and protective effect on mitochondria. It is possible that some if not all aspects of this therapeutically beneficial array involve unique interaction with membranes containing calmodulin, which leads to modification of important enzyme systems. It is likely that Ca^{2+} blockers produce a complex array of actions by interacting with membranes causing perturbations and conformational changes involving lipids and proteins (Fig. 14).

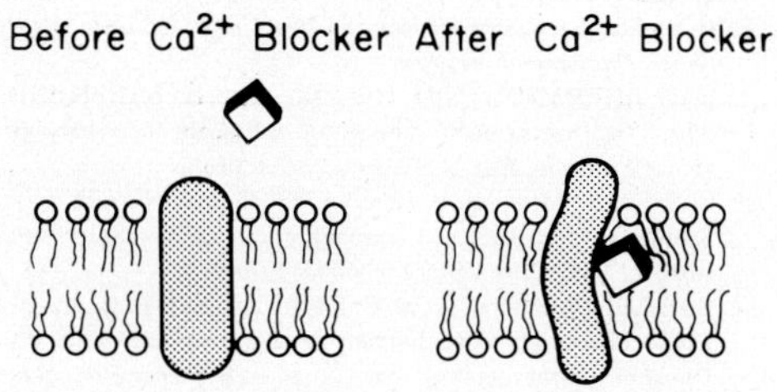

FIG. 14. Conformational changes involving lipids and proteins caused by interactions of membranes and Ca^{2+} blockers.

REFERENCES

1. AMILE, J. P., AND K. LANDMARK. The effect of nifedipine on the sinus and atrioventricular node of the dog heart after beta-adrenergic receptor blockade. *Br. J. Clin. Prac.* 8, Suppl: 9–13, 1980.
2. BELLAMY, R. F. Diastolic coronary artery pressure-flow relations in the dog. *Circ. Res.* 43: 92–101, 1978.
3. CARRIER, G. O., H. LULLMAN, L. NEUBAUER, AND T. PETERS. The significance of a fast exchanging superficial calcium fraction for the regulation of contractile force in heart muscle. *J. Mol. Cell. Cardiol.* 6: 333–347, 1974.
4. CHEUNG, W. Y. Calmodulin plays a pivotal role in cellular regulation. *Science* 207: 19–27, 1980.
5. CHURCH, J., AND T. T. ZSOTER. Calcium antagonists drugs. Mechanism of action. *Can. J. Physiol. Pharmacol.* 58: 254–264, 1980.
6. ESTABROOK, R. W. Mitochondrial respiratory control and the polarographic measurement of ADP:O ratios. In: *Methods in Enzymology. Oxidation and Phosphorylation*, edited by R. W. Estabrook and M. E. Pullman. New York: Academic, 1967, vol. 10, p. 41–47.
7. FLECKENSTEIN, A. Specific pharmacology of calcium in myocardium, cardiac pacemakers, and vascular smooth muscle. *Annu. Rev. Pharmacol. Toxicol.* 17: 149–166, 1977.
8. FRANKLIN, D., R. W. MILLIARD, AND T. NAGAO. Responses of coronary collateral flow and dependent myocardial mechanical function in the calcium antagonist, diltiazem. *Chest* 78: 200–204, 1980.
9. GARLICK, P. B., G. K. RADDA, AND P. J. SEELEY. Studies of acidosis in the ischaemic heart by phosphorus nuclear magnetic resonance. *Biochem. J.* 184: 547–554, 1979.
10. GREGG, D. E., J. J. THORNTON, AND F. R. MAUTZ. The magnitude, adequacy and source of the collateral blood flow and pressure in chronically occluded coronary arteries. *Am. J. Physiol.* 127: 161–175, 1939.
11. GROSS, R., H. KIRCHHEIM, AND K. VON OLSHAUSEN. Effects of nifedipine on coronary and systemic hemodynamics in the conscious dog. *Arzneim. Forsch.* 29: 1361–1368, 1979.
12. HENRY, P. D., R. SHUCHLEIB, L. J. BORDA, R. ROBERTS, J. R. WILLIAMSON, AND B. E. SOBEL. Effects of nifedipine on myocardial perfusion and ischemic injury in dogs. *Circ. Res.* 43: 372–380, 1978.
13. HENRY, P. D. Comparative pharmacology of calcium antagonists: nifedipine, verapamil and diltiazem. *Am. J. Cardiol.* 46: 1047–1058, 1980.
14. HIDAKA, H., T. YAMAKI, T. TOTSUKA, AND M. ASANO. Selective inhibitors of Ca^{2+}-binding modulator of phosphodiesterase produce vascular relaxation and inhibit actin-myosin interaction. *Mol. Pharmacol.* 15: 49–59, 1979.
15. HIDAKA, H., T. YAMAZAKI, T., M. NAKA, T. TANAKA, H. HAYASHI, AND R. KOBAYASHI. Calcium-regulated modulator protein interacting agents inhibit smooth muscle calcium stimulated protein kinase and ATPase. *Mol. Pharmacol.* 17: 66–72, 1980.
16. HIMORI, N., H. ONO, AND T. TAIRA. Simultaneous assessment of effects of coronary vasodilators on the coronary blood flow and the myocardial contractility by using blood perfused canine papillary muscle. *Jpn. J. Pharmacol.* 26: 427–431, 1976.
17. KASS, R. S., AND R. W. TSIEN. Multiple effects of calcium antagonists on plateau currents in cardiac Purkinje fibers. *J. Gen. Physiol.* 66: 169–192, 1975.
18. LOWRY, O. H., M. J. ROSEBROUGH, A. L. FARR, AND R. L. RANDALL. Protein measurement with the Folin phenol reagent. *J. Biol. Chem.* 193: 265–275, 1951.
19. LÜLLMAN, H., P. B. M. W. M. TIMMERMANS, AND A. ZIEGLER. Accumulation of drugs by resting or beating cardiac tissue. *Eur. J. Pharmacol.* 60: 277–285, 1979.
20. McALLISTER, R. E., D. NOBLE, AND R. TSIEN. Reconstruction of the electrical activity of cardiac Purkinje fibers. *J. Physiol. London* 251: 1–59, 1975.
21. MEISHERI, K. D., O. HWANG, AND VAN BREEMAN. Evidence for two separate Ca^{2+} pathways in smooth muscle plasmalemma. *J. Membr. Biol.* 59: 19–25, 1981.
22. NAGAO, T., M. A. MATLIB, D. FRANKLIN, R. W. MILLARD, AND A. SCHWARTZ. Effects of diltiazem, a calcium antagonist, on regional myocardial function and mitochondria after brief coronary occlusion. *J. Mol. Cell. Cardiol.* 12: p. 29–43, 1980.
23. NAKAJIMA, H., M. HOSHIYAMA, K. YAMASHITA, AND A. KIYOMOTO. Effects of diltiazem on electrical and mechanical activity of isolated cardiac ventricular muscle of guinea pig. *Jpn. J. Pharmacol.* 25: 383–392, 1975.
24. NAYLER, W. G., E. FASSOLD, AND C. YEPEZ. Pharmacological protection of myocardial function in hypoxic heart muscle: effect of verapamil, propranolol and methylprednisolone. *Cardiovasc. Res.* 12: 152–161, 1978.
25. NAYLER, W. G., R. FERRARI, AND A. WILLIAMS. Protective effect of pretreatment with verapamil, nifedipine and propranolol on mitochondrial function in the ischemic and reperfused myocardium. *Am. J. Cardiol.* 46: 242–248, 1980.
26. NAYLER, W. G., J. MAS-OLIVA, AND A. J. WILLIAMS. Cardiovascular receptors and calcium. *Circ. Res.* 46, Suppl. I: 161–166, 1980.
27. PADELETTI, L., F. FRANCHI, A. BRAT, R. P. DABIZZI, AND A. MICHELUCCI. The cardiac electrophysiological effects of nifedipine. *Int. J. Clin. Pharmacol. Biopharmacol.* 17: 270–293, 1979.
28. PAYET, M. D., O. F. SCHANNE, E. RUIZ-CERETTI, AND J.-M. DEMERS. Inhibitory activity of blockers of the slow inward current in rat myocardium, a study in steady state and of rate of action. *J. Mol. Cell. Cardiol.* 12:187–200, 1980.
29. PIASCIK, M. T., M. F. PIASCIK, C. L. JOHNSON, R. G. RAHWAN, AND J. D. POTTER. Binding of the calcium antagonist 2-butyl-3-dimethylamino-5,6-methylenedioxyindene to calmodulin. *J. Pharmacol. Exp. Ther.* in press.
30. SCHWARTZ, A., T. NAGAO, AND M. A. MATLIB. The reaction of diltiazem on occlusive-induced regional myocardial damage: a beneficial effect on mitochondria *in situ*. In: *New Drug Therapy With a Calcium Antagonist*, edited by R. J. Bing. Amsterdam: Excerpta Medica, 1979, p. 48–58.
31. SHUG, A. L., E. SHRAGO, N. BITTAR, J. D. FOULTS, AND J. R. KOKE. Acyl-CoA inhibition of adenine nucleotide translocation in ischemic myocardium. *Am. J. Physiol.* 228: 689–692, 1975.
32. TRIGGLE, D. J., AND V. C. SWAMY. Pharmacology of agents that affect calcium. *Chest* 78: 174–179, 1980.
33. VAGHY, P. L., M. A. MATLIB, L. SZEKERES, AND A. SCHWARTZ. Protective effect of verapamil and diltiazem against inorganic phosphate induced impairment of oxidative phosphorylation of isolated heart mitochondria. *Biochem. Pharmacol.* in press.
34. VAN BREEMEN. C., O. HWANG, AND K. MEISHERI. The "Ca^{2+}-antagonist" action of diltiazem on vascular smooth muscle. *Eur. J. Pharmacol.* in press.
35. WIESTER, M. J., P. F. BONVENTRE, AND G. GRUPP. Estimate of myocardial damage induced by diphtheria toxin. *J. Lab Clin. Med.* 81: 354–354, 1973.
36. WOLLENBERGER, A., E.-G. KRAUSE, AND B. E. WAHLER. Uber den tatsachlichen Orthophosphate- und *N*-Phosphorylkreatingehalt des Saugetierherzens. *Pfluegers Arch. Gesamte Physiol. Menschen Tiere* 270: 413–427, 1960.
37. YAMAGUCHI, I., K. OBAYASHI, AND W. J. MANDEL. Electrophysiological effects of verapamil. *Cardiovasc. Res.* 12: 597–608, 1978.

16

Influence of Nifedipine on Regional Myocardial Blood Flow, Myocardial Necrosis, and Segmental Ventricular Function

PETER G. ROAN, CARLOS IZQUIERDO, L. MAXIMILIAN BUJA, HOMAYUN HASHIMI, AND JAMES T. WILLERSON

Departments of Internal Medicine and Pathology, University of Texas Health Science Center, Dallas, Texas

NIFEDIPINE, A SLOW-CHANNEL Ca^{2+} antagonist, increases myocardial blood flow during experimental myocardial ischemia (4), preserves myocardial function after the ischemia-reperfusion injury associated with global ischemia during cardiopulmonary bypass (2), and reduces the accumulation of mitochondrial Ca^{2+} in ischemic rabbit hearts (5). In these studies we evaluated the influence of nifedipine in preserving segmental ventricular function, increasing regional myocardial blood flow, and reducing the extent of myocardial necrosis at specific left ventricular (LV) segmental sites in and around the ischemic region in awake, unsedated dogs with permanent coronary artery occlusions for 24 h.

METHODS

Mongrel dogs weighing 25 ± 3.3 kg were anesthetized with pentobarbital (30 mg/kg iv), intubated, and placed on a Harvard respirator. A thoracotomy was performed in the left 5th intercostal space under sterile conditions, and the heart was suspended in the pericardium. A Konigsberg (P22) catheter-tipped manometer was placed into the LV cavity through an apical stab wound. The LV pressure tracing and LV dP/dt subsequently recorded were used for timing purposes to identify the systolic and diastolic periods for calculations of regional wall thickness from the ultrasonic crystals. A catheter was inserted into the left atrium for pressure monitoring. The proximal left anterior descending coronary artery (LAD) was dissected free of the epicardium, and a balloon occluder device was placed around it. In each of the animals studied, temporary inflation of the balloon produced an area of cyanotic myocardium over the anterior left ventricle. Five-MHz titanate-zirconate piezoelectric crystals 3–5 mm in diameter were inserted through the myocardial wall and positioned near the endocardium. The endocardial crystal was paired with a second crystal sutured to the epicardium after its position was adjusted to obtain an optimal signal by the method of Franklin et al. (3, 9). Ordinarily five to six pairs of such crystals were inserted in or near the cyanotic region, and one pair of crystals was inserted in the lateral portion of the left ventricle. The various catheters and wires were exteriorized between the scapulae, and the incision was closed under sterile conditions. The animals were studied when they appeared to be fully recovered from the instrumentation surgery and were free of infection. Internal jugular and carotid arterial catheters were inserted under light anesthesia 24 h before the beginning of the study.

The aortic systolic and diastolic and the left atrial mean pressures were measured in millimeters of Hg. Segmental LV wall thickness was measured in millimeters, assuming the speed of sound through the myocardium to be 1.5 mm/μs (9). To correct for the variable initial separation of crystal pairs, measurements of wall thickness were expressed as percents of control values. The data from the piezoelectric crystals were digitized manually with a Graf/Pen and stored and processed by a DEC System-10 computer.

The NET systolic thickening was used in these studies as a measure of regional ventricular contractility (7) and was defined as the extent of systolic wall thickening minus the amount of paradoxical systolic thinning (if any). Early diastolic thickening, which occurred during the period of isovolumic ventricular relaxation, was not included in the measurement of NET. Throughout the study, NET was expressed as a percent of control NET. Systolic wall thickening was defined and measured as the maximal wall thickness obtained between the initial rapid upstroke of LV pressure and the peak negative LV dP/dt. Time periods defined as *systole* extended from the initial rapid upslope of the LV pressure tracing to its return to its diastolic level; thus systole included both the isovolumic contraction and relaxation phases of ventricular systole (7).

Dogs were studied from 7 to 14 days after their instrumentation when they appeared to be fully recovered. All dogs were studied in an awake, unsedated state and were restrained lightly in a cradlelike device. Thirty-eight dogs were studied; half of these received placebo and half received nifedipine at 50 $\mu g \cdot kg^{-1} \cdot min^{-1}$. Nifedipine or placebo was administered intravenously beginning 5 min after proximal LAD occlusion, and these infusions were continued for 6 h. Measurements of segmental wall thickening, mean systemic arterial pressure, left atrial pressure, and heart rate were made before and 5 min after proximal LAD occlusion, and at hourly intervals for the first 8 h and again at 24 h after coronary occlusion.

At the end of each study, the animal was killed and the occluding balloon examined. In each case the balloon was fully and tightly inflated around the proximal LAD. Hearts were then placed in formalin. Transmural tissue blocks were obtained from each crystal site to include approximately 0.5 cm of tissue on either side of an imaginary line connecting the epicardial and endocardial crystal pairs. The position of the inner crystal was usually found to be located within 1–2 mm of the endocardium. Tissue removed from each crystal pair was counted, and regional myocardial blood flow was calculated using the reference sample technique (6, 10). Histological sections were cut and stained with hemotoxylin and eosin. Photomicrographs of each histological section were taken. The extent of histological necrosis was outlined, and the total and necrotic areas were measured with a planimeter. The extent of necrosis was expressed as a percent of total area and thus varied from 0 to 100 percent.

Segmental function was correlated with the extent of segmental histological necrosis. Regional myocardial blood flows to the various crystal pairs were also analyzed and tabulated relative to the severity of segmental crystal dysfunction 5 min after coronary occlusion (7).

For analysis of segmental function, crystal pairs were assigned to one of three groups (7). These groups were defined according to the value of the perimeter NET measured 5 min after proximal LAD occlusion. The NET systolic thickening was used for grouping the segments because it provided a measure of overall segmental function. Group 1 included segments with NET values ≥ 67–100% of the control value, group 2 included segments with NET values < 67–0% of the control, and group 3 contained segments with NET values < 0% of the control value for that segment. Thus three groups were defined that ranged in severity of reduction of segmental function with LAD occlusions from minimally reduced (group 1) to severely reduced with predominantly paradoxical wall motion (group 3).

For statistical comparisons of resting function and regional myocardial blood flow and the extent of segmental necrosis and responses to nifedipine or placebo, Student's *t* test was used.

RESULTS

Regional Myocardial Blood Flow

Administration of nifedipine was associated with significant increases in regional myocardial blood flow to group 2 segments (segments with moderate ventricular dysfunction) in the endocardial and epicardial regions (Fig. 1). Nifedipine also increased regional myocardial blood flow at the end of the infusion to group 1 segments (segments with mild dysfunction). However, it did not change regional myocardial blood flow to the most severely dysfunctional group 3 segments, nor did it change the endocardial-epicardial flow ratio at any of the segmental sites analyzed.

Extent of Myocardial Necrosis

Nifedipine significantly decreased the extent of necrosis in the group 2 segments in the endocardial, midwall, and epicardial regions (Fig. 2). Nifedipine also tended to reduce the extent of necrosis in group 1 segments, but it did not alter the extent of necrosis in the group 3 segments.

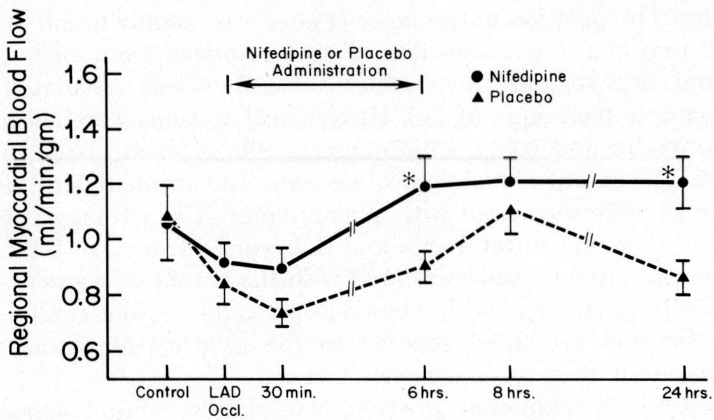

FIG. 1. Regional myocardial blood flow as function of time after experimental coronary artery occlusion. Symbols represent mean values and cross bars show SE. Nifedipine administration was associated with a significant increase in regional myocardial blood flow to moderately dysfunctional (group 2) myocardial sites.

FIG. 2. Percentage of myocardial necrosis at each crystal site as function of various LV segmental sites. Administration of nifedipine was associated with a significant reduction in the extent of myocellular necrosis at moderately dysfunctional (group 2) LV segmental sites.

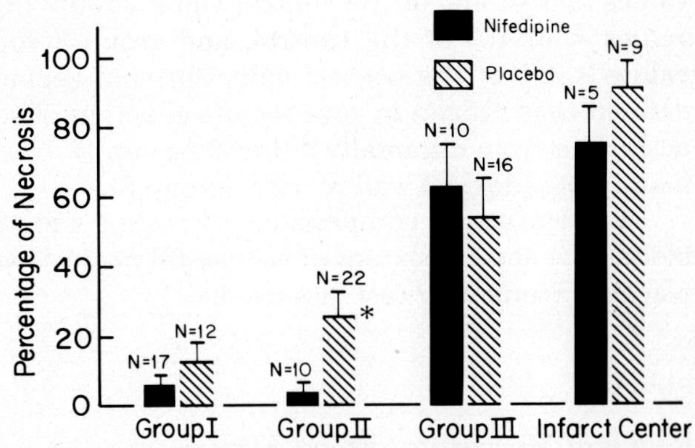

Segmental Ventricular Function

Nifedipine did not significantly alter segmental ventricular function at any of the crystal sites. However, it did tend to increase NET segmental function at the group 2 segments ($P = 0.1$) (Fig. 3).

Influence of Nifedipine on Heart Rate–Blood Pressure Product

Nifedipine did not alter the heart rate–blood pressure product based on values obtained with the placebo during the periods of experimental myocardial ischemia.

DISCUSSION

The data obtained in this study suggest that the administration of nifedipine beginning 5 min after permanent proximal LAD occlusion and contin-

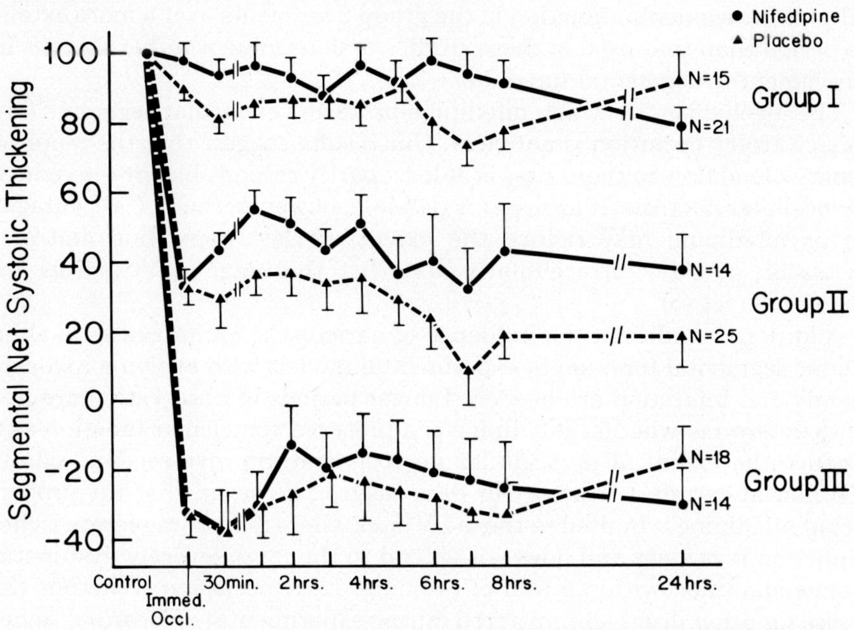

FIG. 3. Segmental net systolic thickening as function of time after experimental coronary artery occlusion. Symbols represent mean values and cross bars show SE. Nifedipine administration did not result in improvement of segmental ventricular function at any crystal site. However, it did tend to improve segmental function at the group 2 crystal sites ($P = 0.1$). Whether nifedipine administration results in significant improvement in segmental function at the group 2 crystal sites must be evaluated further with longer periods of nifedipine administration and with longer periods of observation after experimental coronary artery occlusion.

uing for 6 h results in an increase in regional myocardial blood flow to all layers of ventricular segments with moderate ischemic dysfunction as well as to segments with mild ischemic dysfunction. Nifedipine also reduces the extent of histological necrosis at the moderately dysfunctional sites. At the most severely dysfunctional sites, however, nifedipine does not increase regional coronary blood flow nor does it reduce the extent of myocellular necrosis. These results are consistent with our previous observations suggesting that segmental ventricular dysfunction at those sites with moderate dysfunction 5 min after coronary artery occlusion is dynamic and variable over the following 24 h and that considerable contractile reserve exists at these sites for 24 h (7). At the most severely dysfunctional sites, however, contractile reserve exists for only 2 h after proximal LAD occlusion in this experimental model (7).

Despite the increase in regional myocardial blood flow and the reduction in the extent of myocellular necrosis, we were unable to demonstrate that nifedipine also preserves ventricular function at the moderately dysfunctional sites. Our previous studies (6, 7, 10) and research by other authors (1, 8) suggest that the extent of myocellular necrosis is one important factor influencing subsequent alterations in segmental contractility but that important ventricular dysfunction may exist in the absence of important necrosis. This observation emphasizes the need to evaluate the potential protective effect of nifedipine on segmental function in the group 2 segments over a more extended time period than that used in these studies to determine whether there is later improvement in segmental function.

The mechanism by which nifedipine protects ventricular segments during coronary artery occlusion is not clear. Our results suggest that the increase in coronary blood flow to these sites is at least partly responsible for the reduction in myocellular necrosis. It also seems possible, however, that a Ca^{2+} antagonist such as nifedipine may reduce the extent of Ca^{2+} deposition and/or the distribution of Ca^{2+} intracellularly and that this might play a role in its protective effect (5).

Additional studies of the influence of various Ca^{2+} antagonists in altering ischemic segmental function in experimental models with evolving myocardial ischemia and infarction are needed. Longer periods of observation are necessary to determine whether nifedipine can preserve ventricular function as well as reduce the extent of myocellular necrosis and improve regional coronary blood flow at certain LV sites. Our data suggest, however, that the protective effect of nifedipine is limited to those LV sites where mild or moderate ischemic dysfunction is present and does not extend to the most severely dysfunctional LV sites identified within 5 min of proximal LAD occlusion. Whether this is the case for other drugs administered during experimental myocardial ischemia remains to be determined.

The research reported in this chapter was supported by National Institutes of Health Ischemic SCOR HL-17669.

The authors appreciate the expert technical assistance of Janice McNatt, Judy Ober, David Perry, and Gifford Ramsey, as well as the expert secretarial assistance of Laurie Grey.

REFERENCES

1. BALTAXE, H. A., D. R. ALONZO, J. G. LEE, J. PRAT, W. HUSTED, AND J. W. STOKES. Impaired left ventricular contractility in ischemic heart disease: angiographic and histopathologic correlations. *Radiology* 113: 581–585, 1974.
2. CLARK, R. E., I. Y. CHRISTLIEB, P. D. HENRY, A. E. FISCHER, J. D. NORA, J. R. WILLIAMSON, AND B. E. SOBEL. Nifedipine: a myocardial protective agent. *Am. J. Cardiol.* 44: 825–831, 1979.
3. FRANKLIN, D. L., W. S. KEMPER, T. PATRICK, AND D. McKOWN. Technique for continuous measurement of regional myocardial segment dimensions in chronic animal preparations (abstr.). *Federation Proc.* 32: 343A, 1973.
4. HENRY, P. D., R. SHUCHLEIB, L. J. BORDA, R. ROBERTS, J. R. WILLIAMSON, AND B. E. SOBEL. Effects of nifedipine on myocardial perfusion and ischemic injury in dogs. *Circ. Res.* 43: 372–380, 1978.
5. HENRY, P. D., R. SHUCHLEIB, J. DAVIS, E. WEISS, AND B. E. SOBEL. Myocardial contracture and accumulation of mitochondrial calcium in ischemic rabbit heart. *Am J. Physiol.* 233 (*Heart Circ. Physiol.* 2): H677–H684, 1977.
6. ROAN, P. G., L. M. BUJA, C. IZQUIERDO, H. HASHIMI, S. I. SAFFER, AND J. T. WILLERSON. Interrelationships between regional left ventricular function, coronary blood flow and myocellular necrosis during the initial 24 hours and 1 week after experimental coronary occlusion in the awake, unsedated dog. *Circ. Res.* in press.
7. ROAN, P. G., F. SCALES, S. I. SAFFER, L. M. BUJA, AND J. T. WILLERSON. Functional characterization of left ventricular segmental responses during the initial 24 H and 1 WK after experimental canine myocardial infarction. *J. Clin. Invest.* 63: 1074–1088, 1979.
8. STINSON, E. B., AND M. E. BILLINGHAM. Correlative study of regional left ventricular histology and contractile function. *Am. J. Cardiol.* 39: 378–383, 1977.
9. THEROUX, P., D. FRANKLIN, J. ROSS, JR., AND W. S. KEMPER. Regional myocardial function during acute coronary artery occlusion and its modification by pharmacologic agents in the dog. *Circ. Res.* 35: 896–907, 1974.
10. WILLERSON, J. T., J. T. WATSON, I. HUTTON, D. E. FIXLER, G. C. CURRY, AND G. H. TEMPLETON. The influence of hypertonic mannitol on regional myocardial blood flow during acute and chronic myocardial ischemia in anesthetized and awake intact dogs. *J. Clin. Invest.* 5: 892–902, 1975.

17

Calcium-Channel Blockers and Respiratory Smooth Muscle

FRANK R. GOODMAN

Dow Chemical USA, Indianapolis, Indiana

CALCIUM ANTAGONISTS ARE POTENTIALLY USEFUL as therapeutic agents and as tools for delineating both excitation-contraction coupling and stimulus-secretion coupling. Many types of agents can inhibit Ca^{2+} movements and, in turn, alter the flux of Ca^{2+} across cell membranes. The emphasis in this chapter is on a few members of the relatively new group of agents known as *Ca^{2+} channel blockers* that inhibit transmembrane Ca^{2+} fluxes through the voltage-sensitive channel.

It is well documented that Ca^{2+} plays an important role in activating and maintaining the contractile tone of vascular smooth muscle (15). It is also known that various types of smooth muscle differ in their dependence on and use of extracellular Ca^{2+}. The manner in which pharmacological agents interact with Ca^{2+} also appears to vary (2). Thus much of the effort in drug development has been devoted to delineating the sources of Ca^{2+} important for membrane and cellular events. As a result of this research, Ca^{2+}-channel blockers such as verapamil, nifedipine, and cinnarizine have been developed. These agents, which were designed primarily for the treatment of angina and other cardiovascular disorders, appear to induce their clinical effects by their action on the voltage-sensitive Ca^{2+} channel. It has been suggested, however, that these agents can act on sites other than Ca^{2+} channels (3, 14). The universal involvement of Ca^{2+} in many physiological functions allows one to predict, with some certainty, that these agents should have additional therapeutic values. This idea is further supported by the fact that these agents alter Ca^{2+} mobilization in a variety of muscle preparations (14) and biological processes (12). This chapter examines some of the Ca^{2+}-related sequences involved in asthma and reports on some preliminary experimental observations obtained with different Ca^{2+}-channel blockers in respiratory smooth muscle.

ASTHMA AND CALCIUM

Asthma is a complex disease characterized by intermittent, reversible obstruction of both large and small airways. Airway obstruction and other

TABLE 1. *Ca^{2+}-dependent events in asthma*

1) Bronchoconstriction
2) Mediator release
3) Vagally mediated reflexes
4) Secretion of mucus
5) Mucociliary transport

features of asthma (Table 1) are linked to transmembrane Ca^{2+} fluxes and have been discussed in detail by Middleton (10). Since Ca^{2+} plays such a
' ~ole in all these events, it is reasonable to assume that agents that alter
¬ might be valuable in asthma therapy. The currently used
' 'racellular Ca^{2+} levels by several different mecha-
l~~~ of mediators or induce bronchodilation
hanism of action for these agents
~~erally accepted that
hate or cyclic

TAb

Cromolyn
Steroids
β-Agonist
Prostaglandin E_2
Methylxanthine
Atropine
Cinnarizine (Ca^{2+}-channel blocker)*

* Data from Barrett-Bee (1) and E.

are released from the mast cell. For asthma in humans, histamine and SRS-A seem to be the most important mediators because they induce contractions of smooth muscle. It is believed that Ca^{2+} is essential in the contractile mechanism of airway smooth muscle and that the mechanical functions of isolated bronchial and tracheal smooth muscle show characteristics similar to other smooth muscle preparations (13). Therefore, as with other smooth muscles, the mediators released from the mast cell probably act directly or through vagal stimulation to mobilize Ca^{2+} and initiate constriction of the airways. The manner in which various respiratory smooth muscle preparations respond to similar pharmacological agents appears to differ (4), however, in the same way previously described for various vascular smooth muscle preparations (15). This observation is further supported by the fact that verapamil inhibits responses induced by higher concentrations of acetylcholine (ACh) in trachea to a lesser degree than in the parenchyma (Table 3). Whether these differences are due to differential use of intracellular Ca^{2+} stores or other factors (e.g., a possible muscarinic-receptor blocking action of 10^{-4} M verapamil) is not known.

CALCIUM UTILIZATION

One of the main problems concerning the therapeutic use of Ca^{2+}-channel blockers in the treatment of asthma is that the cellular mechanism and use of Ca^{2+} by respiratory smooth muscle is poorly understood. It has been assumed that the development of tension in respiratory smooth muscle is regulated by changes in the cellular Ca^{2+} concentration. It is not known, however, why the airways of asthmatics are characterized by "hyperreactivity" or how the receptor-mediator interaction is linked, if at all, to the voltage-sensitive Ca^{2+} channels. The extent to which extracellular Ca^{2+} is needed to initiate contractile responses to ACh has been previously studied in canine tracheal smooth muscle, and Farley and Miles (6) found that contractions induced by low concentrations ($< 10^{-6}$ M) of ACh depend on extracellular Ca^{2+}, whereas at higher concentrations of ACh both extracellular and intracellular Ca^{2+} are important. These authors further demonstrated that in the presence of 10^{-4} M verapamil (a concentration eliciting a maximal inhibitory response), the degree

TABLE 3. *Comparison of effects of verapamil on tracheal and parenchymal smooth muscle*

Concentration of ACh, M	Inhibition, %	
	Trachea	Parenchyma
10^{-7}	100	100
10^{-6}	94 ± 10	100
10^{-5}	68 ± 14	94 ± 10
10^{-4}	38 ± 18	77 ± 11
10^{-3}	22 ± 13	57 ± 13

Muscles were exposed to 10^{-4} M verapamil for 5 min before the addition of ACh and while ACh was present. Values are means ± SE for 4 tissues.

of inhibition of the ACh-induced response depends on the concentration of ACh used (7). In contrast to the concentration-dependent results obtained with ACh, elevating the K^+ concentration in the bathing solution does not overcome the inhibitory effect of the Ca^{2+}-channel blocker methoxyverapamil (D 600; Fig. 1). The importance of Ca^{2+} influx for pharmacological activity has also been observed in guinea pig tracheal smooth muscle. Etafenone, a Ca^{2+}-influx blocker, inhibits the responses to ACh, histamine, and other broncho-constrictive agents and potentiates the effects of aminophylline and other bronchodilating agents (8).

In vascular smooth muscle, the contracted depolarized muscle is either not affected or affected only to a limited extent by exposure to the various compounds or agents known to inhibit Ca^{2+} influx. However, tension responses induced by agents that promote Ca^{2+} influx through receptor-operated channels are reversed. Similar findings have also been observed in canine tracheal (Fig. 2) and guinea pig parenchymal tissue in my laboratory (unpublished observations). It was found that the concentration of verapamil, cinnarizine, or flunarizine that resulted in 50% relaxation of a parenchymal strip contracted with partially purified SRS-A was approximately 5×10^{-6} M for all three agents. An important conclusion from these findings is that respiratory smooth muscle, like other smooth muscle, apparently possesses two types of Ca^{2+}

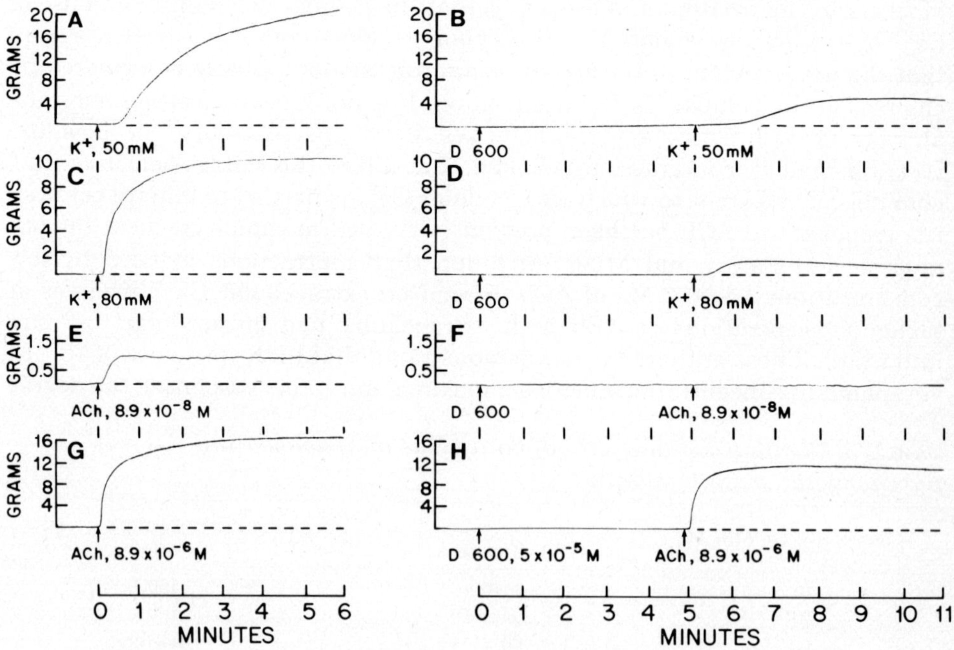

FIG. 1. Inhibition by D 600 of contractile responses elicited by K^+ (*B* and *D*) or ACh (*F* and *H*) in dog tracheal smooth muscle. Contractile responses were obtained before (*A, C, E, G*) and during (*B, D, F, H*) exposure to 5×10^{-5} M D 600. Each agent was added at points indicated by arrows.

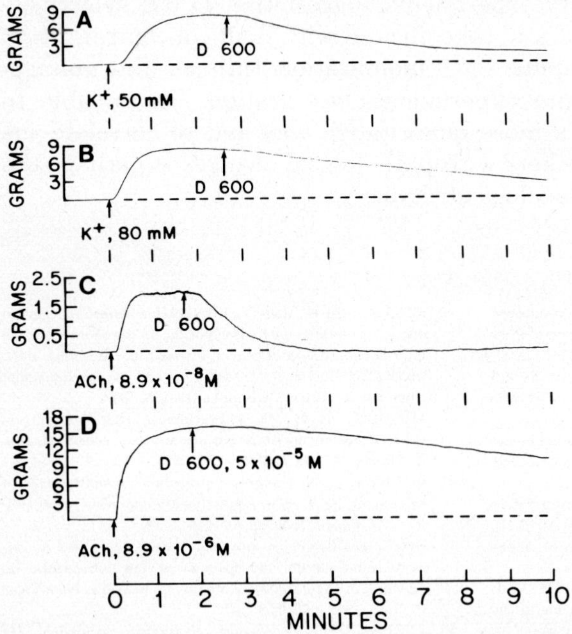

FIG. 2 Effect of 5×10^{-5} M D 600 after initial contraction induced by K^+ (*A* and *B*) or ACh(*C* and *D*) in dog tracheal smooth muscle. Each agent was added at points indicated by arrows.

channels. These channels, as described by Bolton (2), include voltage-sensitive Ca^{2+} channels that open as the potential across the membrane is decreased and receptor-operated Ca^{2+} channels that are controlled by agonist-induced receptor activation. The specific nature and characteristics of these channels in respiratory muscle are less clear. To induce their effect in respiratory muscle, the currently available Ca^{2+}-channel blockers must be used at considerably greater concentrations than those used with vascular smooth muscle. Therefore these channels are either different from or less sensitive than those previously described for the cardiovascular system.

CONCLUSION

Even though the development of newer Ca^{2+}-channel blockers specific for either excitation-contraction coupling or stimulus-secretion coupling is just beginning, this group of agents constitutes a viable approach to the prophylactic treatment of asthma. As with agents currently used for vascular relaxation (e.g., verapamil, diltiazem, and cinnarizine), the success of long-term prophylactic treatment of asthma with Ca^{2+}-channel blockers must await clinical verification. So little is known about the Ca^{2+}-binding properties of the respiratory system (e.g., lung and airway smooth muscle) that nothing absolute can be proposed. For example, the sites of action for inhibition of Ca^{2+} influx and the affinity of the channel blockers for these sites probably vary with the tissues and species being examined. The use of a Scatchard-type plot has indicated the existence of high- and low-affinity Ca^{2+}-binding sites in dog tracheal smooth muscle (G. B. Weiss, personal communication). Furthermore the results of a few preliminary experiments suggest that D 600 selectively blocks increased low-affinity ^{45}Ca uptake elicited with high concentrations of K^{+}. Whether D 600 only inhibits Ca^{2+} mobilization through the voltage-sensitive channels awaits future experimental verification. To be able to describe these phenomena in a more quantitative way and to correlate the effects of the Ca^{2+}-channel blockers with pathological changes is perhaps the most important current pharmacological challenge.

REFERENCES

1. BARRETT-BEE, K. J. Ionophore- and antigen-induced histamine release from mast cells and a comparison of their inhibition. *Biochem. Soc. Trans.* 5: 965–967, 1977.
2. BOLTON, T. B. Mechanisms of action of transmitters and other substances on smooth muscle. *Physiol. Rev.* 59: 606–718, 1979.
3. CHURCH, J., AND T. T. ZOSTĖR. Calcium antagonistic drugs. Mechanism of action. *Can. J. Physiol. Pharmacol.* 58: 254–264, 1980.
4. DRAZEN, J. M., AND M. W. SCHNEIDER. Comparative responses of tracheal spirals and parenchymal strips to histamine and carbachol *in vitro*. *J. Clin. Invest* 61: 1441–1447, 1978.
5. EMANUEL, M. B., J. A. CHAMBERLAIN, S. WHITING, B. G. RIGDEN, AND A. H. CRAVEN. Cinnarizine in the chronic treatment of asthma. *Br. J. Clin. Pharmacol.* 7: 189–195, 1979.
6. FARLEY, J. M., AND P. R. MILES. Role of depolarization in acetylcholine-induced contractions of dog trachealis muscle. *J. Pharmacol. Exp. Ther.* 201: 199–205, 1977.
7. FARLEY, J. M., AND P. R. MILES. The sources of calcium for acetylcholine-induced contractions of dog tracheal smooth muscle. *J. Pharmacol. Exp. Ther.* 207: 340–346, 1978.
8. KITAMURA, S., AND Y. ISHIHARA. Effect of calcium antagonist, etafenone hydrochloride, on the isolated guinea pig tracheal tissues. *Arzneim. Forsch.* 30: 1088–1091, 1980.
9. LICHTENSTEIN, L. M. Mediators and the mechanism of their release. *Chest* 73, Suppl.: 919–925, 1978.
10. MIDDLETON, E., JR. Antiasthmatic drug therapy and calcium ions: review of the pathogenesis and role of calcium. *J. Pharm. Sci.* 69: 243–251, 1980.
11. PUTNEY, J. W. Stimulus-permeability coupling: role of calcium in the receptor regulation of membrane permeability. *Pharmacol. Rev.* 30: 209–245, 1978.
12. ROSENBERGER, L., AND D. J. TRIGGLE. Calcium, calcium translocation, and specific calcium antagonists. In: *Calcium in Drug Action*, edited by G. B. Weiss. New York: Plenum, 1978, p. 3–31.
13. STEPHENS, N. L. *Airway Dynamics*. Springfield, IL: Thomas, 1970, p. 191.
14. TRIGGLE, D. J., AND V. C. SWAMY. Pharmacology of agents that affect calcium. *Chest* 78, Suppl.: 174–179, 1980.
15. WEISS, G. B. Calcium and contractility in vascular smooth muscle. In: *Advances in General and Cellular Pharmacology*, edited by T. Narahashi and C. P. Bianchi. New York: Plenum, 1977, p. 71–154.

18

Psychoactive Drugs as Antagonists of Actions of Calcium

R. ADRON HARRIS

Harry S. Truman Memorial Veterans Hospital and Department of Pharmacology, University of Missouri School of Medicine, Columbia, Missouri

SOME PSYCHOACTIVE DRUGS MAY ACT as antagonists of Ca^{2+} in the central nervous system (CNS). Data that implicate Ca^{2+} in the actions of opiates, alcohols, barbiturates, cannabinols, and phenothiazines on brain tissue are reviewed, but first a brief outline of the regulation and functions of Ca^{2+} in the CNS is presented.

CALCIUM METABOLISM IN BRAIN

Most psychoactive drugs act at the synaptic level, and the dynamics of Ca^{2+} at the nerve ending must be considered. The intraneuronal concentration of free Ca^{2+} is very low (perhaps 10^{-7} M), whereas the extracellular concentration is about 10^{-3} M (32). This segregation allows the influx or release of a small amount of Ca^{2+} to increase dramatically the intracellular concentration of Ca^{2+}. The influx occurs by voltage-sensitive channels, which, though readily blocked by lanthanides (12), are only weakly blocked by compounds such as verapamil, nifedipine, and methoxyverapamil (D 600), all of which are potent Ca^{2+} antagonists in other tissues (37). Inside the nerve ending, Ca^{2+} triggers the release of neurotransmitter by an unknown mechanism (32). A number of processes that can remove or sequester Ca^{2+} assure the transiency of the elevation in intraneuronal Ca^{2+} produced by depolarization (Fig. 1). There is a high-affinity, ATP-dependent uptake of Ca^{2+} by vesicles and other intraneuronal structures that is probably important in the maintenance of low levels of Ca^{2+} (3, 4). Intrasynaptosomal mitochondria also accumulate Ca^{2+}, but the affinity of this uptake is relatively low and likely to be functionally important only when there is a marked elevation of intracellular Ca^{2+}. Calcium may also bind to the cytoplasmic face of the plasma membrane, which contains a number of proteins and acidic lipids. Two processes have been identified for the extrusion of Ca^{2+} from neurons: a Na^+-Ca^{2+} exchange system where the influx of Na^+ is coupled with the efflux of Ca^{2+} and an ATP-dependent Ca^{2+}

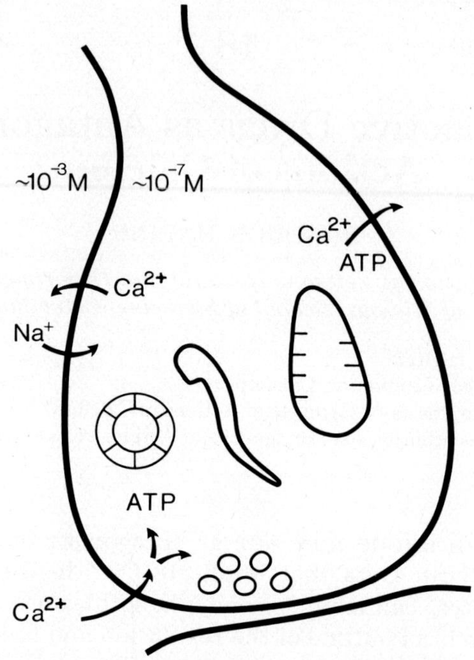

FIG. 1. Synaptosomal Ca^{2+} metabolism.

pump (Fig. 1). Together these data indicate that there are a number of sites at which psychoactive drugs may affect Ca^{2+} availability and thereby alter synaptic transmission. Because the details of neuronal Ca^{2+} transport are beyond the scope of this review, the reader is referred elsewhere (1, 3, 4).

OPIATES AS ANTAGONISTS OF ACTIONS OF CALCIUM

Interest in evaluating opiates as Ca^{2+} antagonists was initially stimulated by the observation that an increase in the availability of brain Ca^{2+} reduces the analgesic effect of morphine in rodents (18, 28). The analgesic effect of subcutaneously administered morphine is completely antagonized by microinjection of $CaCl_2$ into the periaqueductal gray area of the rat brain (Table 1). This area was chosen for investigation because other studies delineated it as an important site for the analgesic action of opiates (27). These in vivo observations have been extended to demonstrate that the effects of opiates on the contraction of isolated tissues such as guinea pig ileum, rat vas deferens, mouse vas deferens, rat colon, and rat aorta are reciprocally related to the concentration of Ca^{2+} in the bath (8, 26). The binding of opiates to their receptors is not affected by physiologically relevant concentrations of Ca^{2+} (40); thus Ca^{2+} must interfere with opiate action by affecting processes that occur after the initial drug-receptor interaction. One opiate effect that is not antagonized by Ca^{2+} is the inhibition of prostaglandin-stimulated adenylate cyclase of neuroblastoma x glioma hybrid cells. This enzyme is inhibited by morphine, but the inhibition is not affected by changes in Ca^{2+} (5).

Reduction of morphine effects by Ca^{2+} raised the possibility that Ca^{2+} antagonists might mimic some of the effects of opiates. In subsequent studies lanthanides were shown to produce analgesia pharmacologically similar to opiate analgesia (Table 2). In animals tolerant to and dependent on morphine, La^{3+} suppresses withdrawal signs and the analgesic effect of La^{3+} is diminished. The cross-tolerance and cross-dependence between La^{3+} and morphine suggest that changes in Ca^{2+} metabolism might be involved in the development of opiate tolerance and dependence. Lanthanum and other Ca^{2+} antagonists also mimic the effects of opiates on neuroblastoma x glioma hybrid cells, and chronic exposure to either opiates or reduced Ca^{2+} concentrations produces similar effects (Table 2).

Biochemical studies of the role of Ca^{2+} in opiate action initially focused on Ca^{2+} localization in brain. It has been reported that acute administration of morphine (or ethanol) reduces the Ca^{2+} content of brain tissue by as much as 50% (43). Subsequent studies, however, have failed to demonstrate an effect of morphine (or ethanol) on Ca^{2+} levels of brain tissue (24). Conversely, studies of Ca^{2+} localization after subcellular fractionation of brain tissue indicate that

TABLE 1. *Antagonism of morphine analgesia by microinjection of $CaCl_2$ into rat periaqueductal gray area*

	% Analgesia	
	T-F	H-P
NaCl-morphine	100	100
$CaCl_2$-morphine	16 ± 4	5 ± 4
$CaCl_2$-saline	1 ± 1	3 ± 3

Rats were injected bilaterally with either 0.2 μmol of $CaCl_2$ in 0.5 μl of buffer or 0.3 μmol of NaCl in 0.5 μl buffer into periaqueductal gray area. Immediately after microinjection, rats were given either saline or 5 mg/kg of morphine sulfate subcutaneously, and tail-flick (T-F) and hot-plate (H-P) responses were assessed 60 min later. Values are means ± SE. [Data from Iwamoto et al. (27).]

TABLE 2. *Similarities between La^{3+} and opiates*

	Ref.
1) La^{3+} produces analgesia (antinociception) in rats and mice	19
2) This analgesia is reduced by Ca^{2+} or naloxone	19, 27
3) The periaqueductal gray area is the neuroanatomical site of analgesic action	27
4) Rats and mice tolerant to morphine are tolerant to La^{3+}	17, 19
5) La^{3+} suppresses morphine withdrawal signs	19
6) La^{3+} inhibits the prostaglandin-stimulated adenylate cyclase of neuroblastoma x glioma hybrid cells	5

TABLE 3. *Effects of acute injection of morphine or ethanol on Ca^{2+} content of rat synaptosomes*

Treatment	μg Ca^{2+}/mg Protein, n = 10	
	L synaptosomes	H synaptosomes
Saline	1.10 ± 0.04	1.23 ± 0.04
Morphine	0.89 ± 0.04*	1.10 ± 0.03†
Ethanol	0.99 ± 0.05‡	1.24 ± 0.03‡

Rats were injected intraperitoneally with saline, ethanol (3 g/kg), or morphine sulfate (25 mg/kg) 30 min before killing. Whole brain homogenates were fractionated by sucrose gradient centrifugation to yield light (L) and heavy (H) synaptosomes (22). Calcium and protein contents were determined as described elsewhere (24). Values are means ± SE. * $P < 0.001$. † $P < 0.05$. ‡ NS.

morphine produces a reliable and selective decrease in Ca^{2+} content of the synaptosomal fractions (Table 3; 6, 24, 50). This observation suggests that opiates inhibit synaptosomal uptake or storage of Ca^{2+}. It has been reported that morphine exposure, both in vitro and in vivo, inhibits the ATP-dependent sequestration of Ca^{2+} at intrasynaptosomal sites (14). These effects are reversed by naloxone, a specific narcotic antagonist. Morphine also reduces the spontaneous accumulation of Ca^{2+} by synaptosomes incubated with ATP (13), but interpretation of this result is difficult because the extent to which the measured Ca^{2+} accumulation represents membrane binding, mitochondrial Ca^{2+} uptake, or nonmitochondrial sequestration is not known. These experiments (13) also fail to rule out the possibility that morphine affects Ca^{2+} efflux rather than Ca^{2+} uptake. It has been reported that morphine inhibits the K^+ depolarization-stimulated uptake of Ca^{2+} by brain synaptosomes (42). My results did not confirm this (Fig. 2). There is, however, electrophysiological evidence that opiates block the voltage-sensitive Ca^{2+} channels of sensory neurons (dorsal root ganglia) grown in cell culture (36). These conflicting results might be reconciled if some, but not all neuronal Ca^{2+} channels were coupled to opiate receptors.

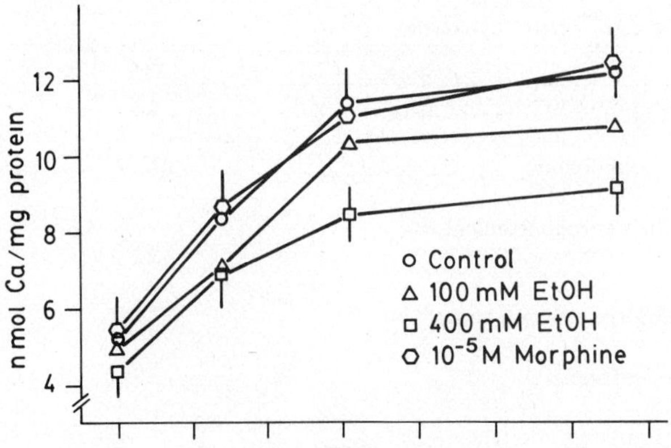

FIG. 2. Effects of morphine and ethanol on synaptosomal Ca^{2+} uptake stimulated by an elevated K^+ concentration. Synaptosomes were isolated from mouse brain and exposed to ethanol (100 or 400 mM) or morphine hydrochloride (10^{-5} M) in vitro. Accumulation of Ca^{2+} was initiated by addition of various concentrations of KCl (substituted for NaCl) and terminated 1 min later, as described by Harris and Hood (16).

Chronic administration of morphine affects Ca^{2+} localization in a manner opposite that of acute treatment. Several studies demonstrate that opiate tolerance is associated with an increased sequestration of Ca^{2+} by synaptosomes (14, 20, 41). This may represent a homeostatic mechanism that counteracts the acute opiate effects and produces opiate tolerance.

Many of the physiological effects of opiates (and endorphins) may be due to an inhibition of neurotransmitter release. This is the mechanism by which opiates and endorphins affect several peripheral tissues (21, 29, 45), and it may be important for their effects on the CNS (36, 44, 48). The above data suggest that opiates inhibit neurotransmitter release by reducing the availability of Ca^{2+} in the nerve terminal. It is possible that opiate receptors are coupled with Ca^{2+} channels, but alternative hypotheses must also be considered.

ALCOHOLS AND BARBITURATES AS ANTAGONISTS OF ACTIONS OF CALCIUM

Although much evidence exists for the role of Ca^{2+} in opiate action, behavioral or physiological evidence that Ca^{2+} reduces the effects of alcohols or barbiturates is limited. Calcium antagonizes the blockade of nerve conduction produced by alcohols and barbiturates (47), but potentiates other effects of these drugs (10, 15).

There is considerable evidence, however, that these drugs inhibit synaptosomal Ca^{2+} transport. In vitro addition of barbiturates or alcohols inhibits the depolarization-stimulated accumulation of Ca^{2+} by nerve endings (Fig. 2; 2, 16, 23). The inhibitory effects of both pentobarbital and ethanol are due to a change in the apparent K_m for Ca^{2+} uptake (2; R. A. Harris, unpublished observations). The effects of these drugs on Ca^{2+} transport display selectivity for various brain regions, with cerebellum and brain stem being more sensitive than striatum and cortex (Table 4; 9). The efflux of Ca^{2+} from synaptosomes is not affected by ethanol but is enhanced by pentobarbital. This appears to be due to a stimulation of the Ca^{2+}-ATPase of the synaptosomal plasma membrane (49). These results suggest that the inhibition of Ca^{2+} accumulation produced by barbiturates is at least partly due to enhancement of Ca^{2+} efflux. In contrast alcohols apparently inhibit Ca^{2+} influx but not Ca^{2+} efflux (16). Although alcohols and barbiturates inhibit intrasynaptosomal (ATP-dependent) sequestration of Ca^{2+}, large concentrations are necessary to produce this

TABLE 4. *Drug effects on synaptosomal Ca^{2+} transport*

Drug	Regional Sensitivity of Ca^{2+} Uptake	Effect on Ca^{2+} Efflux
Ethanol, 50 mM	CBL > BS > STR > CTX	None
Pentobarbital, 0.2 mM	CBL = BS > STR = CTX	Stimulates
Δ^9-THC, 10^{-7} M	BS > CBL > STR = CTX	None

Synaptosomes were prepared from tissue dissected from rat brain and exposed to drugs in vitro. Depolarization-dependent uptake of Ca^{2+} and efflux of Ca^{2+} were determined as described by Harris and Hood (16). CBL, cerebellum; BS, brain stem; STR, striatum; CTX, cortex; THC, tetrahydrocannabinol.

action (25). The neurochemical findings are supported by electrophysiological evidence indicating that ethanol and barbiturates inhibit Ca^{2+} influx. In cultured dorsal root ganglia, ethanol, like opiates, reduces the voltage-sensitive Ca^{2+} current (38). Ethanol also reduces the Ca^{2+} currents of neurons from *Aplysia* and goldfish brain (11), and pentobarbital reduces the Ca^{2+} current of the squid giant synapse (35). The effects of alcohols, barbiturates, and fatty acids on the physical properties of membrane lipids have been compared with their effects on synaptosomal Ca^{2+} transport, and the compounds studied showed a correlation between their ability to disorder or fluidize membrane lipids and their ability to inhibit depolarization-dependent Ca^{2+} uptake (R. A. Harris, unpublished observations). These results suggest that voltage-sensitive Ca^{2+} channels may be regulated by the physical properties of the surrounding lipids and that drugs may affect Ca^{2+} transport by perturbation of these lipids.

The effects of alcohols and barbiturates on the binding of Ca^{2+} to brain membranes have not been studied in detail. Low concentrations of ethanol enhance the binding of Ca^{2+} to brain membranes, but the effect of other alcohols and barbiturates has not been determined (34). Recent data demonstrate that alcohols, barbiturates, and other local anesthetics displace Ca^{2+} from the cytoplasmic surface of the erythrocyte membrane (33); however, this effect was noted only when drug concentrations were very high. Conversely, other studies indicate that alcohols and anesthetics enhance the binding of Ca^{2+} to erythrocyte membranes (46) and cardiac plasma membrane lipoproteins (39). The physiological importance and molecular basis of changes in Ca^{2+} binding remain to be explored.

Barbiturates and ethanol inhibit neurotransmitter release (7, 23), and this action may be secondary to their effects on Ca^{2+} transport. After chronic administration of these drugs, tolerance develops to their effects on both Ca^{2+} transport and neurotransmitter release (9, 16). This adaptation suggests that alterations in Ca^{2+} flux may be one mechanism underlying the cellular tolerance produced by these drugs.

CANNABINOLS AND PHENOTHIAZINES AS ANTAGONISTS OF ACTIONS OF CALCIUM

Even though cannabinols have many of the pharmacological characteristics of alcohols and barbiturates, little attention has been given to their effects on neuronal Ca^{2+} metabolism. In vitro addition of Δ^9-THC (tetrahydrocannabinol, the major psychoactive component of marijuana) inhibits the depolarization-dependent uptake of Ca^{2+} by synaptosomes (Table 4). This effect occurs at low, pharmacologically relevant concentrations. The selectivity for brain regions displayed by this compound is like that displayed by pentobarbital and ethanol. Like ethanol, Δ^9-THC apparently inhibits Ca^{2+} influx rather than enhancing Ca^{2+} efflux (Table 4).

Chlorpromazine, a phenothiazine antipsychotic drug, reduces the depolarization-stimulated accumulation of Ca^{2+} by synaptosomes without affecting the intrasynaptosomal (ATP-dependent) sequestration (25, 30). Chlorpromazine and other phenothiazines have been shown to displace Ca^{2+} from erythrocyte membranes (46), and Ca^{2+} reverses the nerve blockade produced by chlorpromazine (47). The Ca^{2+} effects on some enzymes are mediated by a Ca^{2+}-binding protein, calmodulin. The phenothiazines bind to calmodulin and block its action, thus functioning as indirect Ca^{2+} antagonists (31). These results suggest that several mechanisms are involved in the antagonism of Ca^{2+} effects by phenothiazines.

CONCLUSIONS

The findings presented in this chapter demonstrate that chemically diverse psychoactive drugs inhibit neuronal Ca^{2+} transport. These drugs reduce voltage-sensitive Ca^{2+} movement in the nerve ending, which likely leads to inhibition of neurotransmitter release. The behavioral and physiological effects of opiates, sedatives, and phenothiazines, however, are distinct, and are probably not due to similar indiscriminate inhibition of synaptic function by each of these drugs. The behavioral diversity of these drugs may reside in selective effects on different synapses. It is likely that Ca^{2+} flux is affected by opiates only at synapses containing opiate receptors. In addition barbiturates, alcohols, and cannabinols have been shown to display selectivity for the Ca^{2+} transport of different brain areas. These regional differences in drug sensitivity may reflect the various types of neurons (e.g., different neurotransmitters) found in these brain areas.

An alternative hypothesis is that there is no relationship between the behavioral effects of these drugs and their effects on Ca^{2+} transport. For opiates there is a good deal of evidence (albeit indirect) that changes in Ca^{2+} availability are important in behavioral and physiological effects such as analgesia. Regarding the other drugs, evidence is less compelling. For example, alcohol, barbiturate, and phenothiazine concentrations necessary to alter Ca^{2+} transport in vitro are generally several times higher than those required to produce intoxication in vivo. Further studies are needed to determine which, if any, of the diverse effects of these drugs on Ca^{2+} metabolism are physiologically significant.

This work was partly supported by funds from the Medical Research Service of the Veterans Administration and by National Institutes of Health Grant DA-02855.

REFERENCES

1. BAKER, P. F. The regulation of intracellular calcium in giant axons of *Loligo* and *Myxicola*. *Ann. NY Acad. Sci.* 307: 250–268, 1978.
2. BLAUSTEIN, M. P., AND A. C. ECTOR. Barbiturate inhibition of calcium uptake by depolarized nerve terminals *in vitro*. *Mol Pharmacol.* 11: 369–378, 1975.
3. BLAUSTEIN, M. P., N. C. KENDRICK, R. C. FRIED, AND R. W. RATZLAFF. Calcium metabolism at the mammalian presynaptic nerve terminal: lessons from the synaptosome. In: *Society for Neuroscience Symposium*, edited by W. M. Cowan and J. A. Ferrendelli. Bethesda, MD: Soc. Neurosci. 1977, vol. II, p. 172–194.
4. BLAUSTEIN, M. P., R. W. RATZLAFF, AND N. K. KENDRICK. The regulation of intracellular calcium in presynaptic nerve terminals. *Ann. NY Acad. Sci.* 28: 195–212, 1978.

5. BRANDT, M., C. BUCHEN, AND B. HAMPRECHT. Relationship between the actions of calcium ions, opioids, and prostaglandin E_1 on the level of cyclic AMP in neuroblastoma x glioma hybrid cells. *J. Neurochem.* 34: 643–651, 1980.
6. CARDENAS, H. L., AND D. H. ROSS. Calcium depletion of synaptosomes after morphine treatment. *Br. J. Pharmacol.* 57: 521–526, 1976.
7. CARMICHAEL, F. J., AND Y. ISRAEL. Effects of ethanol on neurotransmitter release by rat brain cortical slices. *J. Pharmacol. Exp. Ther.* 193: 824–834, 1975.
8. CHAPMAN, D. B., AND E. L. WAY. Metal ion interactions with opiates. *Annu. Rev. Pharmacol.* 20: 553–579, 1980.
9. ELROD, S. V., AND S. W. LESLIE. Acute and chronic effects of barbiturates on depolarization-induced calcium influx into synaptosomes from rat brain regions. *J. Pharmacol. Exp. Ther.* 212: 131–136, 1980.
10. ERICKSON, C. K., T. D. TYLER, AND R. A. HARRIS. Ethanol: modification of acute intoxication by divalent cations. *Science* 199: 1219–1221, 1978.
11. FARBER, D. S., AND M. R. KLEE. Actions of ethanol on neuronal membrane properties and synaptic transmission. In: *Alcohol and Opiates: Neurochemical and Behavioral Mechanisms*, edited by K. Blum. New York: Academic, 1977, p. 41–63.
12. GODDARD, G. A., AND J. D. ROBINSON. Uptake and release of calcium by rat brain synaptosomes. *Brain Res.* 110: 331–350, 1976.
13. GUERRERO-MUNOZ, F., K. V. CERRETA, M. L. GUERRERO, AND E. L. WAY. Effect of morphine on synaptosomal Ca^{++} uptake. *J. Pharmacol. Exp. Ther.* 209: 132–136, 1979.
14. GUERRERO-MUNOZ, F., M. GUERRERO, AND E. L. WAY. Effect of morphine on calcium uptake by lysed synaptosomes. *J. Pharmacol. Exp. Ther.* 211: 370–374, 1979.
15. HARRIS, R. A. Alteration of alcohol effects by calcium and other inorganic cations. *Pharmacol. Biochem. Behav.* 10: 527–534, 1979.
16. HARRIS, R. A., AND W. F. HOOD. Inhibition of synaptosomal calcium uptake by ethanol. *J. Pharmacol. Exp. Ther.* 213: 562–568, 1980.
17. HARRIS, R. A., E. T. IWAMOTO, H. H. LOH, AND E. L. WAY. Analgetic effects lanthanum: cross-tolerance with morphine. *Brain Res.* 100: 221–225, 1975.
18. HARRIS, R. A., H. H. LOH., AND E. L. WAY. Effects of divalent cations, cation chelators and an ionophore on morphine analgesia and tolerance. *J. Pharmacol. Exp. Ther.* 195: 488–498, 1975.
19. HARRIS, R. A., H. H. LOH, AND E. L. WAY. Antinociceptive effects of lanthanum and cerium in nontolerant and morphine tolerant-dependent animals. *J. Pharmacol. Exp. Ther.* 196: 288–297, 1976.
20. HARRIS, R. A., H. YAMAMOTO, H. H. LOH, AND E. L. WAY. Discrete changes in brain calcium with morphine analgesia, tolerance-dependence, and abstinence. *Life Sci.* 20: 501–506, 1977.
21. HENDERSON, G., AND J. HUGHES. The effects of morphine on the release of noradrenaline from the mouse vas deferens. *Br. J. Pharmacol.* 57: 551–557, 1976.
22. HITZEMANN, R. J., AND H. H. LOH. On the possible role of brain protein synthesis in functional barbiturate tolerance. *Eur. J. Pharmacol.* 40: 163–173, 1976.
23. HO, I. K., AND R. A. HARRIS. Mechanism of action of barbiturates. *Annu. Rev. Pharmacol.* 21: 83–111, 1981.
24. HOOD, W. F., AND R. A. HARRIS. Effects of pentobarbital, ethanol, and morphine on subcellular localization of calcium and magnesium in brain. *Biochem. Pharmacol.* 28: 3075–3080, 1979.
25. HOOD, W. F., AND R. A. HARRIS. Effects of depressant drugs and sulfhydryl reagents on the transport of calcium by isolated nerve endings. *Biochem. Pharmacol.* 29: 957–959, 1980.
26. ILLES, P., W. ZIEGLGANSBERGER, AND A. HERTZ. Calcium reverses the inhibitory action of morphine on neuroeffector transmission in the mouse vas deferens. *Brain Res.* 191: 511–522, 1980.
27. IWAMOTO, E. T., R. A. HARRIS, H. H. LOH, AND E. L. WAY. Antinociceptive responses after microinjection of morphine or lanthanum in discrete rat brain sites. *J. Pharmacol. Exp. Ther.* 206: 46–55, 1978.
28. KAKUNGA, T., H. KANETO, AND K. HANO. Pharmacologic studies on analgesics. VII. Significance of the calcium ion in morphine analgesia. *J. Pharmacol. Exp. Ther.* 153: 134–141, 1966.
29. KONISHI, S., A. TSUNOO, AND M. OTSUKA. Enkephalins presynaptically inhibit cholinergic transmission in sympathetic ganglia. *Nature London* 282: 515–516, 1979.
30. LESLIE, S. W., S. V. ELROD, R. COLEMAN, AND J. K. BELKNAP. Tolerance to barbiturate and chlorpromazine-induced central nervous system sedation—involvement of calcium-mediated stimulus-secretion coupling. *Biochem. Pharmacol.* 28: 1437–1440, 1979.
31. LEVIN, R. M., AND B. WEISS. Binding of trifluoperazine to the calcium-dependent activator of cyclic nucleotide phosphodiesterase. *Mol. Pharmacol.* 13: 690–697, 1977.
32. LLINÁS, R. R. Calcium and transmitter release in squid synapse. In: *Society for Neuroscience Symposium*, edited by W. M. Cowan and J. A. Ferrendelli. Bethesda, MD: Soc. Neurosci., 1977, vol. II, p. 139–160.
33. LOW, P. S., D. H. LLOYD, T. M. STEIN, AND J. A. ROGERS. Calcium displacement by local anesthetics: dependence on pH and anesthetic charge. *J. Biol. Chem.* 254: 4119–4125, 1979.
34. MICHAELIS, E. K., AND S. L. MYERS. Calcium binding to brain synaptosomes: effects of chronic ethanol intake. *Biochem. Pharmacol.* 28: 2081–2087, 1979.
35. MORGAN, K. G., AND S. H. BRYANT. Pentobarbital: presynaptic effect in the squid giant synapse. *Experientia* 33: 487–488, 1977.
36. MUDGE, A. W., S. E. LEEMAN, AND G. D. FISCHBACH. Enkephalin inhibits release of substance P from sensory neurons in culture and decreases action potential duration. *Proc. Natl. Acad. Sci. USA* 76: 526–530, 1979.
37. NACHSEN, D. A., AND M. P. BLAUSTEIN. The effects of some organic "calcium antagonists" on calcium influx in presynaptic nerve terminals. *Mol. Pharmacol.* 16: 579–586, 1979.
38. OAKES, S. G., AND R. S. POZOS. Electrophysiologic effects of ethanol on dissociated sensory ganglia in vitro. *Alcoholism: Clin. Exp. Res.* 4: 225, 1980.
39. OHNISHI, S. T., D. M. OBZASKY, AND H. I. PRICE. The increase in calcium binding of cardiac plasma membrane lipoprotein caused by general anesthetics and alcohol. *Can. J. Physiol. Pharmacol.* 58: 525–530, 1980.
40. PASTERNAK, G. W., A. M. SNOWMAN, AND S. H. SNYDER. Selective enhancement of (^{3}H)-opiate binding by divalent cations. *Mol. Pharmacol.* 11: 735–744, 1975.
41. ROSS, D. H. Calcium content and binding in synaptosomal subfractions during chronic morphine treatments. *Neurochem. Res.* 2: 581–593, 1977.
42. ROSS, D. H. Effects of opiate drugs on the metabolism of calcium in synaptic tissue. In: *Calcium in Drug Action*, edited by G. B. Weiss. New York: Plenum, 1978, p. 241–259.
43. ROSS, D. H., M. A. MEDIAN, AND H. L. CARDENAS. Morphine and ethanol: selective depletion of regional brain calcium. *Science* 186: 63–65, 1975.

44. SANFACON, G., AND G. LABRECQUE. Acetylcholine antirelease effect of morphine and its modification by calcium. *Psychopharmacology* 55: 151–156, 1977.
45. SCHAUMANN, W. Inhibition by morphine of the release of acetylcholine from the intestine of the guinea pig. *Br. J. Pharmacol. Chemother.* 12: 115–118, 1957.
46. SEEMAN, P. The membrane actions of anesthetics and tranquilizers. *Pharmacol. Rev.* 24: 583–655, 1972.
47. SEEMAN, P., S. S. CHEN, M. CHAU-WONG, AND A. STAIMAN. Calcium reversal of nerve blockade by alcohols, anesthetics, tranquilizers, and barbiturates. *Can. J. Physiol. Pharmacol.* 52: 526–534, 1974.
48. SUBRAMANIAN, N., P. MITZNEGG, W. SPRUGEL, W. DOMSCHKE, S. DOMSCHKE, E. WUNSCH, AND L. DEMLING. Influence of enkephalin on K^+-evoked efflux of putative neurotransmitter in rat brain. *Naunyn-Schmiedebergs Arch. Pharmacol.* 299: 163–165, 1977.
49. WILLOW, M., AND G. A. R. JOHNSTON. Barbiturates and calcium-activated adenosine triphosphatase. *Neurosci. Lett.* 14: 361–364, 1979.
50. YAMAMOTO, H., R. A. HARRIS, H. H. LOH, AND E. L. WAY. Effects of acute and chronic morphine treatments on calcium localization and binding in brain. *J. Pharmacol. Exp. Ther.* 205: 255–264, 1978.

INDEX